Owner's Bible™

By Pat Braden

A Hands-on Guide to Getting the Most From Your Alfa

VELOCE PUBLISHING PLC
PUBLISHERS OF FINE AUTOMOTIVE BOOKS

ROBERT BENTLEY
CAMBRIDGE, MASSACHUSETTS

Table of Contents

Page

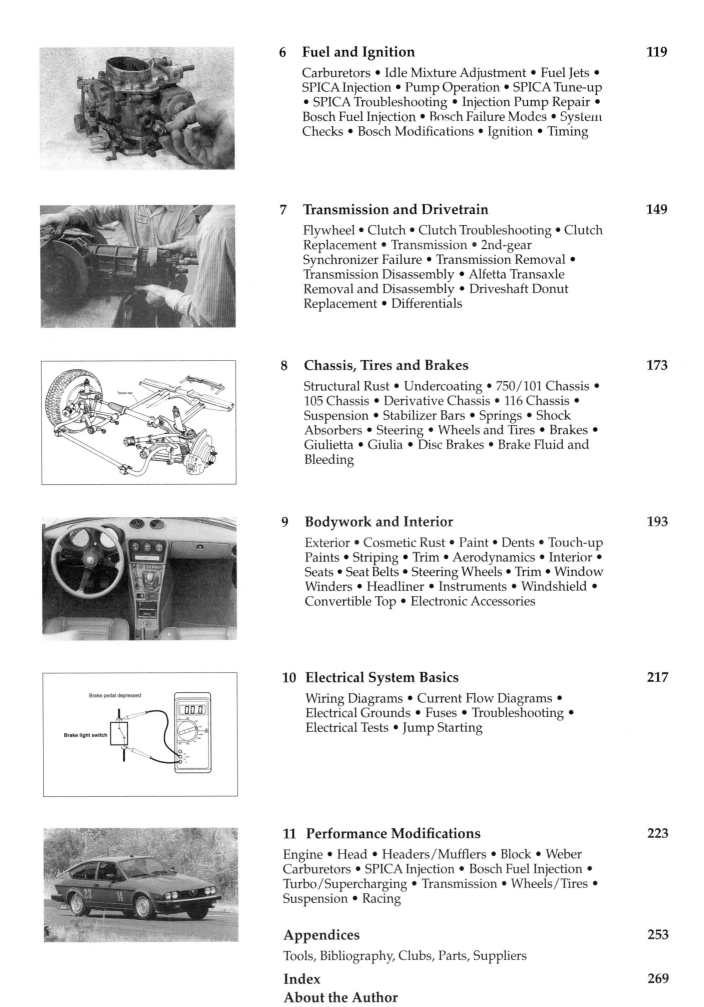

Prepared, Published, and Distributed by:

Robert Bentley, Inc., Publishers
1000 Massachusetts Avenue
Cambridge, Massachusetts 02138

Sole U.K., Australia, New Zealand Distribution:

Veloce Publishing Plc.
Godmanstone, Dorset DT2 7AE, England
ISBN 1-874105-45-6

The publisher encourages comments from the readers of this book. These communications have been and will be considered in the preparation of this and other manuals. Please write to Robert Bentley Inc., Publishers, at the address listed on the top of this page.

This book was published by Robert Bentley, Inc., Publishers. Alfa Romeo has not reviewed and does not warrant the accuracy or completeness of the technical specifications and procedures described in this book.

Library of Congress Cataloging-in-Publication Data

Braden, Pat, 1934–
    Alfa Romeo owner's bible : a hands-on guide to getting the most
from your Alfa / by Pat Braden.
       p.   cm.
    Includes bibliographical references and index.
    ISBN 0-8376-0707-8 : $29.95
    1. Alfa Romeo automobile.   I. Title.
TL215.A35B733 1994
  629.222'2--dc20   94-11886
                                  CIP

British Library Cataloguing in Publication Data – A catalogue record for this book is available from the British Library

Bentley Stock No. GALF

Editorial closing 04/94

96 95 94   10 9 8 7 6 5 4 3 2 1

The paper used in this publication is acid free and meets the requirements of the National Standard for Information Sciences-Permanence of Paper for Printed Library Materials. ∞

Alfa Romeo Owner's Bible™: A Hands-on Guide to Getting the Most From Your Alfa Romeo

Foreword

by Don Black

Don Black has been intimately involved with Alfas for over 30 years, for 26 as Director of Alfa Romeo's U.S. Engineering Office, and currently as Deputy Director of Fiat Auto R&D USA. He also develops and races his own Alfa-powered cars.

ALFA ROMEO ENTHUSIASM borders on being a religion, so I feel the title *Bible* might very well be appropriate. During my first reading of the draft, I didn't light any votive candles, but I was bobbing in waves of nostalgia fed by the wonderful collection of photos and sketches. With more than three decades service with Alfa Romeo behind me, there are plenty of memories.

What I remember more than the cars are the people of Alfa Romeo, who are so important, because they are seen in their product. Using a product can tell you so much about its creators and producers. I learned very early that the Milanese are an intelligent, hardworking, creative, and open people. Above all, they have a wonderful sense of humor. You see these attributes especially in the classic Giulietta. Only "Meneghini" could have created such a delightful personality in an automobile.

I am privileged to have been able to be actively involved in projects with the likes of such great minds as Satta, Garcea, Busso, Surace, Chiti, Chirico, Sirtori, as well as the "Young Turks" Piccone, Guelfi, Satta 2, who continue the Alfa Romeo tradition into the next century.

One of the chassis sketches here reminds me of an early "research" for Satta, to confirm if his "slingshot" 159 G.P. car predated that of Micky Thompson's dragster. That memory faded into another, where I had taken this great man to see his first major league American baseball game…but, that is another short story in itself. These memories seem to cascade one upon another once you are "hooked up" (like a 770-hp sprinter with side bite in the cushion).

When I was drafted by Arturo Reitz to work for his yet unincorporated Alfa Romeo (Inc.), I had a compelling desire to dispel myths and to remove the prevailing cloak of mystery surrounding things technical. I began writing, rewriting, publishing, and distributing as much technical information as our mimeo machine and offset printer could handle. This was done pre dry copiers, pre fax, pre speedy print, and all done by hand. I published the Competition Advisory Service for the racers, making drawings by cutting wax film stencils with a hot knife. I was forced to become my own translator. I was driven to provide accurate information to anyone who wanted it.

To me, Alfa Romeo pistons went up and down like any other pistons; Alfa Romeo wheels rotated like any other wheels, hopefully in the same direction. So why the mystery? Why the intrigue and mysticism for a collection of hardware not much different from other less "exotic" badges? The dearth of technical information caused owners and dealers so much grief. It had to be eliminated, and was, in less than two years. Pat Braden helped to maintain the information stream through several of his publications, as well as being a permanent contributor to the *Alfa Owner*, the Alfa Romeo Owners Club national publication. My personal critique of Braden's work (as he well knows) is that his wonderful sense of humor does not infiltrate his writing.

As the factory's U.S. Engineering Office Director, I have most likely read every book ever written, in any language, about Alfa Romeo. While reading the manuscript of this book, I kept thinking that it is long overdue, a practical reference manual that is sure to provide Alfa Romeo enthusiasts everywhere with a reliable source of expertise. On a more selfish aspect, its pages provide a non-electronic memory for those of us having difficulty with the minutiae we never wrote in our own diaries. It is a terrible thing to forget, but like a computer's hard disk, our "wetware" is also capacity limited and tends to slow with age. Paper is permanent and no batteries are required.

Since this book is very likely to be referred to by Alfisti as the "Braden Bible," it seems only fitting that it be kept in the reader's bedside nightstand as a ready source of inspiration. Having finished my third and final reading of the galleys, I am inspired to complete a very old project of my own: a votive alcohol lamp, blown by a Venetian gaffer in the shape of a 159, and burning the same exotic and politically correct (i.e. green) G.P. fuel, *Dynamin*, that was custom (and secretly) blended by Sandro Sirtori for the race cars of the 30s.

I would like to leave you, good reader, with this advice. Don't take your Alfa Romeo too seriously. Take time to enjoy it by driving it as often as you can. It was made by real drivers for real driving enthusiasts.

"Keep the Shiny Side Up"

BRIGHTON, MI
MARCH 1994

Over three-quarters of a century of Alfa experience: Guido Moroni (left) and Don Black dispute a point at the Montery Historic Races. Moroni was Alfa's chief test driver for 50 years.

Author's Preface

Author Pat Braden holds a GTV6 air-flow sensor.

I F YOU HAVE NEVER driven an Alfa Romeo, you may be stunned to discover a pleasure that is simply unavailable in many passenger cars. It's not that Alfa is significantly different in its basic functions: if you turn its wheel and press its brakes it turns and stops. As with any thoroughbred, however, it is the quality of the performance that sets it apart. Not only are there no mistakes, but all the moves are executed to a level of perfection that casual observers may not appreciate.

For those who think a car is supposed to swerve out of control when you slam on the brakes, or roll over if you turn a corner at 25 mph, the initial Alfa experience is likely to be transcendental. Indeed, its absence of unpredictable manners is one of the great driving revelations of the first few miles behind the wheel. Alfa's even temperament is more than just enjoyable. In an emergency, it can mean, literally, the difference between life and death.

If there ever was an affordable exotic, Alfa is it. To get similar performance from a similar car, one might have to double his investment. For many, however, it is not impeccable road manners that are the source of Alfa's allure, but the raw engineering genius of the small mechanical parts. They do more than work; they adorn. The Alfa engine has always been a work of beauty. And the beauty is more than skin-deep. Artfully crafted to their severe essentials, Alfa's exquisite bits and pieces work unseen beneath fine aluminum and steel castings.

There is another dimension to Alfa's allure, and that is its tradition of being a technology-driven company. Road-going Alfas have been powered by twin-cam engines with hemispheric combustion chambers since 1926. Alfa introduced independent rear suspension in 1938 and a variable intake camshaft in 1980. Unit bodies were introduced as a regular production item in 1951.

For most Americans, this is the car that started it all—the Giulietta, here seen in coupe form and racing livery.

In the U.S., Alfa switched to mechanical fuel injection in 1969, in anticipation of the stringent emission controls that devastated the driveability of other makes beginning in 1971.

For these pleasures, many people expect Alfa to prove a troublesome mistress. Indeed, Alfa's styling, performance and Italian heritage puts it in a natural class with two exotic Italian sport cars: Ferrari and Maserati. Of the three, Alfa is the absolutely lowest-priced way for most people to buy into the Italian car mystique.

Many years of experience with the marque have convinced me that Alfa is an unquestionably durable car. Of the Alfas I own, at least four have over 100,000 miles on their odometers and both my wife and I drive Alfas for everyday transportation. Members of the U.S. Alfa Romeo Owners Club have a plaque for +100,000-mile cars. The mark, it turns out, is not especially notable nor unusual: one owner of an Alfa sedan reports 450,000 actual miles.

Facts notwithstanding, Alfa continues to be generally regarded as finicky. My experience suggests that it is the owner or mechanic, and not the car itself, that caus-

The Alfa engine is a thing of mechanical beauty on which owners lavish much care. This spit-and-polish engine compartment *on a coupe features two large Weber carburetors for top performance.*

One of the ultimate Alfa collectibles is the TZ2, a no-frills racer of undoubted beauty. There are only two in the U.S., both owned by the same lucky Alfista.

es the most problems with reliability. Alfa is not a casual item to possess, but then neither is a Ferrari or Maserati.

Frequently, raw enthusiasm prompts the owner into acts that he might otherwise avoid: rebalancing or rejetting functioning carburetors, "improving" properly timed ignition systems, or realigning camshafts when the engine runs perfectly well. There is an oft-repeated adage about mechanical things: "If it ain't broke, don't fix it." The admonition applies especially to a car as sophisticated as Alfa.

The intent of this book is to supply all the information necessary to enjoy owning an Alfa. Recognizing that many owner/enthusiasts want to do their own maintenance, I've included all the basic maintenance and repair procedures in how-to sections throughout the book. If I have erred at all in this, it is in detailing some repairs which are beyond the skill of the casual enthusiast. In a very real sense, this book is for adults only: please use the information with discretion.

On the other hand, this book is not a shop manual. If the solution to your problem appears to be missing from the book, the odds are that (1) it's an exceptionally unusual condition, or (2) the fix involves skills beyond those which can be conveyed in a book. For example, ring and pinion failures are exceedingly rare for Alfas. As a result, I've said almost nothing about them in the

book. Moreover, even if I did show how, the odds on setting up a ring and pinion for the first time, unaided, are very slim indeed.

Finally, there is the matter of safety which needs to be acknowledged here. Most manuals contain a disclaimer paragraph somewhere which admonishes the reader on behalf of safety and common sense (see the inside front cover). My version:

We are conditioned by advertising to think of our cars as friendly puppies or comfortable armchairs. The truth is almost exactly the opposite. Cars, by their very nature, continuously threaten injury or death not only to those who drive, but also to those who work around them. A car runs on an explosively flammable fluid, and the slightest spark required to set it off is easily available from a battery which holds enough power to weld pieces of metal together. The engine offers surfaces which are hot enough to inflict serious burns, and the cooling system is a pressure cooker waiting to blow off steam. Fire and explosions are very real threats around a car, and a fire extinguisher is an absolutely necessary tool.

A running engine is a gauntlet of spinning parts, all of which can easily dismember whatever part of your body gets in the way, or turn a dropped tool into a lethal projectile. The edges of most sheet-metal parts will cut as effectively as the best steak knife.

A lot of work is done beneath the car: while it should be obvious that there is a constant danger of being crushed, it's amazing how many people dive beneath a car with only its jack in place. I never put any part of my body under a car unless it is on jackstands and a floor jack is positioned under the car so it will keep me from being crushed if the car should come off the jackstands. I never crawl under a car unless someone is nearby who knows how to use the jack.

The conclusions to all these warnings are almost self-evident: wear tight-fitting protective clothing and eye protection, work with jackstands, and avoid working on a running car or doing electrical work on a car from which the battery has not been disconnected.

This is a guide for happy Alfa ownership. If you need to know it, it's in this book.

> *For my buddy,*
> *Lee Patrick Braden*
> *4/29/87 – 7/9/93*

Chapter 1

Alfa—A Brief History

VIRTUALLY EVERY BOOK ON ALFA contains a history, so why include yet another one? According to reader surveys of the Alfa Romeo Owners Club, historic information is one of the most desired subjects for the club magazine. Either the factual information is not getting through, or it is so fabulous it bears almost limitless retelling. I prefer the latter theory, and think it is worth recounting the high points here again.

From the late 1920s to the early 1950s, Alfa was indisputably the preeminent sporting marque in the world. That halcyon era gave it a shove that has kept it rolling for years on the basis of its traditional strengths.

In the Beginning...

When the automobile was being developed in Europe, the companies that became involved in its early manufacture were already making horse-drawn carriages or bicycles. A French firm, the Gladiator cycle company, had been founded in 1891 by Alexandre Darracq. In 1896, Darracq sold his bicycle factory to a British firm and began producing an electric-powered car that was exhibited at the Paris Salon. Two years later, he purchased the manufacturing rights to the Bollee four-wheeler voiturette, a light car with a single-cylinder in-

Fig. 1-1. *Via Gattemalata 45, on the old Portello road, is now very much in urban Milano. This was the original factory entrance. Only the gate and part of the building on the right remain today.*

CHAPTER 1

ternal combustion engine. In 1900, Darracq produced his own version of a light-weight car and in the following year produced a 2-cylinder race car. Passenger car production was enlarged to include 2- and 4-cylinder vehicles and the company prospered enough to attract British funds. Darracq's firm was reorganized under British registry in 1905 and the company broadened its range to include larger passenger and sport cars.

In 1907, Darracq decided to open a plant in Italy to produce a line of light, 9-hp taxis that he hoped would become popular in the crowded Italian cities. He selected a suburban site in the Portello section of Milan. Apparently, things did not go outstandingly well. Darracq produced cars there until the firm was taken over, in January 1910, by an Italian group called the Lombardy Automobile Manufacturing Company (Anonima Lombarda Fabbrica Automobili, or A.L.F.A.).

A.L.F.A. and the Cars of Merosi

The new A.L.F.A. directors retained engineer Giuseppi Merosi, a 38-year old engineer who had worked both at Fiat and Bianchi, to design a pair of touring sport cars. These two 1910 models, 4-cylinder cars with displacements of 2.4 and 4 liters, were the first A.L.F.A. automobiles. Both cars, absolutely typical of the era, featured artillery wheels, minimal bodies, L-head monobloc engines with thermosiphon cooling and fuel priming cups, magneto ignition, and a four-speed gearbox with straight-cut sliding gears. The drive was by enclosed shaft with a single universal joint housed just behind the transmission, a feature retained until the adoption of independent rear suspension in the late 1930s.

Both engines went through several series and remained in production until 1920. The 4-liter engine was called the 24-HP originally, and after 1914, the Model 20-30. This engine was adapted for aero use in 1910 and also appeared as the 1917 "ROMEO" air compressor (the first ALFA product to have engineer Romeo's name attached; Alfa's modern trucks are also named Romeo). The smaller engine went through permutations as the 12HP, the 15HP, and the 15-20HP.

In 1913, A.L.F.A. introduced a car with serious sporting potential. A Merosi-designed 6-liter four, its

The Alfa Badge

It was stylish in the early days of the automobile to use the company's initials to name the car (Fabbrica Italiana Automobili Torino was already well established making F.I.A.T.s in Turin) and the acronym A.L.F.A. was a natural choice for the new Milanese company. The name of the car is in no way related to the first letter of the Greek alphabet (which Vincenzo Lancia's company had already appropriated in 1907 for its product line).

The Alfa badge is arguably the most remarkable emblem associated with any marque. For it, the owners combined two symbols that had been associated with the city of Milan since the Crusades: a red cross and a serpent swallowing a man. The red cross is an obvious Christian symbol of medieval heraldry. The man in the serpent's mouth is a Saracen (a Muslim), so the serpent motif recalls the Christian Crusaders' defeat of the infidels, a fact that is probably not detailed in certain of Alfa's current marketing areas. The serpent motif became an honorary sign awarded to a prominent Milanese family. It was eventually permanently attached to the Sforza family.

Originally, the acronym ALFA and the word MILANO appeared on the badge separated with two square knots. The knots were associated with the di Savoia family; they and the blue encircling field symbolize royalty. The company was purchased by Niccola Romeo in 1915, and, in 1920, his name was appended to ALFA by hyphenation. It is only on the badge that the name has ever been hyphenated.

The Alfa badge.

The wreath around the badge commemorates Alfa's World Championship of 1925. During the early 1930s, a few Alfas were assembled in France and "PARIS" replaced "MILANO" on about two hundred badges. "MILANO" was deleted when the Alfasud factory went on-line in 1971 (while "PARIS" fit, "POMIGLIANO DEL'ARCO" would have been a bit much).

Fig. 1-2. *Merosi and Santoni in a "racing" version of a 1911 A.L.F.A. 24-HP. This style of non-bodywork is typical of the era and could be easily replaced with a sedan body.*

Fig. 1-3. *Another 24-HP car, this time with less sporting bodywork. This is in the Alfa museum in Arese.*

Fig. 1-4. *Original 1910 4-liter A.L.F.A. on display at the factory museum.*

cylinders were cast in pairs and external push-rods actuated the overhead valves. Though the valves worked absolutely vertically, the combustion chambers were hemispheric. This model remained in production until 1922 and was the basis for the fantastic "Ricotti Egg" constructed by Castagna.

In 1912, Peugeot had revolutionized the racing world with an engine of astoundingly modern specifications: twin overhead camshafts working four valves per cylinder in hemispheric combustion chambers.

Two years later, A.L.F.A. introduced its own version of this design in a 4-cylinder 4.5-liter version that produced 88 hp at 2950 rpm. After the war the car was developed to put out 102 hp at 3000 rpm. These speeds seem extremely tame now, but bear in mind that the contemporary metallurgy severely limited the speeds of even the most ambitious engines. Also note that, within four years, another Alfa engineer, Vittorio Jano, would introduce a racing engine capable of 5500 rpm and, in the next year, a passenger-car engine that developed its maximum power at 4200 rpm.

Fig. 1-5. *Six liters was not an enormous displacement for 1912, but it was enough to give this sporting A.L.F.A. 70 hp. Car was developed spanning the Great War until, in 1920, it produced 82 hp for a top speed of about 95 mph.*

Fig. 1-7. *Very typical touring bodywork is fitted to this 1920 20-30 ES Sport, a 4.2-liter 4-cylinder.*

In 1921, Alfa decided to enter the luxury car market with a Merosi-designed 6-cylinder 6.3-liter chassis called the G1. This car was intended to carry luxury bodies that would compete in the Isotta Fraschini, Rolls Royce, and Hispano Suiza league, but it proved uncommercial. Only 52 were produced and the sole surviving G1, now in Australia, is the oldest Alfa in private hands.

The RL Series

Until 1920, Alfa Romeo was a largely unremarkable firm producing a fairly conservative line of sporting cars in small volumes. In that year Niccola Romeo (Romeo had bought into the company in 1915, and in 1920 added his name to the A.L.F.A. badge) asked Merosi to begin work on an entirely modern car that would compete in the "3000 Formula" of 1921. Merosi's

Fig. 1-6. *G1 limousine was a monstrous Alfa with a 3.4-meter wheelbase and 6.3-liter engine.*

Fig. 1-8. *If you had invested in Romeo's company in 1918, this is what you would have received. Stock certificate is for "Engineer Nicola Romeo and Company."*

design of a 3-liter, 6-cylinder overhead-valve power-plant in a light chassis became his masterpiece and the largest-selling Alfa Romeo until the 1900 series of 1951.

In fact, the 3-liter Formula lasted only a single year, but Merosi's outstanding design became the basis of a varied series of very sporting vehicles, with racing successes that spawned sales successes.

These RL-series cars are (excepting the G1) the oldest Alfa Romeos still in private hands. For the modern owner accustomed to derivatives of the Giulietta, the RLs are surprisingly large vehicles and their pushrod engines are remarkably spare. These cars were created when the basic mechanics of the internal combustion engine were well understood and engineers were beginning to reach beyond utility to art. The aesthetic standard of the day was an engine as devoid of appurtenances as possible. The organic whole was most beautiful. As a result, the intake manifold on the RL is an

Fig. 1-9. *This RL-series Alfa Romeo toured U.S. as part of Carrozeria Italiana exhibit in the 1980s. Its body is engine-turned aluminum, and was owned by an Indian maharajah. This Merosi-designed car was really quite large, but its efficient engine gave it a top speed of about 80 mph.*

abbreviated pipe, while the twin carburetors of the RLSS bolt directly to the head, sans manifold.

Indeed, "elegant" is an appropriate description of these Merosi cars. They are also powerful for their years. Peter Hull drove an RLSS for everyday transportation for many years and reported to me that the only real drawback of the car was its appetite for camshaft lobes.

There was the RL at 56 hp, the RLS at 71 hp, and the RLSS at 83 hp. The RLSS-TF (Targa Florio) developed 95 hp at 3800 rpm and had a 3.6-liter engine with a seven-main bearing crankshaft (the other RLs had five-main cranks). The RLSS-TF was the first Alfa to carry the "Quadrifoglio," the four-leaf clover painted on a triangle by some unnamed pit crew member as an omen of good luck.

A 2-liter 4-cylinder version of the RL appeared as the RM, and a longer-wheelbase version was called the RMU. While much more could be said about these cars, their rarity and the greater importance of subsequent models requires that we press on.

Probably on the basis of the RL's immediate success, Alfa decided that participation in the Grand Prix would further enhance its sporting image. As a result, in 1922, Merosi designed a 2-liter car to compete in the 1923 season. It was not a disaster as a design, but an unfortunate accident with the car killed Alfa's premiere driver, Ugo Sivocci. Moreover, Fiat fielded a car of clearly superior design and won the 1923 championship. Merosi found himself suddenly in disgrace—and out of work.

For an Italian car designer, there are only two cities of employment: Turin and Milan, and fierce loyalties keep the rate of itinerancy low. Merosi must have stayed in Milan doing what little he could to earn a living. In Griffith Borgeson's book "The Alfa Romeo Tradition" we learn that a destitute Merosi returned to Alfa in the late 1930s to a menial job, an incredibly insulting turn of events.

The Cars of Vittorio Jano

Within days of Sivocci's death, Romeo raided Fiat for one of the young engineers involved with the Fiat Grand Prix car, Vittorio Jano. Jano was from Turin, and the offer of a quadrupled salary convinced him to move to Milan.

Grand Prix P2

Jano designed a completely new car for Alfa that first appeared in June 1924. The P2 won its first race convincingly and went on to win the championship for

Fig. 1-10. RLSS was a most sporting car. V-shaped radiator and twin badges are characteristic of series.

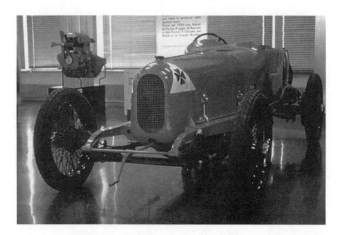

Fig. 1-11. A few made-for-racing RLs featured crankshafts with seven main bearings. These cars are referred to as Targa Florio, RLSS-TF, or just RLTF models. It was Ugo Sivocci's RLTF, legend has it, which first bore the Quadrifoglio.

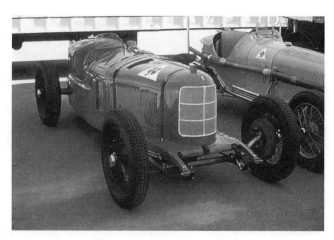

Fig. 1-12. Jano's first design for Alfa was this 2-liter P2 race car. Its two narrow seats permitted customary riding mechanic. Front end shows some appreciation of aerodynamics for a car capable of nearly 125 mph. Note friction shock absorbers, easily tightened by hand. In fact, the major suspension component on cars of this era was the chassis itself and not the very-rigid suspension. Not seen in this angle: asbestos wrapping on the high exhaust pipe to protect the driver's arm.

Fig. 1-13. *P2 engine. Installation in car is very tight, all but hiding supercharger. Updraft Mimini carb sits atop manifold that carries pressurized air from supercharger (bottom). Spring-loaded valve on carbs is pressure blow-off to prevent* *backfires from damaging supercharger. Note brace just in front of forward carb to reduce manifold flexing. Fork-shaped collector at front of block routes oil vapor from crankcase back to supercharger.*

Alfa in 1925. Jano's design remained competitive into the early 1930s, an astonishingly long life for a competition car. With the work of Vittorio Jano, Alfa became the most desirable sporting car in the world.

In 1927, a 6-cylinder Jano design was introduced for passenger cars, a design still reflected in the modern 4-cylinder Alfa engine: twin cams, hemispheric combustion chamber, in-line with main bearings flanking each rod throw. It is a classic design not unique to Alfa, since it is also found in virtually all high-performance engines, including the Miller engine used at Indianapolis, the Duesenberg, XK-series Jaguars, Maserati, and the most sporting Ferraris. Alfa, however, has developed this design over so long a period that if any marque "owns" the design by appropriation, it is Alfa Romeo.

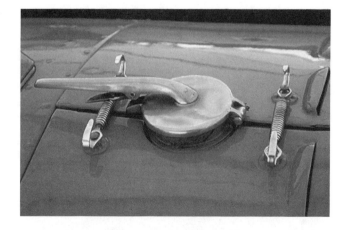

Fig. 1-14. *Exquisite detail: P2 oil-filler cap.*

Fig. 1-15. *Jano's 6-cylinder design with Zagato bodywork is one of the all-time, all-marque classic cars. This is a twin-cam supercharged model.*

6C1750

The original Jano passenger car design was a 1927–28 trio of 1.5-liter engines: single-cam, twin-cam and twin-cam supercharged. In 1929, the basic displacement was enlarged to 1752 cc. This larger engine, the 6C1750, is one of the great technical and aesthetic achievements of automotive design. It is an absolutely reliable engine with remarkable power output: roughly 1 hp per cubic inch for the supercharged unit.

I alluded to the spareness of Merosi's design. Jano's technique was more dramatic, and his finned intake manifolds and superchargers are stunning in their

Fig. 1-17. *The author's 6C1750 Gran Sport Zagato: 100 cu. in., 100 hp, 100 mph. Photograph was taken in 1961.*

Fig. 1-16. *Racing practice translated to a road car: the Gran Sport 1750 offers many features of P2 engine, including a su-* *percharger. Intake manifold finning is one of the most beautiful acts of craftsmanship found on an automobile.*

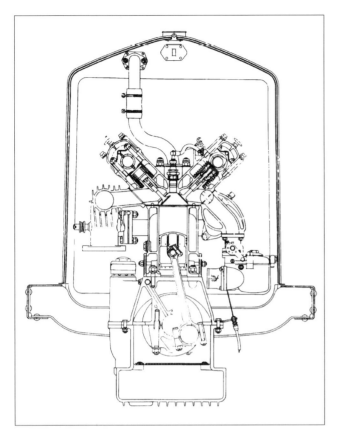

Fig. 1-18. Cross-section of Gran Turismo engine, un-super-charged. Note the long, slender connecting rod.

graceful beauty. Also beautiful were the various bodies lavished on the chassis by Farina, Touring, Castagna, Brianza, and Zagato.

8C2300

In 1932, Jano created an 8-cylinder version of this engine, the 8C2300. While he retained the same bore and stroke, he actually formed two 4-cylinder versions into a single straight-8. The crankshaft was bolted to-

Fig. 1-20. Accessory-drive side of Jano's 8-cylinder engine. On this "passenger" side are located water and oil pumps, as well as generator. Supercharger is driven from the same central gear set on the other side of the engine.

Fig. 1-19. Arcangeli and Bonini in 1931 8C2300 Mille Miglia entry. Riding mechanic had very little room, and appears most uncomfortable.

gether at its middle to two large gears (of opposite angle, to give offsetting end-thrusts) that served to drive all the accessories, including the supercharger. (The center four cylinders of the design made up one 4-cylinder engine. The two outboard cylinder pairs formed the other 4-cylinder engine, and the two pairs were displaced 90 degrees to obtain the maximum number of power strokes per revolution.)

Fig. 1-23. This is ex-Norm Miller/Keith Hellon Brianza-bodied 8C2300, a four-seat tourer.

This engine was used extensively in sports car racing, and finally appeared in "Monza" form with magneto ignition and abbreviated (if any) fenders. Several Monza Alfas were entrusted to Enzo Ferrari, who enlarged them to 2.6 liters. Ferrari was first retained as Alfa's racing manager and then his company, Scuderia Ferrari, actually took over the Alfa racing effort until 1937 when it returned in-house as "Alfa Corse."

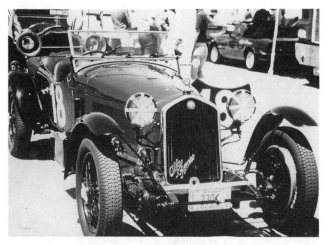

Fig. 1-21. An 8 cylinder at speed during 1985 Monterey vintage races. Note slotted "Monza" grille.

Fig. 1-22. Many unfamiliar controls greet modern motorist who tries this Brianza-bodied 8C2300. Large knob beneath dash center is shock-absorber control. Fuel mixture is controlled by lever on dash directly behind the steering wheel; ig- *nition timing is adjusted with lever on steering wheel. Not quite visible is accelerator pedal located between brake and clutch.*

Fig. 1-24. Prancing horse is a legitimate emblem on this 8-cylinder Alfa, for it is one the of cars taken over by Scuderia Ferrari when Alfa's racing effort was transferred to fledgling Ferrari organization. Ferrari enlarged displacement of some cars to 2.6 liters. Note tight fit for driver.

At this point, the story breaks into two, passenger cars and racing cars, for Alfa began committing a major portion of its resources to winning races. To keep the sudden proliferation of models as manageable as possible, I'll track the racing cars first and then come back to the passenger cars.

Alfa's Racing Effort

Several manufacturers at this time were attempting to put more horsepower into a chassis by fitting more than one engine. Both Maserati and Alfa fielded side-by-side twin-engine race cars. In Alfa's case, the Tipo A consisted of twin 6C1750 supercharged engines complete with two side-by-side transmissions. Two separate driveshafts were used to drive stub axles at the rear. While we can today immediately seize the idea that this layout begs to be the first independent-rear suspension on a race car, Alfa missed the opportunity

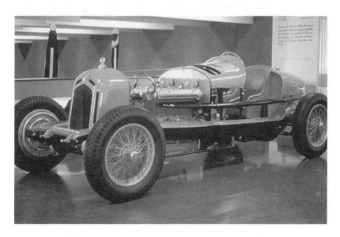

Fig. 1-25. Tipo A used two side-by-side 6C1750 engines to develop 230 hp. Two gear boxes were used, with shift levers connected by a sturdy link. Car won one race in 1931.

and put the two stub axles on a solid beam. Though the power developed by the Tipo A was not remarkable, its traction under acceleration was. The idea of a split driveshaft was retained in the next car, the Tipo B.

Instead of two 6-cylinder engines and associated transmissions, the Tipo B, the first single-seater Grand Prix automobile, utilized an enlarged version of Jano's 8-cylinder engine. It was equipped with twin superchargers and, to accommodate the two driveshafts, a differential was fitted to the tail of the transmission. The two driveshafts splayed rearward from it and drove the two stub axles on a common beam that had served so well in the Tipo A. The Tipo B had essentially the same success as the P2: that is, it was unbeatable.

Fig. 1-26. At first glance Tipo B looks very much like race-trimmed 8C2300. Its most significant exterior difference is its single seat.

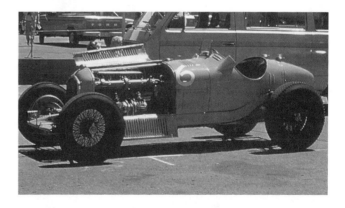

Fig. 1-27. Tipo B has intake and exhaust sides switched compared to 8C2300. Twin superchargers feed front and rear cylinder banks independently. These cars used superchargers to increase horsepower and gain greater output from poor-grade fuels, so proper fuel blending was an important element of a winning race team. Distinctive smell of old racing blends was a memorable feature of every race. Sandro Sirtori is remembered as Tipo B fuel blender.

Fig. 1-28. Cockpit of Tipo B reveals differential just behind gearbox.

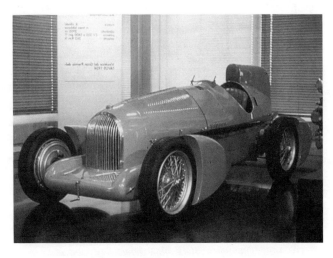

Fig. 1-29. A Tipo B variant: this car has extensive aerodynamic refinements, including a full belly pan. This style car won the Avus Grand Prix in 1934. The flowing lines make aerodynamic Tipo B a true work of art.

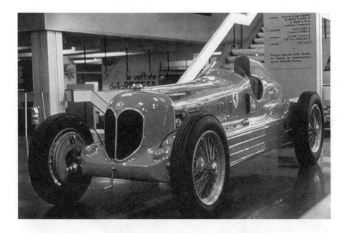

Fig. 1-30. Scuderia Ferrari built two twin-engine cars utilizing components from Tipo B. This "Bimotore" maintained a narrow profile by putting one 8-cylinder engine at front and another at rear. Almost 180 mph was possible, thanks to total of 540 hp from two 3.2-liter engines.

Keeping Up With the Germans

The Germans arrived at the racecourses beginning in 1934. From that point on, the story of Alfa's racing effort becomes muddled: unsuccessful attempts to catch the Mercedes and Auto Union cars, the loss of their top drivers to the Germans, and only occasional solace in lesser-class races that the Germans eschewed.

Fig. 1-32. *Dubonnet independent front suspension of Tipo B looks more like solid tubular axle. Suspension was added in an effort to improve its performance against Mercedes and Auto Union cars. Not much free wheel travel was gained—nor any advantage. Note hydraulic brake line, mandated by the independent suspension.*

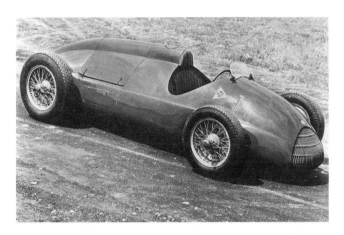

Fig. 1-31. *This model 512 flat-12 would have been a footnote in Alfa history had not Luigi Fusi saved it for factory museum. It is a 1939 experiment in aping Auto Union layout, but for voiturette races in which type 158 was proving successful.*

Essentially, Alfa's response to the German successes was to enlarge the P3 engine to 3.2 liters, then, as the Tipo C, to 3.8 liters. Other ineffective designs included a V-12 with both high- and low-chassis configurations

Fig. 1-33. *This 1937 12-cylinder Alfa is the reason Jano lost his job. A taller model preceded it in 1936, but neither high nor low versions could keep up with German cars.*

and a V-16. Other designs were discussed, including an Auto-Union inspired rear-engined flat-12 that was actually built as a prototype and is in the Alfa museum.

All these designs were exciting but feckless. Alfa could not match the money poured into the German cars. Nor could they match the Germans' attention to detail. In a larger sense, Alfa was stuck trying to upgrade an obsolescent design while the Germans forged into next-generation technology with independent suspensions and exceptional lightness. When the formula was weight-limited, Alfa simply upgraded its standard fare, while Mercedes struck with a monster engine from the proverbial clean sheet of paper.

The same worldwide depression that brought Hitler into power also caused Alfa Romeo to be absorbed into the Italian government under the Institute for Industrial Reconstruction (IRI). Why anyone thought a bureaucracy could win races is beyond belief, but then much of what happened during those years strains credulity anyway. Winning races was definitely on Mussolini's agenda, but his Fascist bureaucracy simply couldn't marshall the skills to match those of his northern "friend."

After several disastrous races in 1937, Jano—like Merosi—was fired and Enzo Ferrari was asked to develop a voiturette-class race car that might win some races. Alfa sent Colombo to help with the design, which for some time was regarded as one bank of the 16-cylinder engine being developed by Alfa. In fact, work on the smaller engine established the design of the larger, and not vice-versa. The voiturette model, the Type 159, would eventually regain back-to-back world championships for Alfa in 1950 and 1951.

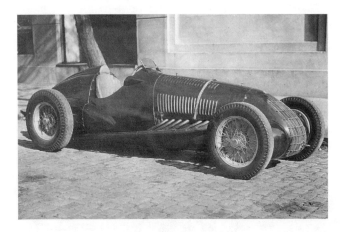

Fig. 1-34. Wrong-way engineering. Essentially an upgraded 8C2900B, with enlarged engine and longer track, 308 of 1937 was a track version of a road car. It was developed by Alfa Corse after Jano's firing, but proved uncompetitive.

Fig. 1-36. An absolutely memorable combination: Fangio in the 159 with which he won the 1951 World Championship. Photograph taken at Monterey in 1985.

Passenger Cars

Prewar Cars

To return to passenger-car production: With Jano's sophisticated 8-cylinder passenger car design of 1932, it was clear that Alfa was pricing itself out of the market. To correct the error Alfa decided to take its first step downscale. Jano was asked to design a cheapened version of his 6-cylinder engine for a new passenger car. The result appeared in 1934 as the 6C2300, an engine with similar displacement to the 8-cylinder car, but producing only 68–95 hp (depending on tune) compared to the 142 hp of the supercharged eight.

This line of 6-cylinder cars became Alfa's standard offering during the Fascist years. They were stylish, lively cars with twin cams and a 4-speed transmission. The series was revised in 1935 to include synchromesh

Fig. 1-35. By late 1930s, engineers understood very well how to package mechanical components inside constraints of an aerodynamic body. Note large and very graceful carburetor air intake that begins on the sill just in front of windshield and curves downward to 3-throat Weber. This is a postwar version of the Alfetta, a 1.5-liter straight-8 with two-stage supercharging. Maximum power output reached 450 hp.

Fig. 1-37. 6C2300 was much less sporting than its 8-cylinder cousin, but still provided a level of performance and style few other marques could match. Top speed neared 90 mph.

on the top two gears and independent rear suspension. In 1939, the engine was enlarged to 2.5 liters.

In 1937, a very few road-going Alfas carrying essentially a detuned version of the P3 engine were sold. These cars, the 8C2900A and 8C2900B, are the subject of Simon Moore's great book. The 8C2900 Alfa Romeo is clearly the most advanced automobile of the prewar

years and its design and performance foreshadow the greatest grand touring cars of the 1950s and 1960s. In retrospect, we can see that this model car was the greatest Alfa of them all. The possible exception is the Type 158-9, which dominated the racecourses of the postwar world until 1952. That car, however, couldn't get you down to the local grocery for a loaf of bread.

From the Italian standpoint, the years just before 1940 were good indeed, even if they weren't winning races. Great plans were being made for a reordered world, and Alfa was caught up in the mood by designing several futuristic, aerodynamic cars that would be scheduled for production through the 1940s. Models included the S 10, a 3.5-liter V-12, and the S 11, a V-8 sedan with four overhead camshafts and a body that owed much to the Chrysler Airflow. A race car, the Type 163, was manufactured to enter the races scheduled for 1941. Its supercharged 3-liter mid-engine produced 190 hp at 7400 rpm. A 2-liter postwar design, the Gazelle, was sketched by the Alfa designers who had moved from Portello to Lake Orta in 1943.

To jump ahead just a bit, an upscale follow-on to the Gazelle was sketched in 1950 as the 6C3000. This was a very aerodynamic car that borrowed much from the large greenhouse made famous by the Zagato "Pan-

Fig. 1-38. Absolutely the most desirable road-going Alfa, this Touring-bodied 8C2900 Corto combined beauty and brute force most elegantly. Bodywork plans for this car are illustrated in Fusi's second edition.

oramic" prototypes. Though a body was never built, one running 6C3000 engine was created. It was this engine that Colombo put into an all-new competition coupe, the 6C3000 C50, for Consalvo Sanesi to drive in the 1950 Mille Miglia. By all reports the car was a disaster.

With the departure of Jano in disgrace during the 1937 race season, Alfa used the skills of Ferrari, Gioacchino Colombo, and Wifredo Ricart to design race cars and Bruno Trevisan to develop passenger cars. Jano's co-worker, the draftsman Luigi Fusi, remained with the company helping to develop the great designs of Alfa's golden years. As noted above, it was Fusi's singular dedication that saved all the prewar records and much of the material of which modern Alfa histories such as this are concocted.

And then the war. Many of the prized Alfas were dismantled and hidden against the day when, win or lose, they would surely race again. The Portello factory was decimated near the end of the conflict, but enough material was saved to allow Alfa to re-enter production rather quickly after the war, with the 6C2500.

Postwar Revival

It seemed only days after the war that the first races were run. The observation is not a slight to the horrors of war so much as it is a testimony to the ineluctable resilience of the human spirit. Alfas came out of the gate ahead. For numerous reasons, the sight of an Italian car carried fewer negatives than a German one (remember, they tried to give Volkswagen away just after the war and no company, including Ford, would have it). In the span 1946–52, Alfa absolutely dominated the racecourses as it had in the years between 1925 and 1937.

6C2500

Largely as a result of its racing dominance, Alfa became an object of desire of the most wealthy and/or os-

Fig. 1-39. Alfa as art: the wonderfully integrated lines of this 6C2500 coupe was a highlight of the Carrozeria Italiana exhibit.

Fig. 1-40. Not fast, but elegantly beautiful, the 6C2500.

Fig. 1-41. Postwar 6C2500 was available with a 3-carburetor Super Sport engine. These cars were comfortable tourers, not sports cars, and even multiple carburetors failed to produce performance much over 100 mph. Racing versions of 6-cylinder cars existed since 1939, eventually inspiring Colombo's ill-performing 6C3000 C50 of 1950.

tentatious personages of the world, including the Prince Aly Kahn and screen star Rita Hayworth. Body builders competed to shower Alfa with their most dramatic innovations. It was the 6C2500 that inspired a whole generation of "shaved" American hot rods beginning in the late 1940s.

Most readers will have seen one of these 6-cylinder Alfas. They are surprisingly large, absolutely comfortable long-distance tourers. They are not notably quick to accelerate, but their aerodynamic style gives them a grace and top speed that was decades ahead of their time. They have incredibly light steering. Some were available with left-hand drive, but many are right-hand drive with a 4-speed column shift and a dash that can only be likened to a jukebox. Bench seats were all the vogue in the early 1950s, so it is very rare to find a passenger 6C2500 that did not have a virtual slab to sit on.

1900

As the postwar world economy began to sort itself out and median European incomes increased, Alfa management decided that it should build a car that would appeal to an even broader range of buyers. This was, I believe, the single most pivotal decision made in the history of the company. Alfa had always been a technology-driven company. With no sales department, it depended entirely on the excellence of its product to attract buyers.

The decision to enter mass production would eventually force Alfa to court buyers using the same techniques as Ford and General Motors. That Ford and GM were better at the game should not come as a surprise; neither should the fact that an Italian bureaucracy would be slow to learn essential lessons of business.

The design selected to enter the mass market was certainly in keeping with Alfa's high levels of engineering. While BMW was struggling to stay afloat producing egg-shaped micro-cars under license from Isetta, In 1950 Alfa introduced the 1900, a sedan unlike anything ever before presented to the public. The appearance of the car was unremarkable, but its power and agility were unmatched by any other car. It was the first true sport sedan. Power enough for very nearly 100 mph came from a 2-liter engine with classic Alfa elements:

Fig. 1-43. The most popular coupe built on the 1900 floorpan was by Touring. This early version was built specifically for racing and has participated in modern Mille Miglia.

twin overhead camshafts and hemispheric combustion chambers. Initially, a 4-speed column-mounted shifter was available; later, a 5-speed fully synchronized unit was optional. The 1900 utilized unit frame construction in that external body panels were stressed and there was no separate frame. The design proved to be extremely durable and served even as taxis—and police cars—for at least a decade.

Fig. 1-42. A benchmark of automobile engineering, this 1900 sedan was Alfa's first car designed for mass production. It was advertised as the "family car that wins races." With a level of performance and reliability few cars have been able to match, the 1900 invented the sport sedan category.

Though conceived as a sedan, the 1900 was rebodied by Touring, Farina, and Zagato into very sporting coupes. Bertone created a trio of aerodynamic bodies on the 1900 that explored the use of tail fins.

Alfa created a follow-on to the 6C3000 C50, the completely redesigned 6C3000 CM engine, which looked like a 6-cylinder version of the 1900, and installed the powerplant in a space-frame chassis. Experimental bodywork by Touring earned the car the name "Flying

Saucer." The six-cylinder 6C3000 CM displaced 3.5 liters and was quite successful. The Pininfarina "Superflow" Spider on a 6C3000 CM chassis in fact became the basis for the styling of the Giulia Duetto.

The 1900 engine was updated in 1958 to the 2-liter that was dressed in thoroughly modern bodywork. With the 2-liter, Alfa began producing cars in trios of Sedan, Coupe, and Spider. The process continued when

Fig. 1-44. Disco Volante borrowed 1900 engine design though not actual engine parts. Built exclusively for racing, cars nevertheless carried some very trendsetting sheet metal. This is original bodywork that gave Flying Saucer its name. Later versions would range to 3.5 liters, with significantly different sheet metal, but Flying Saucer nickname remained.

Fig. 1-46. Two-liter sedan offered crisp lines and a luxurious interior. Later 2600 sedan differs externally in minor details, including a smooth horizontal line where hood breaks downward toward grille.

Fig. 1-45. Don and Margaret Black about to set off in museum's 6C3000 CM Disco Volante, at Meadowbrook in 1992. This is the most successful body style of Disco series. Next to

it is a 6C1750 Gran Sport, followed by a Type 33 and racing-modified Giulietta.

the 2-liter was replaced in 1960 by the 2600 6-cylinder engine. The bodies for these large 4- and 6-cylinder cars are virtually identical.

Until the 164, the 2600 series was the last of the large Alfa Romeo production cars for the U.S. designed for luxury long-distance touring in the same vein as the RL, the 8C2300, 8C2900B, and 6C2500. The appearance of the Type 164 car in 1990 may herald Alfa's successful return to this arena.

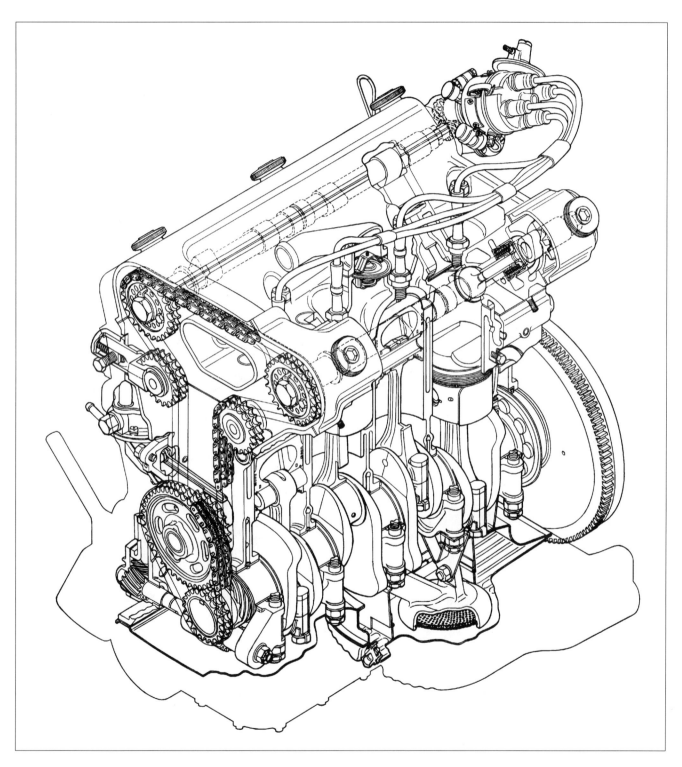

Fig. 1-47. *1900 and 2-liter shared very similar engines (their head gaskets interchange). This 2-liter engine has valve adjusting shims like Giulietta.*

Fig. 1-48. Bertone coupe version of 2-liter foreshadowed Giulia Sprint GT by a year or so. It is much larger and less agile than smaller car—but also more comfortable.

Fig. 1-50. Osi built a few luxury sedans on 2600 floorpan. This design is a bit more understated than the factory production sedan.

Fig. 1-49. 2600 coupe is differentiated from 2-liter model by its hood scoop. A true grand touring car, it is capable of covering long distances reliably without tiring driver or passenger.

Fig. 1-51. Intended for limited production, this Zagato-bodied 2600 prototype features hood scoops and recessed headlamps that (mercifully) did not make it to assembly line. A very limited number of these desirable cars was produced.

Satta's Giulietta Dynasty

The 1900 was such a remarkable sales success that Alfa ventured another step downscale in 1954 with the Giulietta designed by Orazio Satta. This car was essentially a miniaturized 2-liter with a slightly more interesting 1300 cc engine. The engine featured an aluminum block with slip-in steel cylinders and its aluminum head had pressed-in valve guides and seats.

This is where most of us Alfa enthusiasts came in. Suddenly, one of the giants of the automobile world was producing a sports car the average enthusiast could hope to own. It was small, relatively powerful, and very comfortable. Furthermore, it had a twin-cam engine when its competition (mostly British) was still using pushrods. Just lifting the hood on an Alfa was explanation enough for ownership.

It's quite apparent that Alfa was completely unequipped to capitalize on the Giulietta's popularity. The Giulietta sedan, coupe, and Spider were all jewel-like and perfectly covetable. The suspension allowed a

bit more body roll than was expected for the time, but the roadability of the car and its charming nature made it a clear favorite of the automotive press. The Giulietta reconfirmed Alfa's preeminence as a manufacturer of sporting cars, and while its image was not so imposing as the 1750 or 2900 cars, it was loved every bit as much.

Since many Giuliettas are still available, a bit more technical detail is appropriate. The original series of the Giulietta was given serial numbers beginning with 750 (in recognition of the fact that the first prototype engine displaced 750 cc). The choice of serial number was meaningless until 1959, when some strange-looking Giuliettas appeared (unannounced; Alfa still didn't have any marketing sense) with detail differences in the engine and body, and a completely new transmission; the mechanical parts carried numbers beginning 101. It became clear only slowly that Alfa had done a subtle redesign of the entire car and, as old parts were exhausted from the warehouse, the new pieces began appearing.

Fig. 1-52. *One of the all-time great designs and the owner of millions of hearts, Giulietta Sprint coupe proved to be an almost impossible act to follow. Flat-bottomed taillights show that this is a 101-series version.*

In 1959, the mish-mash of 750 and 101 parts was confounding; 750 engines appeared in 101 bodies with either 750 or 101 gearboxes.

The most obvious change to the new series was that the 101 Spider body had a longer wheelbase and was able to carry a small fixed window on the door. More significantly, the camshaft and crankshaft diameters were enlarged to permit (eventual) larger displacements. The mechanical fuel pump of the 750 engine had been located on the head while the 101 located it on the block near the distributor. The 750 gearbox was a cylindrical aluminum casting and gears were assembled and then inserted through its end; on the 101, the gearbox was split so the main and layshafts could be lifted out individually. Further, the 101 gearbox had a collapsing shift lever to lock out reverse gear.

(Owners of the 750 series cars should avoid fitting 101-series replacement units. An unmodified 750-series Giulietta is already a solid financial investment and the remaining cars shouldn't be modified if at all possible.)

Soon after the introduction of the Giulietta, Alfa released a hot-rodded version with extensive modifications which included special pistons, cams, and dual Weber carburetors. The designation of this series was initially "Veloce," which means fast in Italian. Later on the term Veloce was dropped in favor of "Super" for the U.S. market, but the Veloce designation has stuck, and 101 Spiders with dual Webers are no longer called Super.

It was popular for many years to hang Webers from later Alfas on the Giuliettas to try to make them into Veloces. A proper 750 Veloce has a separate cold-air intake in the front grille on the driver's side, a tachometer that reads to 8000 rpm, and a 4:10 rear axle ratio. Also, the original Veloce carburetors were 40DCO3; the later 101 cars used 40DCOE carburetors. If there is a plaque on the firewall indicating body style, 750B indicates a Sprint coupe, 750C is a sedan, 750D is a Spider, 750E is a Sprint Veloce and 750F is a Spider Veloce.

Today, we use the term "Normale" to refer to Giuliettas with the 35APAIG Solex carburetor. This is a term of convenience; it was never a contemporary designation.

Fig. 1-53. *The author's collection of two generations of sedans and coupes. At front are Giulietta/Giulia series cars (perhaps the best pair of postwar Alfas offered). Just behind (third) is a 1750 coupe, then a 2-liter sedan. While there were minor structural differences between Giulia and 2-liter sedans, character was completely different—in favor of Giulia.*

The Giulietta was so successful that it redefined Alfa Romeo's image for the generation that grew into maturity in the 1960s. For these people, myself included, Alfa was a jewel-like and superbly engineered small car that ranked below Ferrari, Maserati, and even Lancia on the scale of desirable Italian cars, but it made up for all that because it was affordable. In the early 1960s, most Alfa owners were ignorant of Alfa's history, and so considered it only in terms of the Giulietta, as a maker of small, mid-range sport cars.

As engaging as it is, the Giulietta runs down the road at uncomfortably high engine speeds if one is accustomed to big American V-8s, and it has a serious problem starting in deep cold if everything is not in exact tune and the battery is not fully charged. Beyond those annoyances, it is one of the most delightful cars you're ever likely to own.

Fig. 1-54. Variation of Bertone-bodied Giulietta coupe was this Bertone-bodied Sprint Speciale. Noted for its aerodynamics and effortless high-speed cruising abilities, this car is now one of the most collectible Alfas of postwar era.

Post-War Alfa Passenger Cars (seen in U.S.)

6C2500 (1946–1953)

1900
Berlina (1950–1954)
T.I./T.I. Super (1952–1955)
Sprint/Super Sprint (1951–1958)
Cabriolet (1952)

2000 (102)
Berlina (1958–1962)
Sprint (1960–1962)
Spider (1958–1961)

Giulietta (750)
Berlina (1955–1960)
Sprint/Sprint Veloce (1954–1959)
Spider/Spider Veloce (1955–1959)

Giulietta (101)
Sprint (1959–1962)
Sprint Veloce (1959–1962)
Sprint 1300 (1963–1964)
Spider (1959–1962)
Spider Veloce (1959–1962)

Giulietta (specials)
Sprint Speciale (1957–1962)
Sprint Zagato (1959–1961)

2600 (106)
Berlina (1962–1968)
Sprint (1962–1966)
Spider (1962–1965)
2600 SZ (1965–1967)

Giulia (101)
Sprint (1962–1964)
Spider (1962–1965)
Spider Veloce (1964–1965)

Giulia (105)
T.I. (1962–1967)
Super (1965)
Sprint GT (1963–1966)
GTV (1965–1968)

Duetto (1966–1967)

Giulia (specials)
Sprint Speciale (1963–1965)
TZ (1963–1967)
GTA (1965–1967)
GTC (1964–1966)
4R Zagato (1965–1967)
T.I. Super (1963–1964)
GTA 1300 Junior (1968–1972)
Junior Z 1300 (1970–1972)
Junior Z 1600 (1972–1975)
TZ 2 (1965–1967)

1750 (105)
Berlina (1967–1972)
GTV (1967–1972)
Spider (1967–1972)

2000 (105/115)
Berlina (1971–1974)
GTV (1971–1974)
Spider Veloce (1971–present)

Alfetta (116)
Sedan (1972–1978)
Sport Sedan (1979–1985)
GTV (1975–1980)

GTV6 (1981–1986)

Special
Montreal (1971)

Milano (1987–1989)

From Giulia to 1750 to Alfetta

Though the Giulietta was a hard act to follow, Alfa trumped its own ace with the Giulia, which appeared in the U.S. in 1965. In retrospect, we know now that it was the planned Giulia engine that prompted the change of dimensions in the 101-series Giulietta. The Giulia added about 300 cc displacement to the Giulietta engine without having to increase bearing journal diameters or any other significant dimensions except block height. The extra displacement added both power and reliability and the Giulia must rank as one of the most reliable Alfas ever produced.

The Giulia engine first appeared in bodies that were generally indistinguishable from Giuliettas. Within a couple of years, however, new 105-series body styles were introduced, to the despair of Alfa enthusiasts. The new Bertone Sprint body for the Giulia was a notchback that looked quite awkward in comparison to the Giulietta. The new Spider was called the Duetto, but could not be given that name officially because it would

Fig. 1-57. Giulia Super was one of the most durable cars ever—of any marque. It spawned more variations over a longer period of time than any other model Alfa. Engines included both gasoline and diesel.

Fig. 1-55. While most recognize the notch-backed Giulia coupe, few have seen its convertible counterpart, the GTC.

Fig. 1-56. Duetto was a movie star as Dustin Hoffman's wheels in The Graduate. Its powertrain was better received than its styling.

have been a copyright infringement on Volvo's PV445 "Duett." The Duetto was derided for a rear end that didn't seem to know how to end. Now, both cars are cherished classics, but were an exceedingly hard initial sell. The new sedan was bearable and was, of the three new body styles, the most enthusiastically received.

Alfa virtually owned the sport sedan category in 1967. There was nothing that could touch the Giulia Super sedan for speed, comfort, or reliability. But through a series of uninspired designs and the oil crisis, Alfa stood stock-still in the market while other manufacturers more skilled in marketing, most notably BMW, moved ahead. In the U.S., the Giulia replacement, the 1750, simply never caught on, in part because of its new SPICA fuel injection and the lack of service information about it, but also because the design lacked the charm of the earlier Alfas. The 2000 sedan, which appeared in 1971 as a 1972 model, regained some of the Giulia's charm, in part because it had a more sporting dash.

In 1967, Alfa had displayed a dream car at the Montreal World's Fair. It was such a hit that, almost five years later, Alfa released a limited production run of the 1750-based car. The Montreal had a V-8 engine and two 4-cylinder SPICA fuel injection pumps bolted end-to-end. The rear axle was essentially a stock Giulia unit with a large oil sump appended to keep things cool. This car recalled the 2600 in attempting to be a luxurious cross-country runner. It was both beautiful and capable, but perhaps four years too late, a detail of deadly significance then but of added attraction today.

The 2-liter Spider introduced in 1971 was a "Kamm tail," which means it lost the Duetto's rear taper. The Spider has become a modern Morgan: few changes, but none needed anyway. It is a classic and very desirable vehicle regardless of the year. In the U.S., the Spider engine went through many changes, with the 1750 engine with SPICA fuel injection in 1969, a 2-liter engine in 1971, a variable intake camshaft in 1981, and Bosch EFI

Fig. 1-58. Alfa V-8: the Montreal. A hit as a show car at the 1967 World Fair, it entered production in 1971.

Fig. 1-59. Alfa's flagship model, after the 90, was the Alfetta. First series had hoods that hinged at front, and protruding door handles. Second series, here in European trim, was more luxurious, with front-opening hood and flush handles. This model was sold in U.S. as the Sport Sedan.

Fig. 1-60. Alfa won't let a great name die. Giulia continued in production long after it ceased coming to America. One of the last, this 1976 diesel, shared lineup with Alfasud and Alfetta.

Fig. 1-61. Alfa made many forgettable show cars and only a few memorable ones. This is an Alfetta-based version of Fiat X1-9 from Pininfarina, shown at 1975 Paris Salon as proposed replacement for venerable Spider.

in 1982. A single-throttle mechanical fuel injection system was offered in 1980, although some of those cars were finally sold in 1981. The interior has become more and more lavish, but the essential grace of the car has not diminished over its (very long) life.

A major driveline redesign (recalling the Type 158/159 race car) appeared in the Alfetta coupe and sedan in 1975. The transmission was moved from just behind the engine to just in front of the differential. This chassis, typically referred to as the 116 (following the 750/101/105 logic of part numbering) offered refined weight distribution and a de Dion rear suspension. Technically, the de Dion rear is still a solid rear axle, but it is a very light design that makes the transmission and differential sprung weight. The Alfetta driveline was an expensive design and the driveshaft, with several large rubber joints, proved a special chore to maintain. The handling of the chassis was superb and proved that, after years of showing everyone else how well a solid rear axle could handle, Alfa had retained its mastery of the rear suspension for passenger cars.

Alfasud

The attempt to become a mass producer of automobiles caused a large proliferation of Alfa models for Europe and elsewhere in the 1970s. A plant was opened near Naples to produce the Alfasud, a flat-opposed 1.3-liter sedan that eventually appeared as a sporty coupe.

Because of governmental control, Alfa had to fill certain social functions that are not normally a part of business. If one draws a line across Italy at Rome, all the country to the south is referred to collectively as the Mezzogiorno, a relatively impoverished region that the

Fig. 1-62. *When first introduced, Alfetta coupe received same negative reception as all new Alfa body styles. And, like all new styles, it slowly became appreciated for its undoubted* virtues. *It is Alfa's most refined coupe, here seen in GTV6 form.*

government wanted to improve. A number of social programs were instituted for the Mezzogiorno, and Alfa was tapped to make a significant contribution to its economy. During World War II, Alfa had constructed an aircraft engine plant at Pomigliano d'Arco, just outside Naples. The plant was thoroughly destroyed toward the end of the war but was rebuilt during the reconstruction. In the late 1970s it seemed a great idea to bring back another Pomigliano d'Arco plant to produce a new line of inexpensive Alfa Romeos. The near-disaster of trying to industrialize an essentially agrarian work force needs a whole book of its own. It's enough to say here that the new Alfasud was beleaguered with labor and quality problems.

The 'Sud was planned for introduction in the U.S. (the federalized car now belongs to Tom Zat), but it was never brought in. In other parts of the world, however, the 'Sud became a favorite Alfa. The coupe version looked very much like a little brother to the Alfetta coupe. The pairing has a historic parallel: the 'Sud was to the Alfetta as the Giulietta was to the 2000.

In retrospect, Alfa should have brought the 'Sud to the U.S. It would have proved that a car can have front-drive and excellent handling and still be a genuine Alfa. Instead, the point had to be made with the 164 at just the time that Alfa ownership changed hands. There are many people in the U.S. who feel that the 164 just isn't a real Alfa. The 'Sud would have relieved, if not eliminated, the scepticism.

Various body styles were tried around smaller displacements of the 4-cylinder twin-cam engine and Alfa is one of the few manufacturers to reuse a name: the New Giulietta and the New Giulia were both attempts to capitalize on past glories. But then, Alfa even marketed its own replicar, the Quattroruote Gran Sport, a Giulia commissioned by Quattroruote magazine and built by Zagato to look (distantly) similar to the 1929 Alfa Gran Sport.

The Modern Era

In 1981, Alfa introduced to the U.S. market its first new sporting powerplant since the 1956 Giulietta. This Surace-designed 2.5-liter V-6 engine was actually designed in 1971, but the project was shelved for lack of funds for tooling. It retained many Alfa characteristics, including replaceable steel cylinders in an aluminum block, but it introduced the first pushrods in an Alfa engine since 1926. The V-6 infused new life into the current coupe body style for a few years but the coupe was then quietly retired from the U.S. market. The Milano (known as the 75 in Europe) appeared in the U.S. in 1987 with the V-6 engine. Fast and agile, the controversial styling of the rear kick-up probably did more harm than good for sales. The Milano was determined to be too expensive to produce by new Fiat management and was killed in 1990, when the front-wheel drive 164 sedan appeared in American trim.

Fig. 1-63. *A study in ergonomic mis-design, Milano still appeals as an essentially solid, if quirky car. Leaking head gaskets and power steering pumps marred car's early years. Rear* *ends of other manufacturers' subsequent designs (notably VW Jetta) minimized visual uniqueness of tall trunk and made Milano seem one of the crowd.*

Alfa's history in the late 1980s was not centered on its product line-up, however. The most significant events of the last several years concern control of the company. Years of bureaucratic wrangling and 13 years operating in the red made Alfa an albatross to the Italian government that owned it, and inspired Ford to offer to buy the company. It was widely reported in the press that BMW was initially approached as a possible purchaser. Some of Alfa's prototype design and testing facilities were of interest, but the bulk of Alfa's old facilities weren't. BMW refused, and, for a while, it appeared that Alfa would be a Ford subsidiary.

The prospect that Ford would produce cars in Italy terrified Fiat management, but pleased Alfa. In fact, very little of Alfa was of use to Fiat. Much of Alfa's production capacity was old and the Alfa model line-up significantly overlapped Fiat's. Fiat used all its considerable political clout not so much to obtain Alfa as to keep a competitor out of its own backyard. The prospect of Fiat gobbling up Alfa just to keep Ford at bay terrified Alfisti, for it seemed that Fiat had little enthusiasm for Alfa as a marque. Further, while Fiat is a worldwide giant with undoubted marketing and manufacturing skills, its track record in the U.S. has been dismal.

There was also terror within the Alfa organization. When it took over Lancia, Fiat slashed the workforce and it was expected that Fiat would do the same to Alfa. Thousands of Alfa employees had continued to receive paychecks while homebound when there was no work for them to do. All these people were clearly vulnerable under Fiat management. The Italian union preferred Ford in the hope they might be more benign.

There's an old saying about not being able to beat the locals at their own game: Ford lost, Fiat won. In the Fiat organization, Alfa and Lancia share manufacturing resources and their product lines share components.

Alfa clearly lost identity under Fiat. Fiat's initial actions were to approve a redesigned 164 and number the Milano's days, both actions intending to put Alfa on a profit-making path. As a result of these two actions, Alfa has a much more attractive and profitable new car in place of a car that was not profitable to produce nor especially pleasant to look at.

Under Fiat, Alfa hoped to gain the marketing acumen it so badly needed. This hope also probably explains the agreement between Fiat and Chrysler. But it seemed Chrysler failed to understand what it was marketing, and the liaison failed to weather the hard times of 1991. Chrysler backed out of ARDONA at the end of two years, leaving Alfa completely in the hands of Fiat management.

In retrospect, the two most significant events to have shaped Alfa's current condition were its incorporation into IRI in 1931 and the decision to mass produce the 1900-series cars in 1950. The wisdom of both decisions will probably be debated well into the 21st century.

Chapter 2

Your Alfa—A Guided Tour

V ERY FEW OWNERS are ever properly introduced to a car. If you buy one new, the salesperson is supposed to handle the introduction. However, since his only goal in life is to sell you something, he tends to feel that anything that happens after you sign the sales contract is an avoidable diversion from his next sale. If you buy a used Alfa, the previous owner may know nothing about the car, or worse, be filled with inaccurate information which he conveys to you as gospel.

Owner's manuals typically provide only the nitty-gritty details, such as where to find the spare tire and fuses. Clearly, you bought your Alfa to do more than discover how to rotate its tires.

May I introduce you to Alfa?

1. In General

The single most important feature about an Alfa, I'm convinced, is its long and brilliant history. This won't help you humble Corvettes from a stop light, but it will make you feel better about losing a stupid drag race.

Alfa's pedigree goes back almost 90 years. If you don't think such an intangible can be important, con-

Fig. 2-1. *Two-liter coupe engine bay illustrates what you'll see if you open most U.S. 4-cylinder Alfa hoods. Starting at lower right corner and working toward firewall: oil vapor separator, air cleaner, and ignition coil. Behind air cleaner you can just make out intake manifold, as well as parts of the SPICA* *fuel injection system. On far side, beginning with firewall and working forward: dual brake cylinder with reservoir, clutch reservoir behind it, voltage cutout relay, radiator coolant overflow bottle, horn relay, and battery.*

Fig. 2-2. *A Weber-carbureted 2-liter engine is nonstandard for U.S., but many SPICA injection systems have been converted to carburetors. Webers are bolted high up on engine on passenger's side (top of photo). Shiny U-shaped casting on top of engine covers two camshafts and their connecting chain. On driver's side is exhaust system manifold.*

sider the fact that it was the central message for the 164's introduction in the U.S. People who buy cars such as Alfa are looking for more than transportation. Indeed, they tend to be quite emotional about their cars. Owning a car with an honored past is important, and Alfa has one of the finest pedigrees you can buy.

Next, new Alfas are always on the cutting edge of styling. The great Italian body builders routinely choose Alfa chassis to showcase the avant-garde. Because they are the cutting edge, new Alfas frequently look strange to us. Again and again, our initial reactions to these strange new designs have mellowed and they become acknowledged classics.

Alfas are also on the cutting edge of technology. The superiority of the older Alfas over their contemporaries was dramatic: Alfa was the twin-cam, cast-aluminum standout in a world of pushrod-operated, cast-iron engines. The advantage is no longer so clearly defined, but evidence of it persists. Of four sedans (Saab 9000,

Fig. 2-3. *A U.S. Milano V-6 opens wide. This engine visually is quite different from typical Alfa in-line four. Further, Milano has an electronically controlled fuel deliver system, so there are no carburetors. Large shiny stamping at far left is air filter. Attached to it is a black-capped aluminum casting, part of electronic fuel injection system, which measures air volume flowing into engine. Ribbed, black L-shaped hose carries intake air to finned aluminum plenum located dead-center on top of engine. This car has self-leveling suspension and anti-lock brakes, so several items under hood will mystify many Alfa owners with their newness.*

Fig. 2-4. *A detail of the V-6 engine bay. At front of engine is thermostatic housing that carries three (two are visible) temperature sensors. Two have typical Bosch rectangular snap connectors for wires attached to them. Along right side of engine from top: square brake fluid reservoir, top of cruise control unit with a long control cable leading to air intake plenum, coolant overflow bottle, dome-shaped reservoir for anti-lock brake system, ignition coil, and battery. Camshaft cover has a script Alfa Romeo logo cast into it. Three black circles along bottom of script are grommets that seal spark plugs from water and dirt.*

Lancia Thema, Fiat Chroma, and Alfa 164) built on a common floorpan, Alfa is acclaimed the best.

Because of its technical excellence, Alfa is a better car than most people are drivers—that is an enormously important fact. Alfas are designed to be driven at 10/10ths without breaking or misbehaving. Moreover, if you should overcook it and lose control, an Alfa won't bite you back by suddenly changing its handling characteristics.

All Alfas have character. Some of it comes from the wonderful sounds of a free-breathing engine. Some comes from intelligently placed controls, carefully selected gearing, and finely tuned suspensions. There is a special friendliness in all of this. If you must go fast and far, Alfa will get you there feeling more refreshed—if not exhilarated—than most other cars on the road.

An honored past, pacesetting style and technology, wrapped up in a car with an endearing personality. Why would anyone not want to buy an Alfa?

Because, I need to caution, Alfa can be a demanding car to own. Its superior capabilities are not a license to explore the limits of wanton destruction. Like any athlete, an Alfa must be exercised knowledgeably and maintained attentively.

2. Engine

2.1 4-Cylinder

Most Alfa 4-cylinder engines are a sandwich of four aluminum castings: the camshaft cover, head, block, and oil pan. (Early Giulietta engines had stamped steel oil pans.) Each casting is sealed to its mating casting with a gasket. Aluminum has the advantage of good heat dissipation and light weight, so Alfas have an excellent power-to-weight ratio. Further, because all the wearing surfaces of the Alfa engine are replaceable, the engine is infinitely rebuildable to new specifications.

Camshaft Cover

The camshaft cover serves as little more than a cap for the head. It attaches directly to the head with six large, decorative nuts. Two bolts at its forward edge were added early in the design to help reduce the tendency to leak oil which is thrown at high speed from the camshaft chain.

The cam cover is one of the most decorative parts of the engine and is frequently polished to a mirror-like finish which, though not "stock," is very attractive. The U shape of the cover announces, at a glance, that there is a twin-cam engine underneath.

Fig. 2-5. *A very stock 750 Giulietta Veloce engine compartment, with polished camshaft cover. On left fender wall, front to back, fuel filter and Weber air box. Air filter occupies most of firewall, but to its right is brake fluid reservoir. Along right fender, in front of intake hose, Marelli voltage regulator and horn relay. Note in-line thermostat in the upper radiator hose: it's identified by two clamps in middle of hose.*

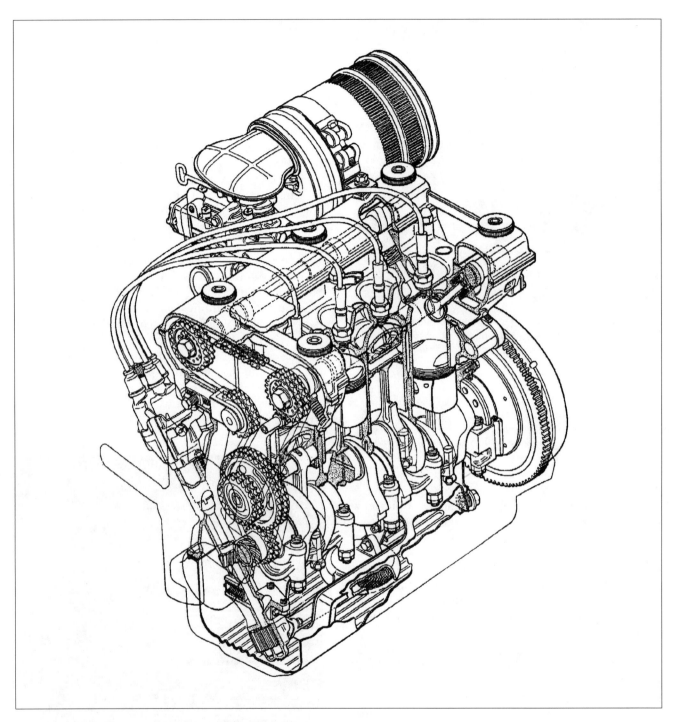

Fig. 2-6. *Basics haven't changed since 1955, when this 750-series Giulietta engine first saw production. Only relatively minor details distinguish basic engine from current model. Note that this drawing shows stamped steel sump.*

Head

The head carries the camshafts and spark plugs as well as the intake and exhaust manifolds. The basic purpose of the head is to control the flow of gasses through the engine. The combustible intake gasses are a mixture of air and fuel supplied by the carburetor or fuel injection system. The spark plug ignites the intake gasses in a controlled explosion which is the engine's source of energy. The exhaust gasses are the hot residue of the combustion process.

There's a classic exercise in self-identity that begins: "If I cut off one of your legs, would you still be you?" The somewhat grisly process continues until some crit-

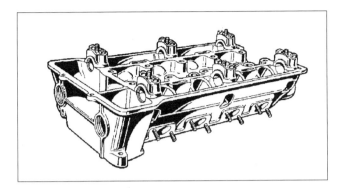

Fig. 2-7. *Bare Alfa twin-cam head. Camshaft bearing surfaces are easily visible.*

Fig. 2-9. *A Veloce owner's life is not easy. Compare cramped confines of this 101-series Giulietta Veloce with its Solex-carbureted counterpart. Relocated air filter has displaced shiny Bosch voltage regulator to firewall, seen just above cast aluminum air intake horn.*

Fig. 2-8. *101 Giulietta Spider engine compartment has more room than Veloce. Ignition coil is only item attached to left fender wall. At firewall, brake fluid reservoir sits just above steering column. On right fender wall, back-to-front: horn relay, Lucas voltage regulator, and headlamp relay.*

Hemispheric Combustion Chamber

You make a hemisphere by lopping a sphere in half, ending up with a cavity which resembles a salad bowl.

The primary modern advantage of the hemispheric combustion chamber is that you can fit larger valves. Initially, the design was popular because the gas flow follows a virtually straight line. The absence of turns in the intake and exhaust path eliminates flow restrictions that rob the engine of its breathing efficiency. In order to get the straight-through flow, Alfa splays the intake and exhaust valves at an included angle of about 80°. This arrangement also allows you to put the spark plug dead center in the hemisphere. But the centrally located plug makes it hard to operate such widely splayed valves from a single camshaft. Thus, twin overhead camshafts are only a way of getting the optimum combustion chamber design; the chamber is not a side benefit of the cam design.

There are additional advantages. The hemispheric combustion chamber puts the spark plug virtually dead center in the bore and promotes smooth burning of the fuel. Any tendency to knock (pre-ignition) is reduced. Also, since there are two camshafts, there's the opportunity to adjust the relationship between the intake and exhaust cams for maximum performance. And, with Alfa's direct-acting cam layout, there is very little inertia in the valve train. Since valve float (from excessive inertia) is the main limiting factor in engine speed, Alfa's twin-cam design allows much higher en-

ical point is reached where lopping off that part (the head, for instance) no longer leaves "you." The analog is that a lot of parts of the Alfa can be removed without losing its Alfa identity. But, take away the twin-cam head and you've probably removed the item that, more than anything else, makes an Alfa an Alfa (V-6 owners must suffer the fact).

It makes sense. Virtually every Alfa built between 1926 and 1981 is powered by an engine with twin camshafts. While nifty, twin cams are only a side benefit of the basic design. The real goal of Alfa's design was to obtain hemispheric combustion chambers.

gine speeds and therefore higher output for a given displacement.

When the single-cam Alfasud four and V-6 engines appeared with only one camshaft per bank, the immediate question was: "Is that really an Alfa?" If one focuses on the combustion chamber design and not the way of actuating the cams, then the answer is an easy "yes" because both have hemispheric chambers.

In theory, the twin-cam layout is an almost perfect design for the internal combustion engine. In the 1920s and '30s, when fuel was of very poor quality, the advantages of a centrally placed spark plug and straight-through gas flow were undeniable. Modern engines live on much better fuel and the need to control emissions has made the hemispheric chamber less desirable than a cleaner-burning wedge-shaped chamber. Though it is now obsolescent, the fact that the twin-cam, hemispheric chamber design has survived so long speaks volumes for its basic excellence.

Excepting the flat-4 and the new V-6, only minor changes to the basic Alfa twin-cam specification have occurred since 1926. Valve adjustment was simplified from threaded caps to shims in 1954, although early 'Sud engines reintroduced threaded adjustment. The ratio between bore and stroke has narrowed over the years as fuel quality has improved. Camshaft drive has evolved from gears and shafts to chains, and now to rubber belts.

Since aluminum is a soft metal, replaceable inserts are used for the long-wearing valve seats and guides. Alfa's valve seats are durable; the guides, less so. If the guides wear excessively, lubricating oil is drawn through them to be burned in the combustion chamber, causing the engine to smoke. To reduce the possibility of excess oil leaking past the guides, rubber seals are placed around the tops of the guides.

The camshafts turn on bearing surfaces machined in the head's aluminum. In spite of the fact that the aluminum is not a good bearing material, the demands on the

Head Casting Marks

To assist in engine model identification, for several years Alfa cast an identifying symbol on the front surface of the cylinder head. The symbol is enclosed in a circle:

1300	▮
1600	✚
1750	△
2000	■

Beginning with Alfetta models, the size of the engine is cast into the rear surface of the head. On the V-6, the displacement is cast into the front of the left head.

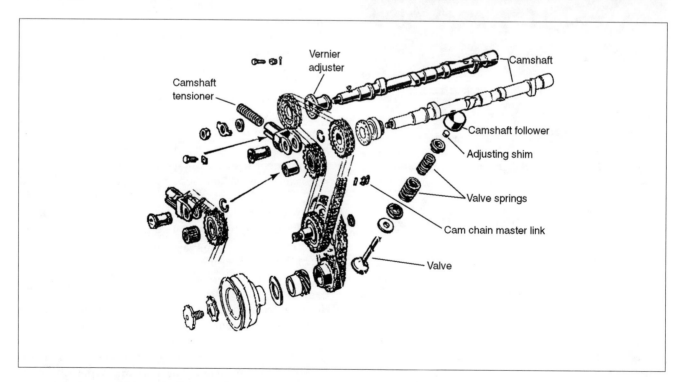

Fig. 2-10. Alfa's valve layout is traditionally double overhead camshafts, driven by two chains. This illustration shows valve adjusting shim, spring-loaded chain tensioner (above crankshaft pulley), and tachometer drive (lower right hand corner).

Fig. 2-11. Giulia Sprint GT engine bay holds same mechanicals as Giulietta Veloce, but with room to spare. Infamous fuse block is located at extreme top right of this view. Battery joins engine compartment and a plastic brake fluid reservoir appears in lower right.

six camshaft bearing surfaces are so minimal that no special bearing inserts are required.

The cam bearings are pressure-lubricated through oil passages drilled in the head. These passages are fed by mating passages in the block and the joint between the passages is sealed by the head gasket. Oil falls from the bearings to troughs that run beneath each cam. Oil in the troughs bathes the camshaft followers to lubricate them; a very small amount of oil is drawn past the followers and rubber valve seals to lubricate the valves as they work in their guides.

On Giulietta-derived engines, the camshafts are driven from the crankshaft through two chains. See Fig. 2-10. The primary chain runs off a small sprocket at the front of the crankshaft and turns a larger intermediate sprocket at half-speed, which is the speed at which the camshafts turn. A long chain connects the intermediate sprocket to the two cams. A spring-loaded tensioner keeps the chain from bellying-out at higher speeds.

The camshaft sprockets are mounted to the cam using a vernier adjustment so each cam can be timed independently—and at whatever position the owner wishes. The opportunity for mischief (and costly mistakes) is obvious. On Alfa engines built in 1981, a hydraulic system on the intake camshaft adjusts that cam's timing to provide improved responsiveness without increasing exhaust emissions. Later models added solenoid control.

The seal provided by the head gasket has proved to be the weakest feature of the Alfa engine. If there is an oil leak at the head gasket, it's usually noticed because the oil runs down the outside of the engine, most usually near the rear. If the oil leaks into the engine's internals, then the coolant becomes contaminated with emulsified oil and starts to look like a chocolate malt. In

severe leaks, enough coolant mixes with the oil line to destroy the oil's ability to lubricate. Such a failure will destroy an engine in only a very few miles.

Intake Manifold

With only very few exceptions, Alfa's intake manifold has always been located on the left (looking from the front) side of the head. Originally, this location assured that the driver on right-hand drive cars didn't have to contend with the heat from the exhaust pipe.

On carbureted engines, the manifold serves as an adapter between the carburetor and the head. If the carburetor is a downdraft Solex, then the manifold provides a right-angle turn for the air-fuel mixture as it flows from the vertical carburetor to the horizontal inlet passages of the head. On sidedraft Weber-carbureted engines, the manifold is little more than a spacer between the carburetors and the head. Regardless of the carburetor setup, the manifold is heated to help keep the fuel atomized on its way into the engine. The engine thermostat sits at the front of the manifold. (On early Veloces, the thermostat was in the top radiator hose.)

Coolant Temperature Sensor

The engine coolant temperature sensor is located on the intake manifold of most 4-cylinder Alfas. See Fig. 2-12. This is a screw-in, nonrepairable plug. As it gets hotter, its electrical resistance is reduced, a fact reflected in a rising temperature gauge on the dash.

Fig. 2-12. Coolant temperature sensor for SPICA fuel injected engines is located on intake manifold. Here, connecting wire is removed for clarity.

Exhaust Manifold

The exhaust manifold is a set of tubes that carries the hot, high-pressure exhaust gasses away from the engine. Many 4-cylinder Alfa exhaust manifolds are made up of two pieces: one exhausts gas from the front and rear cylinders and the other handles the exhaust of the center two cylinders. This layout takes advantage of the alternating exhaust gas pulses to help extract the exhaust more efficiently—such a design is called a "header." Typically, Alfa exhaust manifolds have been much more efficient than those fitted to other production cars, even those purporting to be high-performance designs.

Fig. 2-13. *Exhaust manifold side of 1979 SPICA-injected sedan. Hot-air hose (to preheat air on cold days) normally runs from air cleaner (top) across engine to stovepipe bracket over exhaust manifold. Accordion-pleated tube just above exhaust manifold is air injection manifold. It connects through one-way valve to air pump for emissions control.*

Block

Alfa has one of the lightest blocks in the industry, thanks to its wet-sleeve design, aluminum composition, and refined casting techniques. Alfa's block's basic design is so challenging that few manufacturers have even attempted it.

It is an H-section: upper and lower halves separated by a sturdy horizontal web. In its upper half, the block is little more than a container for coolant. The cylinders are machined steel sleeves that slip into the horizontal web. Since coolant touches the cylinder sleeves directly, the engine is called a "wet-sleeve" design. The bottom half of the block contains the sturdy webs that support the main crankshaft bearings.

The advantage of a wet-sleeve engine is that it provides removable cylinders (you never have to "bore" an Alfa block, you just replace the cylinders) and optimum

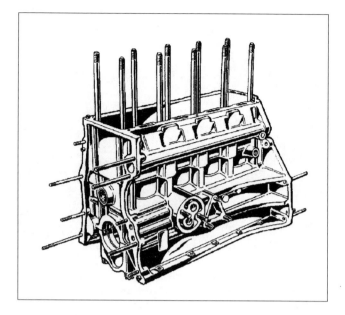

Fig. 2-14. *Bare aluminum block casting. Coolant from pump is distributed along engine through tube cast into block. We see its entrance flanked by two long studs just above C-shaped curve on block's front surface. Follow tube back along block and notice how coolant flushes upward from four rectangular passages on block's top surface. These passages provide a first-priority flow of coolant to exhaust valve seats in head.*

Fig. 2-15. *A close-up of left V-6 cylinder block reveals wet-sleeve cylinders, same design as 4-cylinder Alfa engines. Note water passage cast along left hand edge. Oval passages direct coolant at exhaust valve seats, hottest part of engine.*

cooling. As a result, a wet-sleeve engine tends to run cooler and last longer. It is also infinitely rebuildable because all the parts that wear are replaceable.

The challenge of a wet-sleeve engine is to maintain a gas-tight seal at both the top and bottom of the cylinder sleeves. The sealing problem is exacerbated by the fact that, when heated, steel and aluminum expand at different rates. Thus, it's necessary to compensate for expansion by machining the cylinders and block to slightly different heights. "Slight" is on the order of one or two thousandths of an inch. The job of the cylinder head gasket is made more difficult because it has to accommodate the two different expansion rates while still maintaining a gas-tight seal against very high pressures.

The bottom half of the Alfa block carries the five main bearings in which the crankshaft rotates. Many 4-cylinder engines have only three main bearings, but the five mains of the Alfa engine give it absolute durability. Alfa main bearings almost never give up.

Crankshaft

The crankshaft is a steel forging with drilled oil passages to the hardened bearing surfaces. Soft metal plugs are used to seal the ends of the oil passages. These plugs sometimes work their way free, causing a sudden drop in oil pressure. Typically, this is not a catastrophic failure, but each engine rebuild should include verifying that the plugs are securely in place. Racing Alfa engines occasionally are modified to have threaded plugs, but cutting the threads in a forged crank is not a casual task for the amateur mechanic.

A gear at the front of the crankshaft drives an accessory shaft for both the oil pump and the distributor. On carbureted engines, the mechanical fuel pump is also operated by a cam on the accessory shaft. On 750-series Giuliettas with Solex carburetors, the fuel pump was located on the head and operated off the exhaust cam.

Fig. 2-16. *Hood of this Quattroruote Zagato houses a Giulia Super engine, Webers and all.*

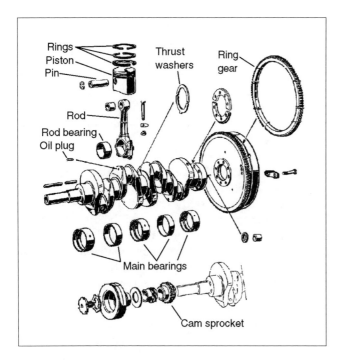

Fig. 2-18. *Oil pressure sending unit is located below intake manifold at rear of engine. Again, connecting wire (and the starter) is removed for clarity.*

Fig. 2-17. *The typical Alfa "bottom end." This illustration shows a Giulia Super crankshaft, rod/piston assembly, and flywheel/ring gear along with crankshaft bearings, piston rings, crankshaft pulley, and camshaft drive sprocket.*

Oil Pressure Sensor

On the Giulietta, oil pressure was sensed mechanically using a Bourdon tube gauge which received pressure from the exhaust side of the engine. Beginning with Giulias, oil pressure was sensed electrically by an oil pressure sensor nestled under the intake manifold toward the rear of the block. It is drumlike, with a single wire connection. See Fig. 2-18. For several years Alfa couldn't seem to get this little item right, and failed sensors were legion. The problem is that they didn't just give up the ghost: they lied a lot. You'd be driving along only to notice that your oil pressure seemed to be about half of what it should be. Then, with the next nervous glance, the pressure would return to normal. Or, worse, they would consistently indicate either virtually no oil pressure or peg the needle.

Alfas carry quite a bit of oil pressure. About 55 psi is normal but the dash gauge isn't accurate enough to use real numbers. If your oil pressure gauge stays somewhere near the middle of the dial, you're OK, especially if it goes up and down with the speed of the engine. If it never strays beyond the first quarter of the dial then you should probably have the pressure checked using a manual gauge. If it is really low, you're into a rebuild either of the pump or the entire engine.

Older Alfas always showed oil pressure even at idle. Modern Alfas show no oil pressure at all at idle with a hot engine, a characteristic I've never appreciated.

Several models of Alfa back up the gauge with a low-pressure warning light. If your Alfa has such a light, rely on it instead of the gauge to tell you when you're in trouble. If it ever lights, shut down the engine immediately.

Coolant Drain

If you have to drain the coolant, either for an annual change or for engine work, you should know that disconnecting the lower radiator hose does not entirely drain the block of coolant. All the older Alfas had a little petcock on the exhaust side of the engine toward the rear which was intended as the block drain. Newer Alfas replaced the petcock with a brass threaded plug.

Oil Filter

Older Alfas had cartridge-type filters which were a mess to replace. The modern spin-on filter as fitted to all 2-liter engines is more convenient. You can't change the filter on any Alfa, it seems, without spilling dirty oil.

Timing Cover

The camshaft chain drive is enclosed by an aluminum casting that also carries the water pump. A thin paper gasket mates the timing chain cover to the block. Typically, this joint leaks just enough oil to keep the engine covered with a grimy combination of road dirt and lubricant. Since the front main oil seal is also carried on the timing chain cover, it's important to verify that oil isn't leaking from the seal.

Fig. 2-19. *On most Alfas, block is drained of coolant by removing this brass plug. It's located at very rear of exhaust side of engine.*

Water Pump

The water pump is carried on the front surface of the timing cover. Prior to the Alfetta, the radiator fan is attached to the nose of the pump. Alfa water pumps are unremarkable, and usually trouble free. If they go bad, they leak water to get your attention.

Tachometer Drive

Just below the water pump is a connection for the tachometer drive cable. If your tachometer doesn't work, this cable is probably broken. A replacement inner cable can be made up easily, but a new cable is relatively inexpensive. Spiders built after 1979 used electronic tachometers, but the front casting wasn't changed and you can retrofit mechanical drive units on some newer engines. When would you ever want to consider such retrograde engineering? Just after you've priced a replacement electronic unit.

Distributor

The other item on the front timing cover is the distributor itself. This is the thing with all the spark plug wires coming from it. The distributor is really two devices in one. The top part routes the electrical energy (spark) from the coil to the appropriate cylinder. The lower half of the distributor contains a set of points which control exactly when the coil discharges its energy. Beginning in the early 1980s, Alfa used electronic, pointless distributors and, finally, the computerized

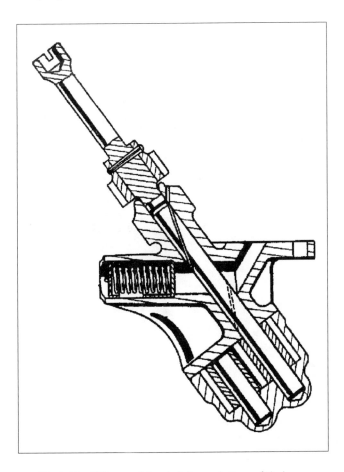

Fig. 2-20. *Oil pump is located at very bottom of timing cover and is composed of two meshing gears. A spring-loaded, high-pressure relief valve is located above gearset. Along shaft above valve is driven gear that meshes with crankshaft. At very top is drive to distributor.*

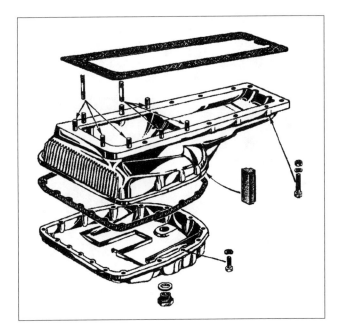

Fig. 2-21. *Alfa's oil pans have been delights of casting. For good reason: the most important single thing you can do for an engine is to keep its oil cool, and a multi-part, finned aluminum sump does it best. You'll find this style on most Alfas prior to Alfetta (except 750 and 101 Giulia).*

Motronic engine management system. Alfa distributors are keyed to go in only one way. In theory, you can't get it wrong. Of course.

The secret to getting it right is the knowledge that the distributor has to be fully seated against the timing cover before you tighten anything up. Why would anyone ever want to remove the distributor? It's easier to gap the points that way.

Oil Pan

Most mass-produced engines have a single steel stamping on the bottom of the engine to serve as a reservoir for the engine's lubricating oil. Only the early 750-series Giuliettas used such a simple stamped-steel pan. Typically, Alfa's oil pans are complex two-piece aluminum castings shaped to help assure cool oil for long engine life. For many models, the bottom of the pan is flared and heavily finned to enhance cooling. The lower section of the pan has a cooling labyrinth for the oil. See Fig. 2-21.

2.2 V-6 Engine

Though it is a completely new design, much of what has already been said applies equally to the V-6 engine, which follows classic Alfa design in most details. The V-6 is an excellent test of those items that are uniquely Alfa. While it doesn't have double overhead camshafts, it does have hemispheric combustion chambers, wet cylinder sleeves, and superb aluminum castings.

The V-6 was introduced in the U.S. in early 1981 as a 2.5-liter engine for the Alfetta Coupe, now called the GTV6. It had been available for a couple of years in the Alfa 6, a luxury sedan not imported into the U.S.. While the V-6 would have significantly revitalized the Spider, its older chassis would not accept the driveline the V-6 was intended for: a de Dion rear suspension with transmission in unit with the rear differential.

The V-6 was clearly a sporting engine, and only a peek inside was enough to reveal ample room for subsequent increases in displacement. Initially, Alfa denied

Fig. 2-22. V-6 head with its cam cover removed. Camshaft is toward top of casting and three pushrods extend down from it to rocker arms for exhaust valves. To left of each pushrod is a deep well for spark plug. Toothed-belt pulleys for cam and dis- *tributor are also clearly seen. Note distributor drive at lower right beveled corner of head, spin-on oil filter mount further down to left, motor mount and then starter at bottom left.*

any plans for a V-6 larger than 2.5 liters. It was not long, however, before a 3-liter version went racing in South Africa. Five years later, a 3-liter V-6 went into production to power the Milano Verde sedan and its successor, the 164.

Head

The most novel feature of the V-6 engine is its use of a transverse pushrod to operate the exhaust valve. This is a design that was made famous by BMW in its 328-series sport cars of the late 1930s and perpetuated in the Bristol engine after the war. The Alfa layout allows the engine to retain hemispheric combustion chambers with only a single overhead camshaft (The BMW/Bristol cam is in the block.). In Alfa's V-6, the inlet valve is opened directly by the camshaft. The exhaust valve is operated from the same camshaft through a rocker and a short push rod. The inertial mass of the exhaust valve train is so low that there is no significant performance penalty. Redline on the V-6 is a healthy 6000 to 6400 rpm, depending on model.

While the 4-cylinder Alfa head has poor torsional rigidity (read: it warps easily), the V-6 head is a much deeper casting, and stiffening webs run between the head bolts and the spark plug wells to help resist warpage. The V-6 head gasket was at first a multipart

Fig. 2-24. Combustion chambers for the V-6 are also hemispheric.

assembly with individual round metal gaskets, but current issue is a single-piece unit.

The inlet path of the V-6 shows a change mandated by modern emission controls. One of the characteristics of the classic hemispheric combustion chamber is that the straight-through gas flow creates very little "swirl," so any pockets of poorly mixed air and fuel tend to stay that way. In the V-6, the inlet passage dumps directly on the top of the piston and then must make a right-angle turn to the exhaust port. This sudden change in direction creates some swirl, promotes more complete combustion, and lowers the amount of unburned hydrocarbons. The design of the combustion chamber in the head is not a true hemisphere. It is more shallow, to reduce the surface area of the piston and lower emissions. The straight-down inlet path also provides a compact induction system, keeping the V-6 a fairly narrow installation.

The camshafts are driven by a toothed, fiberglass-reinforced belt. Though the toothed-belt drive was pioneered by Fiat, Alfa used it to drive the SPICA fuel injection pump on 1969–81 U.S. Alfas, and to drive the cams on the AlfaSud engine. It proved to be quite reliable. The advantages of belt drive are several: it is quiet, requires no lubrication, and is simple to change. A rub-

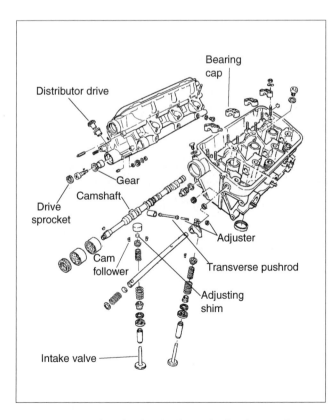

Fig. 2-23. Milano head and valve train showing novel transverse pushrod and rocker-arm. Valve assembly, including adjusting shim for intake valve, is typically Alfa.

ber belt, is, however, not permanent and needs much more regular replacement than a chain.

The single largest source of maintenance for V-6 owners is probably tensioning pulley for the belt. The pulley goes to full tension at start-up, and then slacks off as engine speed rises. This "detensioner" is operated by engine oil pressure. Some leak, spilling oil over the front of the engine.

Block

The cylinders form a V of 60° and so the Alfa V-6 beats in regular rhythm, unlike the early 90° V-6s of American practice, which sacrificed even-firing smoothness to manufacturing convenience (they could be assembled on the same line as the V-8).

Like the fours, head-gasket sealing around the oil passages on V-6 engines has proved to be an early problem, which has now been largely solved by a one-piece head gasket.

There is a fix for early, leaking V-6 head gaskets, a kit that provides external flexible oil lines to bypass the cast-in oil passages (which depend on the head gasket for their seal). This kit is available from several aftermarket sources.

The pistons of the V-6 are significantly flatter than the steeply-domed 4-cylinder pistons. This more level topography reduces the cold surface area of the piston and helps lower emissions. There are small valve pockets cut in the tops of the pistons to improve clearance between the valves and pistons and to provide a little extra margin of safety in case of valve float. Since the V-6 is electronically rev-limited, the valves should float only if the engine is improperly downshifted.

Typical Alfa practice is to use two compression rings and one oil control ring, and that tradition is continued with the V-6. The compression rings are narrow, following racing practice.

Incidentally, Alfa pistons (like all modern pistons) are not truly round, but rather slightly oval. The reason for this is that a piston has a large mass of metal along the axis of the wrist pin and very little metal 90° to that axis. The expansion rates of the piston, then, vary around the piston's circumference. The piston is intentionally machined oval to help offset this differential expansion rate.

The V-6 rods are very similar to those in the 4-cylinder, except that they have a large boss at their bottom that can be ground when balancing the rods. A small squirt-hole near the shank of the rod provides lubrication and cooling to the thrust side of the cylinder wall.

A traditional strength of the 4-cylinder engine was that each rod was supported on both sides by a main bearing. On the V-6, two rods swing between supporting main bearings. The rod bearings themselves form a helix along the crankshaft axis and are spaced 60° apart. Color-coded bearing inserts provide a precise method

Why Alfa Chose a V-6 Design

It's dangerous to estimate why a company makes any kind of decision, but the question naturally arises why Alfa chose a V-6 engine as its next-generation powerplant. The final decision, made by Surace, was easy enough, since the engine had been designed in 1971 and had simply sat on the shelf for want of tooling funds.

There is a kind of classic limit of 500 cc per cylinder (or 2 liters for a 4-cylinder engine), though both Mitsubishi and Porsche have been very successful in producing 4-cylinder engines of 2.6 and 3.0 liters respectively. An enlarged 4-cylinder Alfa engine, of classic design, might have been considered. Alfa also has a precedence of a straight-6 of 2.6-liter displacement with the 2600 cars of the mid-60s. Or, it might have returned to the wonderful V-8 engine of the Montreal, which also displaced 2.6 liters. I should note, also, that both Scuderia Ferrari and Alfa Romeo created 2.6-liter straight-8 engines based on Jano's 8C2300 masterpiece: what a design that would have been to emulate!

Taken in order, a large 4-cylinder engine is probably too much a compromise design for Alfa. Such an engine (including Mitsubishi's and Porsche's) requires a countershaft to cancel the vibrations of the large four, and that adds complexity without real benefit. A 4-cylinder engine simply isn't smooth enough for a luxury car.

A straight-6 offers the drawback of excessive crankshaft torsional vibration, that is, the crank is so long that it tends to twist under load. Alfa's solution to the straight-8 cranks was to split them in the middle with an accessory gear. That solution, while elegant, is simply too expensive today.

A small-displacement 4-cam V-8 would have been perhaps too wonderful. There might have been some problem with image: the V-8 is cursed with the mantle of gas-guzzler. Certainly, the expense of the engine would have priced the car above what we are now accustomed to pay for a modern Alfa. Maybe the folks at Alfa/Lancia Industriale could keep the thought in reserve, however.

If there is a design which is synonymous with modern high-performance, it is the V-6. And, a six still suggests fuel economy to most buyers. More prosaically, I think the most powerful argument for the six was that it fit in the Alfetta engine bay.

of bearing-fitting during rebuild. Alfa crankshafts are traditionally unbreakable. Almost a decade's experience with the V-6 suggests the tradition of reliability is nobly maintained.

Oil Pan

The 7-quart oil pan is deeply finned alloy, another Alfa tradition. The pan is designed to enhance oil cooling: cooled oil lubricates better and improves engine life. The oil pump is a five-lobed eccentric rotor. Its design differs from the 4-cylinder gear-pump design.

2.3 Other Alfa Engines

Alfasud Boxer

The Bossaglia-designed Alfasud was originally a front-drive flat-4 five-place sedan, later joined by a Sprint coupe version. Engine displacements ranged from 1.19 to 1.5 liters and output was from 63 to 95 hp.

The water-cooled flat-4 was based on a cast-iron block that included the cylinders, but its two alloy heads carried single overhead camshafts which were driven by toothed belts. The engine had a number of interesting technical details, including an extremely rugged bottom end and unique valve adjustment.

The main bearing caps were cross-bolted in the block to give extra rigidity and the caps themselves had toothed mating surfaces with the block to further resist deformation during heavy loads. Similar serrations were used to help locate the big end caps on the rods.

Originally, the stems of the valves were externally threaded for an adjusting screw. The cup-type cam follower rested on the screw. A hole in the cam follower and a mating hole through the camshaft lobe itself allowed quick clearance adjustment without removing the camshaft. Because each camshaft lobe had a hole in it, the lobe was cut away so that the cam's two working surfaces bore against the follower near its outer edge. Later versions of the engine used adjusting shims.

The several horsepower variations were obtained primarily by the choice of carburetion. A flat-4 engine suffers from inherent breathing problems when only one carburetor is used (a well-known VW feature). The smallest-displacement 'Sud engine used one single-throat carburetor, while the other engines used a dual-throat carburetor mounted centrally above the block. The most powerful "Quadrifoglio Oro" used two twin-throat carburetors, giving a Venturi for each cylinder.

The AlfaSud flat-4 showed up in the ARNA, a joint Alfa-Nissan venture, as well as in the Type 33, a sedan positioned just below the (new) Giulietta sedan. The 33 was also available in four-wheel drive, with a 1.7-liter four-cam engine.

6-Cylinder 2600 Engine

An enthusiast's first look at the 2600 engine is a study in awe. It is a much larger proposition than the four, and even the V-6. On first view, it seems incredibly long, and the three PHH Solex carburetors only emphasize the mass of the powerplant.

Fig. 2-25. Engine bay of 2600 is dominated by three Solex carburetors and air intake plumbing. Notable in this car is original VDO windshield washer bag on driver's-side fender wall.

CHAPTER 2

The 2600 is a mildly tuned, relatively low-power mill that is built to be rock-solid reliable. It achieves that goal remarkably well. There are a lot of 2600s still running. The water pump is often the weakest part of the engine; fortunately, a stock 105-series water pump rebuild kit can be made to work (the main shaft is shorter, but not so short it can't be used).

There is a potentially catastrophic condition 2600 owners must remember: the PHH Solex carburetors have a vacuum-controlled secondary throttle that is operated by a shaft running through a drilled boss. See Fig. 2-26. The shaft is exposed to the rigors of the engine bay without benefit of lubrication. The fact that it is steel running in aluminum invites electrolytic corrosion. If the shaft seizes, the secondary will stick open, seriously leaning out the air/fuel mixture and possibly melting a piston. Always lubricate these shafts and check them frequently. You should be able to move them easily just by pressing your thumb against them. The 2600 does not enlarge the vocabulary of the 4-cylinder engine. A familiarity with the Giulietta/Giulia is enough to qualify a mechanic on the 2600.

Fig. 2-26. Owners of 2600s take note: exposed, vacuum-operated shaft for secondary throttle plates must be free to move. Test it with your finger, as shown.

Fig. 2-27. An early 1900 coupe engine compartment is exceptionally sanitary. Black box on far side, just in front of two manual heater valves, is Bosch voltage regulator. Fuse box is located behind air cleaner and to left of distributor.

2-Liter and 1900

If you absolutely must have an Alfa with a cast-iron block, then either of these 4-cylinders can be your choice. The original 1900 had exhaust valves that were adjusted with two small gears, with one threaded to the valve stem. This exotic technique was abandoned on the 2-liter. Other than that, there is almost no difference between the two engines (in fact, their aluminum heads are interchangeable). The very earliest 1900 engine carried its oil pump on the front timing cover plate.

The 1900 transmission is unique. The 2-liter gearbox is the same basic unit as the 5-speed 101-series Giulia transmission. Both the 1900 and 2-liter suspensions use proper A-arms and kingpins up front and a solid rear axle that looks like it belongs under a Ford or Chevrolet.

Montreal

Alfa has a refreshing willingness to cast up some low-production models on a whim. The Montreal is a prime example of this. Basically, a Montreal is a Giulia/1750 with a few unique castings, most notably its V-8 engine, which has Giulietta heads and valves, and Junior rods. There was an exchange of letters pub-lished in the Alfa club magazine in late 1988 between an independent technician who worked on Montreals and Alfa's chief American engineer. They disagreed on the true origins of the Montreal engine. The technician claimed it was a 1750 knock-off; the Alfa engineer confirmed it was an independent design. The Montreal is a long-distance tourer in the mold of the 2600. It is not a parking-lot racer, but because of its beautiful body and superb V-8 engine, it is incontestably one of the most desirable of the modern Alfas. Like the 1900, 2-liter, and 2600 engines discussed above, a familiarity with the Giulietta engine or its variants qualifies one to work on the Montreal's V-8, including the two SPICA fuel injection pumps bolted end-to-end.

2.4 Other Things in Engine Compartment

Reservoirs

There are usually two large plastic fluid reservoirs in the engine compartment. One is for the windshield washer fluid and the other is the coolant expansion/recovery reservoir. On newer Alfas (post-Giulia), smaller plastic reservoirs are provided for the brake and clutch master cylinders.

Brake and Clutch Cylinders

The brake and clutch hydraulic cylinders are attached to the firewall, inside the engine compartment and above the steering column. They are black, cast-

Fig. 2-28. Montreal engine bay is dominated by air cleaner. Starting from upper left corner and working clockwise: coolant overflow bottle, fuel filter, windshield washer fluid bottle, two coils with two relays just behind, distributor cap, brake booster with fluid reservoir, clutch reservoir, two small air conditioning relays, and receiver/drier for air conditioning. Large black oval at bottom left is a vacuum reservoir. Peeking out at front of engine are shiny fuel injection lines atop SPICA injection pump.

Fig. 2-29. Brake fluid reservoir for most modern Alfas looks like this. Wires are for brake warning light in dash.

metal tubes which are closed on one end and connected to the pedals on the other end. On modern Alfas, the brake cylinder is mounted on a booster diaphragm which is also black and looks like two cake covers stuck together. The booster uses engine manifold vacuum working on a large diaphragm to provide "power brakes."

Fig. 2-30. Clutch fluid reservoir. You can see stylized "A" cast into the plastic body, identifying this as an ATE reservoir. Note, however, that ATE reservoirs appeared on clutch master cylinders from many different suppliers; the cylinders may have different specifications (internal diameter).

Fig. 2-31. Alfa provided cut-apart cars for service training. This Alfetta sedan shows clearly how air conditioning compressor is mounted atop alternator. Flexible tube which courses over top of engine carries heated air to air cleaner as part of the emission controls.

3. Carburetors

3.1 Solex

Most Giulietta Alfas in the U.S. came with a dual-throat downdraft Solex 35 APAIG carburetor while Giulia TIs used the 32 PAIA dual-throat carburetor. On the 35 APAIG, the second throat was operated by a combination of mechanical linkage and air flow. When you got to about three-quarter throttle, the throttle plate for the second throat was opened mechanically and a counterweighted throttle plate below it was forced open gradually by the inrushing air. The 32 PAIA used a vacuum diaphragm on the secondary.

On either Solex, the two main jets are screwed into the back sides of brass carriers which take a 14 mm wrench. The intermediate jet, which supplies both mid-range and idle fuel for the primary throat, is a press-fit on the backside of a brass 10 mm carrier.

Most Solex-equipped owners dreamed of Webers, and you'll still find Giuliettas running around which have been modified with dual Weber carburetors. The Solex was a perfectly good setup, and is actually preferable if your driving is limited to the city. Whole 35 APAIG Solex carburetors may be hard to find, but the jets are still plentiful because the Volkswagen Solex jets fit just fine. Be sure you get the proper number jet (usually 120 main, and 180 or 190 air correction). If you must have a Weber, several two-throat vertical models from the DGA or DGV families will fit.

3.2 Weber

The two side-draft Weber carburetors take up almost as much room as the engine itself, suggesting a level of power the 4-cylinder Alfa engine never did have. There's no doubt that two dual-throat Webers are the classic setup for a high-performance 4-cylinder engine. With Alfa's twin-cam engine, they make an irresistible combination, mechanically and aesthetically.

Webers have a macho image and have been attributed a mystique which doesn't bear close inspection. A Weber carburetor is quite straightforward, is easy to maintain, and keeps its tune once properly set. You may hear exactly the opposite, but if you need reassurance, I suggest my book on the subject, *Weber Carburetors*, published by HP books in 1988.

Under the thumbscrew and disc sit the jets which control the Weber's main fuel circuits. They're conveniently located together to ease fine-tuning under race conditions. Closer to the engine are the four spring-loaded idle richness mixture screws, while the idle speed adjustment is a single spring-loaded screw located between the two carburetors.

A set of air horns makes a pretty sight, but you should never drive an Alfa engine without an air filter.

Fig. 2-32. *A Weber carbureted box-stock 1300 cc GT Junior. This is a European model, betrayed by shape of the air cleaner. You'll occasionally see a similar view of carbs and air cleaner on a U.S. SPICA-injected model converted to carburetors.*

4. Fuel Injection

In 1969, with the introduction of the 1750 engine in the U.S., Alfa began fitting SPICA fuel injection in order to meet federal emission standards. Since any emission control is anathema to gung-ho performance-car owners, it was fashionable to rip off the SPICA system at first provocation and fit Webers. It is also cheaper to fit Webers rather than buy a factory-rebuilt SPICA fuel injection pump.

Webers offer no significant advantage over a properly tuned SPICA system. I will say that a Weberized 1750 or 2-liter has quite a different character from a car with SPICA injection. It seems to breathe deeper and certainly can take advantage of other tuning modifications more easily. For most applications, though, you should love your less-polluting SPICA system.

The Bosch system, on the other hand, is demonstrably superior in all respects to both Weber and SPICA.

4.1 SPICA Fuel Injection

Fuel Injection Pump

The SPICA fuel injection pump sits low on the intake side of the engine and is driven off the crankshaft by a small toothed belt. See Fig. 2-33. The pump resembles a little engine, with a crankcase and block. Indeed, it contains four pistons in cylinders which force small quantities of fuel through the injectors and into the intake manifold. It also contains an oil filter, which should be changed regularly about every third oil filter change.

The pump is covered in detail in Chapter 6. Below are some components that work with the pump.

Barometric Sensor

This takes care of fuel mixtures when you're flying high. For pre-Alfetta cars, on top of the sensor is a temperature compensating lever with the letters N, C, and F stamped below. This is a rough thermostatic setting

Fig. 2-33. SPICA installation in a 1750 (air cleaner removed). This is an early pump, lacking fuel cutoff solenoid that on later pumps mounts to right of thermostatic actuator.

corresponding, in English, to Normal, Cold, and Freezing. Leave it on N unless you're in near-zero temperatures. Reserve the F setting for truly arctic conditions.

Thermostatic Actuator

The injection pump's automatic choke. This is an expensive and somewhat fragile device. Just make sure the capsule in the middle of the metal tube running to the manifold is firmly mounted. The unit is still available from Alfa Ricambi, or they also sell John Shankle's Sure-Start mechanical replacement which is considerably less expensive but more troublesome to operate.

Cold-start Solenoid

The purpose of the cold-start solenoid is to give an enriched mixture while the starter is cranking. It's energized by a wire which comes directly off the starter solenoid. If you crank and crank and Alfa still won't start, you've probably fouled the plugs with a too-rich mixture. Pull the wire to the cold-start solenoid, set it where it won't ground out, and start the engine. In the process of cranking with the wire removed, you'll be feeding a moderately lean mixture which will probably clean the plugs enough to start. Be sure to get out and replace the

Fig. 2-34. Two-liter SPICA fuel injection pump removed from engine. Four towers are fuel outlets to the injectors. Behind them is pump's logic section, complete with two solenoids (fuel cutoff in front). Oil filter housing is at lower right.

wire, then properly readjust the solenoid as soon as you get home so you won't have to pull the wire again. A properly adjusted SPICA system should start easily.

In spite of this, many owners have spliced in a switch so they can disable the cold-start solenoid from the dashboard. Using insulated spade connectors, run 14-gauge insulated wire from the terminal on the cold-start solenoid to a single-pole single-throw switch mounted conveniently on the dash. Run another length of wire from the switch to the connector on the end of the wire that used to go to the cold-start solenoid. Mark the switch clearly so you can remember when it is closed. Keep it in the closed position unless you're beginning to foul the plugs during a cold start.

Fuel Cutoff Solenoid

Shuts off fuel during deceleration and prevents dieseling when the engine is shut off. It also is the means by which you adjust fuel delivery, by screwing the solenoid in or out.

Long Link

Tells the fuel injection pump how far the throttles are open. This link establishes the air-fuel mixture delivered by the pump.

Short Link

Operates the throttle plates in the manifold.

Injectors

The manifold carries the four injectors that spray a small quantity of fuel directly at the intake valve as it is

Fig. 2-35. On left side (looking from the front), you see FIS-PA main fuel filter. This is a late installation which does not pass fuel return through fittings 2 and 4. Fitting 3 comes from fuel pump and fitting 1 (in shadow) connects to inlet of fuel injection pump.

opening. The injectors themselves are trouble- and maintenance-free.

Set Screws

On Alfas with SPICA fuel injection, there are two little screws sitting on the top of the intake manifold which are just begging to be fiddled with. These screws control the total rotation allowed for the throttle bellcrank. They were set at the factory and should never need to be adjusted. If you have any reason to think that the screws on your car may have been adjusted, their settings should be confirmed using the procedure in Chapter 6.

There is another factory-set screw on the SPICA system, and that one is located on the rear face of the fuel injection pump. The screw is covered by a plastic protector which is safety-wired in place. If that screw is disturbed, there is no field repair and the fuel injection pump has to be returned to the factory for readjustment. If you're thinking about buying a used Alfa and notice that the plastic cover to this screw is missing, you need to ask why it was removed. If the car doesn't run well, and the cap is missing, take about $1000 off the price of the car: that's about what parts and labor will be to have it put right.

Fig. 2-36. If this screw on SPICA pump has been tampered with, much work may be needed to recalibrate injection system. Normally, wired-on plastic cap covers screw.

4.2 Bosch Electronic Fuel Injection

Carburetors and fuel injection systems such as SPICA depend on pressure sensing devices and levers to tell them how much fuel the engine needs. Because these systems are entirely mechanical, they have to correct for such mundane things as inertia, friction, and limited ranges of reliable operation. As a result, and even at their best, they can give only an approximation of the engine's real fuel needs.

In contrast, an electronically controlled system can supply exactly the right amount of fuel under all possible operating conditions. It does not have to contend with approximations or frictional losses. It can hold the injectors open for precisely the right amount of time, and even verify its own operation by monitoring the oxygen content of the exhaust gasses.

In point of fact, an automotive fuel system doesn't have to be so exact just to make an engine run well. If it were not for the stringent emission standards required of modern engines, electronic fuel injection would be overkill and unnecessary complication.

Principles of Operation

The heart of the Bosch Jetronic electronic fuel injection (EFI) system is a small computer (located behind a panel in the front of the passenger compartment) that evaluates the information provided by several engine sensors and then issues commands to operate the fuel pump and injectors. It's an input/output sort of thing. For input, the system uses:

Air-flow Sensor

This is the pretty aluminum casting located just downstream of the air filter. Its electrical signal depends on how much a spring-loaded flapper valve is held open by the flow of air into the engine. Ultimately, this valve measures the mass of air entering the engine.

Throttle Position Sensor

The throttle position sensor is located on the outside of the throttle body, which contains the throttle plate and is bolted to the side of the large intake plenum. This signal varies with the angle of the throttle plate. Most of the time, the throttle position and air volume measured by the flapper valve are directly related: floor the throttle and the engine begins gulping more air. But, when you're coasting down from a high speed with the throttle closed, there is a significant disparity between throttle angle and the amount of air the engine breathes.

Full-Throttle Switch

Wide-open-throttle operation causes the system to go into a kind of damn-the-torpedoes mode. The injectors fire twice, instead of once, every four crankshaft revolutions and the oxygen sensor in the exhaust pipe is disabled so it doesn't try to correct for the over-rich condition.

Temperature Sensor

Located at the front of the engine on the thermostat housing on the GTV6 and Milano, this sensor reports

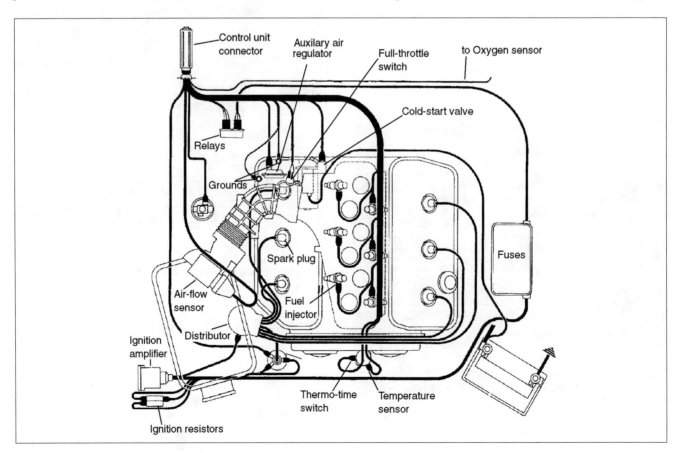

Fig. 2-37. *Schematic view of Bosch fuel injection system.*

coolant temperature. Generally, a colder reading requires slightly more fuel for the engine to run properly.

Thermo-time Switch (Jetronic Systems)

Located near the temperature sensor, this is an electrically heated switch which controls the cold-start injector (see below). It opens, when the engine warms up, to disable the injector. Because it is electrically heated it opens in about three minutes even if the engine hasn't reached operating temperature.

Oxygen (Lambda) Sensor

This is a fragile sensor which looks like a spark plug and is screwed into the exhaust pipe just ahead of the catalytic converter. Truly space-age, its signal is in the 0.5 to 1.0-volt range, and varies with the amount of oxygen in the exhaust gas. The signal is carried in a shielded cable to protect it from electrical interference. If the oxygen content of the exhaust gas changes, the system leans or enriches itself to maintain the desired oxygen strength. The oxygen sensor makes the EFI system self-correcting and so it is called a closed-loop system.

Engine Speed Sensor

In addition to discharging the coil, the regular pulses from the distributor provides a report of engine speed. Thus, the Bosch system collects information on throttle position, airflow volume, engine temperature, engine speed and the composition of the exhaust gas.

All these signals are used by the electronic control unit (ECU, or "brain"), to control:

Injectors

The injectors are simple solenoids which open and close to let fuel spray into the engine. Their duty-cycle is varied by the system. If less fuel is needed, the injectors are opened for a short duration. If more fuel is required, the injectors are opened for a longer duration.

Cold-start Injector (Jetronic Systems)

The cold-start injector is located at the rear of the large aluminum air plenum which sits between the banks of cylinders on top of the engine. It is controlled by the thermo time-switch to create an easier-starting, over-rich fuel mixture by squirting extra fuel into the air plenum when the starter is engaged.

In addition, the Bosch EFI system includes several items which are not directly controlled by its ECU:

Auxiliary Air Device (Jetronic Systems)

When the engine is cold, this device delivers extra air to the engine, causing it to run faster. A heated bimetallic strip senses cylinder head temperature and closes the auxiliary air supply when the engine is warmed up. On Motronic systems, a stepper motor adds air.

*Fig. 2-38. V-6 engine compartment shows Bosch injection components. **Top**: Ribbed rubber hose runs from air-flow sensor to throttle valve on engine. Connectors for throttle switches can just be seen. Bolted to intake manifold is decel valve. Round-topped object below valve is crankcase breather. **Bottom**: Wavy pipe below intake manifold is injector rail. At front of engine with wires attached are temperature sensors and thermo-time switch.*

Decel Valve

Things get a bit upside-down during deceleration: the engine is turning over fast, but the throttle plate is closed. The closed throttle tells the fuel system to deliver only a little fuel, but the engine speed sensor says to deliver a lot. Without some correction, the engine will decide to go lean to the point of backfiring. The decel valve simply opens an auxiliary air passage which causes the system to add more fuel and avoid backfiring. Simply put, the decel valve means there's never a closed-throttle condition on coast-down.

All of the small-tube plumbing on the intake plenum and the bright mushroom-shaped valve to which the plumbing goes represents a method of controlling emissions during deceleration.

Fuel Pump

Two electric fuel pumps are located at the rear of the car, one in the fuel tank, the other external. The external pump supplies a high-pressure source of fuel and cannot be replaced with a low-pressure aftermarket pump such as the kind carried at your local parts emporium.

Pressure Regulator

Gasoline circulates the system at about 40 psi, which is some ten times greater than the system using carburetors. It is this high fuel pressure which actually sprays fuel into the engine. As noted above, the injectors are no more than valves. Along with injector open-time, fuel pressure determines just how much fuel passes the injectors when they open, so a carefully regulated fuel pressure is essential.

Injector Rail

Fuel circulates through the system through a metal tube called the fuel rail and is then dumped back into the fuel tank. The fuel pressure regulator is located on the fuel rail.

Fuel Filter

An injector passes an astoundingly small amount of fuel during each duty cycle. As a result, its orifice is small and the fuel washing through it must be as clean as possible to prevent clogging. A filter is located at the rear of the car, after the fuel pump. There is no service life suggested for this filter, but if it clogs it may be a cause of low fuel pressure.

5. Distributor/Coil Ignition

The distributor is divided into three parts. The top part, which is made of an insulating material and has all the thick wires coming from it, "distributes" the spark to the proper cylinder. Electrical energy from the coil enters through the center wire and is rerouted by a rotor just beneath the cap to the proper spark plug wire.

The second part of the distributor makes and breaks a ground circuit which controls the charging of the coil. Typically, the circuit is made and broken using a set of points which require occasional maintenance. In modern Alfas, electronics have replaced the mechanical points, and the need for maintenance.

The third part of the distributor adjusts the exact moment the points open so that the fuel has its greatest opportunity to burn completely. Although the speed of an engine changes, the amount of time it takes for the fuel to burn remains constant. Thus, higher engine speeds need a spark which goes off slightly earlier than for lower engine speeds. This change in spark timing is taken care of by a weight which spins around the distributor shaft but is restrained from flying outward by a

Fig. 2-39. Eight spark plug leads instead of four: the Marelli dual-plug distributor and cap as fitted to GTA Alfas.

spring. At higher speeds, the weight pulls against the spring with greater force and so moves outward. The motion of the weight is used to advance the spark. On Giuliettas with Lucas distributors, a vacuum mechanism was also used to modify spark timing.

Late-model Alfa engines use Bosch "EZL" electronic ignition with Jetronic injection or the integrated Bosch Motronic fuel injection system.

The coil-and-distributor ignition system dates back at least to the 1920s. That may suggest that there is something wrong with the system, but that is not exactly true. Properly maintained, the Alfa system is adequate for everyday driving. Enough so that I have never used any of the aftermarket "hot-shot" systems on any of my Alfas. But the stock system operates on the ragged edge of acceptability. If you don't maintain the ignition system in good condition, or do something to the engine which requires a "hotter" spark, then a high-voltage aftermarket system may solve some of your starting problems.

In my experience, most hard-starting problems come either from poor starting technique or the use of old or improper spark plugs.

6. Emission Controls (U.S.A.)

Unlike most American cars, Alfas do not carry a large complement of emission controls. Early on, they got away with simplicity because a significant degree of combustion cleanness was engineered into the engine, especially the fuel system. Alfas built prior to 1969 had no emission controls worth mentioning because they were maintenance-free. Beginning in the mid-'70s, emission restrictions began to challenge even the SPICA system, a situation which got progressively more severe until 1982 when Bosch EFI became standard.

I need to add a few words of caution to the majority of Alfa owners who live in states that do not currently require regular emission-control checks: don't throw the parts away. It's popular to discard emission controls in the hope of improving performance. That is a very short-sighted attitude.

There is a probably unstoppable movement toward uniform emission-control regulations. California and a few other states already require that all the original emissions-control items be physically present on the car before it can be licensed. Junkyards make a lot of money selling used emission-control pieces now, and the problem is only going to get worse because the factories are not going to make new, but obsolete, emission control pieces.

I know of several Alfas in California which can't be registered legally because they have been stripped of emission controls. The biggest problem is the Alfa which has been converted from SPICA to Webers. Most of those cars are now only good for parts.

It's in California now—coming to your state soon.

Intake Air Heater

Beginning in 1976, Alfas piped some hot air from around the exhaust manifold into the air cleaner's intake. The heat helped keep the fuel vaporized and so lowered emissions, especially when the engine was cold. This is not a power-robbing system. It is identified by the accordion-fold tube running between the exhaust manifold "stove" and the air cleaner intake. Also, there's a flapper valve on the front of the air cleaner that controls cold-air intake. It's operated by a bimetal strip.

Air Pump

An air pump looks like a miniature generator, is belt-driven from the front of the engine, and has two large rubber hoses attached to its rear. See Fig. 2-40. One of the hoses goes to a manifold which feeds pressurized fresh air into the exhaust. The other hose sucks fresh air from the air cleaner. The fresh air helps burn any fuel not consumed in the engine itself. Alfa Spiders had the air pump mounted above the generator; Alfetta sedans and coupes mounted it outboard of the SPICA fuel injection pump. Because of increasingly severe emission standards, you will likely see the return of the air pump.

Incontestably, an air pump robs power from the engine. It also contributes to the tendency to backfire.

Fig. 2-40. You'll be hard pressed to see it, but air pump on Alfettas rides outboard of SPICA fuel injection pump. Even skilled mechanics, new to Alfa, will miss it. Location makes changing SPICA oil filter a chore. Air pumps were fitted 1975 through 1981.

Fig. 2-41. You may never see one of these: infamous "monofarfalla" intake plenum on 1981 SPICA-injected Alfas. EGR pipe attaches front (right) of plenum and runs across cam cover to exhaust manifold. Hat-shaped object just in front of EGR pipe is vacuum-controlled EGR valve.

Exhaust Gas Recirculation (EGR)

Alfas produced in 1980 and 1981 were fitted with an exhaust gas recirculation (EGR) system. You can identify an EGR system by the pipe which connects the intake and exhaust manifolds. These systems have a vacuum-operated valve on the exhaust manifold which opens to dilute the intake charge with nonflammable exhaust gas. EGR is not only power-robbing, but it also degrades a car's driveability. It is frequently disabled even though it reduces nitrous oxides. Like the air pump (above) Alfas will regain EGR as emission standards are raised.

Vapor Separator

The vapor separator traps and condenses oil fumes from the crankcase and returns oil via a small-diameter rubber hose which runs in front of the engine, from the intake to the exhaust side, and ends up at a nipple low on the dipstick holder. The only reason I mention this is that the hose needs a little bit of oil in it for the system to work properly, and the separator needs regular cleaning.

Fig. 2-42. *Canister shown dead center in this photo is a vapor separator, used to recycle crankcase fumes. Small rubber hose at its base (not shown) leads back to engine oil sump. If your engine stalls, the small hose may be plugged.*

Catalytic Converter

This isn't in the engine compartment, but I include it here because it's an integral part of the emissions control system on some Alfas. The catalytic converter looks for all the world like an extra muffler. On Alfa, it's covered with a perforated heat shield. Catalytic converters are a little high-temperature chemistry set which turns one kind of pollution (covered by federal legislation) into another (not covered). They are expensive, can be relatively short-lived (60,000 miles) if the engine runs rich all the time, and smelly like sulphur or rotten eggs.

Fig. 2-43. *Stock catalytic converter is usually housed in a perforated-metal heat shield which frequently rattles itself off car. That's what's happened to this one. If your Alfa simply won't pass a smog inspection, even after a careful tune-up, change the cat.*

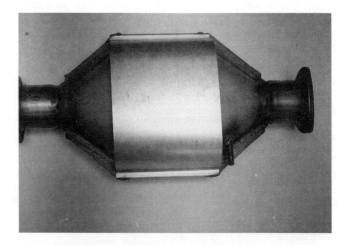

Fig. 2-44. *This aftermarket catalytic converter looks different but works exactly like round, stock unit. On some cats, mounting flanges are rotated 180° to each other: on others, flanges are oriented the same on both ends.*

7. Transmission and Drivetrain

Come with me for a moment beneath your Alfa. Jackstands, please.

7.1 Pre-Alfetta Cars

I had initially named this section "Giulietta to 2-Liter Coupes and Sedans and all Spiders" only to conclude that we still haven't come to terms with a neat way of identifying what is clearly a large family of virtually identical designs. I do think in the future we will call this group "Giulietta-Series" but the model variations are still so fresh they resist collection into such a neat class.

Perhaps as an alternate, we could just say "Classic Chassis," for the solid-rear-axle layout of the Giulietta-series cars dates back to Alfa's beginnings in 1910. Alfa has paid for its adherence to the solid rear axle by being labeled "antique" and other less-laudatory adjectives.

But the Alfa design has been refined to the point that it has an incomparably forgiving nature and no real performance penalty.

For all Alfas which predate the Alfetta, toward the rear of the mass of aluminum castings that comprise the engine (and things bolted to it) is a very large bulge which tapers into a roughly cylindrical casting. At the rear of this cylinder is a three-pronged yoke which is connected to the driveshaft by a large rubber donut. Close inspection will reveal that the large bulge, called the bell housing, bolts up to the rear of the engine and is not a part of the engine casting. Similarly, the cylindrical casting is the transmission, which is bolted firmly to the bell housing.

Because all the components are cast from aluminum, the entire assembly—engine, clutch, and transmission—is manageable enough to be pulled out as a unit for repairs. The flywheel and clutch mechanism is located inside the bell housing, where it is protected

Fig. 2-45. Four-cylinder engine and transmission removed from the car. Tapering bulge to rear (left) between engine and transmission is bellhousing, which encloses the clutch. At far left end of transmission is the driveshaft "donut."

from dirt and oil. The clutch is unremarkable in its design and absolutely conventional. In fact, the clutch disc for Giuliettas is the same as used on the MG TD.

The Giulia offered the 5-speed gearbox originally introduced in the 2-liter/2600 cars. The five forward gears enhanced Alfa's image of exclusivity. Many Alfa gearboxes suffer from weak second-gear synchronizers. Enthusiasts have learned to live with the characteristic.

The rubber donut just behind the gearbox works as a universal joint to allow some variation in the driveshaft angle and also helps absorb driveline shocks. On the cars we're discussing here, the donut is very reliable (read on).

The middle of the driveshaft is splined so it can change length and there is also a steady-bearing supporting the shaft near the spline. In wintry climates it's common to wash the underside of an Alfa to help keep salt from rusting the body. It's important not to spray a strong stream of water directly at the steady-bearing for fear of flushing out its lubricant. At the rear of the driveshaft, on what is called the companion flange of the differential, there is a conventional universal joint.

I don't want to launch a discussion of suspension dynamics, but I do need to explain what is meant by "unsprung weight," because Alfa does so much to keep it low. The road springs act as a dividing line between sprung and unsprung weight: the body rides smoothly over the dirt road while the wheels are bouncing crazily. Simply put, unsprung weight is everything that goes up and down with the wheels: hubcaps, wheels, tires, brakes, axle shafts, differential gears, and housing. It's important to keep unsprung weight low so the wheels

Fig. 2-46. *Giulietta rear axle and its suspension is typical layout of traditional solid rear axle drivetrain.*

can follow the road surface as intimately as possible. Low unsprung weight also makes getting a smooth ride much easier.

Alfa's solid rear axle assembly (all of which is unsprung weight) is quite light, even though only a small part of it is cast aluminum. The axle makes intelligent use of lightweight tubes and properly proportioned geometric shapes to achieve strength and lightness at the same time.

Part of Alfa's excellent roadholding comes from the lightness of the rear axle, and part from the precision of its location. At its top center, the axle is located by a triangular link which controls rotational and side-to-side movement. On the 750- and 101-series cars the upper locating link is a triangle fabricated of tubes. On 105-series and later cars, a sturdy T-shaped steel forging serves the same purpose. Lower trailing links on each end of the axle act as torque-reaction rods.

Beginning with the Giulia Super, the rear axle carried disc brakes. On earlier cars, lightweight radially finned-aluminum drums were used.

7.2 Alfetta and Later

The Alfetta series cars (and Milano) have one of the most exotic drivelines ever created for a passenger car. Their transaxle design and de Dion rear suspension form what is arguably the most sophisticated driveline possible for a front-engined car. This is not a new configuration. Ferrari, Pegaso, and Lancia all employed transaxles.

In the layout, the small bellhousing at the rear of the engine encloses only a rubber donut, which is connected to the driveshaft; the clutch and transmission are located at the rear of the car and are part of a large alloy casting which also carries the differential gears. This casting is bolted to the body. Jointed driveshafts drive the rear wheels, which are held solidly in position by a lightweight but sturdy tube.

This style of rear suspension is named de Dion after its inventor (of de Dion Bouton fame) and dates to the first years of this century. The lightweight tube which carries the rear wheels provides the same positive location as a solid rear axle, but since the driveshafts and

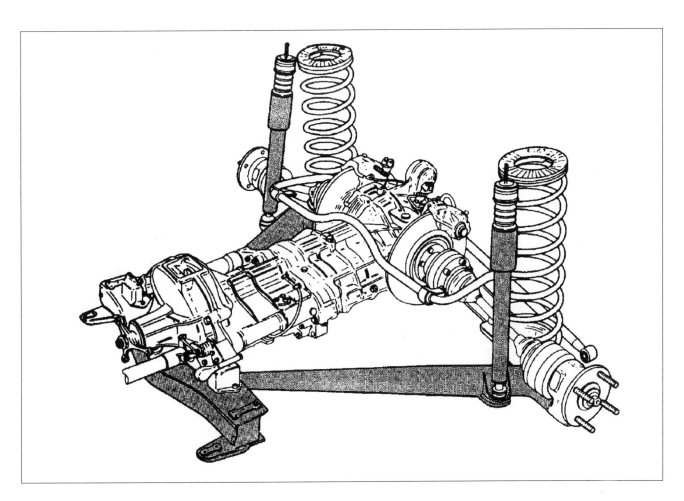

Fig. 2-47. Alfetta transaxle combines clutch, transmission, differential, and rear brakes in a single unit.

axle gears are not part of the tube, the unsprung weight is much less.

Placing the transmission at the rear of the car also helps equalize the fore/aft weight distribution of the major mechanical components. Most cars, with both transmission and engine up front, are nose-heavy and tend to understeer. Cars with the transmission and rear axle gears together (a transaxle) at the rear can have an ideal 50/50 weight distribution without compromising passenger room. The result is a better-handling car that steers predictably over a very broad range of driving situations.

The de Dion rear suspension and transaxle layout were used on the world-championship 1950–51 Alfetta formula car, so Alfa attached the Alfetta name to its new line of passenger cars for 1975.

The Alfetta driveline has proven a mixed blessing. It certainly offers all the advantages of excellent road-holding and good fore/aft weight distribution. But the driveshaft donuts which proved so trouble-free on earlier cars have become an Achilles' heel for the Alfetta. There are three in the Alfetta driveline, and the one which gives the most problem is located just behind the engine.

It's not exactly clear why there should be a problem. Several factors may contribute. The Alfetta driveshaft turns at engine speed rather than road speed, which means it's going a lot faster most of the time. Also, there's no clutch up front to help absorb driveline shock. But both problems should be solved by the fact that the Alfetta front donut, especially, is a much larger part than its predecessors. There may be additional pressure exerted on the front donut by a slightly out-of-alignment condition caused by heavy acceleration with weakened rubber motor mounts.

No one, in fact, knows why. But the front donut on Alfettas has clouded the virtues of an otherwise excep-

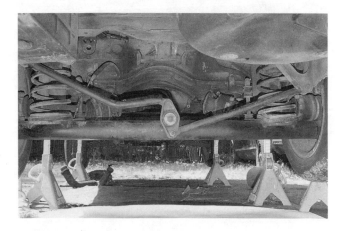

Fig. 2-48. *A de Dion rear end with transaxle and driveshaft removed. Z-shaped Watts link dead-center locates lightweight rear axle tube.*

tional design. Owners of Alfettas should consider the front donut as consumable as brake pads and, probably, in need of as-frequent changes. You should look at it regularly and replace it when it begins to tear (surface cracks are generally OK).

7.3 Front-Wheel-Drive

Beginning with the 164 in the U.S. (and the Alfasud elsewhere), Alfa became a front-wheel-drive (FWD) car. Early FWD cars had several annoying, if not actually dangerous, characteristics which have mostly been engineered out of the modern versions.

While there is no need to dwell on old troubles, owners of front-drive automobiles should be cautioned that the cars handle inherently differently from rear-drive vehicles. There is a natural tendency for a front-drive car to steer itself under acceleration in a tight turn. This phenomenon, known as torque steer, means that the driver sometimes has to fight a bad-handling front-drive car for control in turns. A light throttle through a turn is the fastest way around, providing the speed going into the turn is initially not too fast.

The 164 is remarkably well civilized, but the owner should remember that the rear wheels are not powered, and are provided mainly to keep the back end from dragging on the ground.

7.4 Chassis

Cars originally followed the construction of horse-drawn vehicles in that they had a body sitting on a chassis. The chassis was a significant suspension member on the old cars. Its primary function was to absorb the forces created by the suspension as it worked its way over the road. A flexible chassis spread the stresses of the suspension over the length of the car. Also, a longer chassis helped smooth the ride by averaging out the irregularities of the road.

With a flexible chassis, the road springs can be quite stiff to control lean during hard cornering and the shock absorbers can be quite rudimentary since the chassis itself works to damp wheel motion on rough roads. Alfas used separate chassis until the 1900 series of cars was introduced in 1951.

In 1927 Vincenzo Lancia introduced the Lambda, a car which did away with the chassis entirely. Suspension forces were spread over the entire tublike body structure and local sheet-metal reinforcements were used to strengthen the suspension attachment points. This chassis-less design was called monocoque or unit body construction. See Fig. 2-49. The advantage of the unit body is that, since it does away with the chassis, it is both cheap to manufacture and lightweight. All production Alfas from the 1900 model of 1951 have unit bodies.

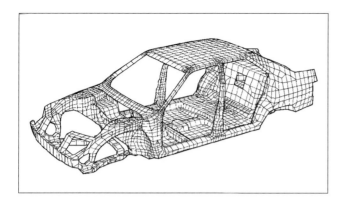

Fig. 2-49. A finite-element exercise in computer graphics from Alfa. Subject is unit body of 164. Alfa has a large computer-aided design capability, which was progressively applied beginning in the 1980s.

Body strength on unit-body Alfa convertibles is obtained by contouring the floor pan deeply and including large-section, welded-up sheet metal members which help form the door sills. Additional strengthening members run just inboard of the sills and also help support the lower front suspension A-arm and the motor mounts. At the rear, the trailing arms and the upper control link are mounted directly to the floor pan's sheet metal.

On the Giulietta convertibles, the dash was a structural part of the body. Beginning with the Duetto, the dash was removable, and so ceased to contribute to chassis rigidity. The 1750 and later Alfa convertibles all suffered from cowl shake, a side-to-side motion of the dash as the car traversed a rough road.

Unit bodies are vulnerable to rust. It took Alfa several decades to learn how to design a rust-resistant unit body, so you should inspect an Alfa carefully for rust before buying. Typically, rust forms along the joint where two layers of sheet metal are lapped together. On all Alfas, the floor pan should be inspected from beneath the car to identify rust which compromises the car's structure. Pay special attention to the area around the longitudinal reinforcing members. Small rusted-through holes on the floor, especially where the driver puts his right heel, are common but they do signal the need to search more closely underneath for significantly-rusted areas.

Cosmetic rust is much more visible but less important to body strength. Rocker panels used to rust quickly on the Giulietta convertibles and there was some concern that a car with rusted rockers was not structurally sound. In fact, the rocker panels themselves are cosmetic and not structurally significant, though the sills inboard of the panels are. Beginning with the Duetto, however, the rocker panel is a structural member.

Early Alfas rusted in the trunk around the battery, and, from the trunk, outward around the rear wheel wells, while the Alfetta coupes are prone to rusting around the front and rear glass and the shock towers in the engine bay. Nothing short of heroic efforts can slow rust propagation over the wheel wells or around the glass, so you simply accept the fact that all Alfas will have some rust, and the older they are the more rust they'll have. Even brand-new Alfas may have some rust, thanks to the salt-water bath they got while sitting dockside waiting for delivery to dealers.

There is a fabric or rubber "wiper" which seals the outside of the side windows as they go up and down. If this lip is damaged, rain will run down the window and fall into the door. Modern Alfas have drain holes along the bottom of the door, but if they are clogged, the bottom edge of the door will begin to rust away.

In the 105-series Giulia, Alfa used rubber bushes in the suspension. The through-bolts seize to the inner steel insert over time to the point that the suspension will seize, and worst-case seizures will actually tear the front cross member from the body. This failure is announced well in advance by squeaks from beneath the car as it travels over bumps. It is cured by replacing the offending bushes.

On all Giulia and later cars, the frontmost suspension pivot point (located near the back of the headlamp) should be checked for wear. If you hear a solid "clunk" every time you go over a speed bump or rough road, the odds are that the front trailing-arm socket requires replacement.

Fig. 2-50. This front trailing arm bushing (at right end of adjustable link) is ripe for replacement.

Beginning with the Milano, Alfas sold in the U.S. had rack-and-pinion power steering. Early SPICA units leaked rather quickly, and for some time there was no good cure except to replace the system, a very expensive repair for an only annoying problem. I know several Alfa shops which are currently rebuilding the leaking rack, but this is an item you must check if buying a used late-model Alfa.

7.5 Tires

How to Read a Tire

U.S. federal laws have created a new pastime: deciphering all the codes that must be printed on a tire's sidewall. If you're stuck in a parking lot waiting for your spouse to finish shopping, you might want to spend some time reading what's printed on your tires. They have load ranges, speed ranges, aspect ratios, belly sizes, rim sizes, inflation pressures, and an indication of whether they are of radial or bias construction. Some are measured in metric and others in English units. You can safely remain ignorant of the details since it's illegal for a store to sell you a tire that does not meet the minimum needs of your car.

Fig. 2-51. Tires are required by law to identify their construction. This radial tire has three tread plies and a single sidewall ply. Two tread plies are steel, and this is a Michelin tire (on a Campagnolo alloy rim).

The significant codes on a tire vary depending on whether they are radial or bias ply tires. For radials, a typical code is 225/50VR14. That means a belly width of 225 mm, a 50% aspect ratio (the height and width dimensions of the tire's cross-section are approximately square), Very high-speed Radial with a 14-in. rim diameter. The values in all these categories must match on each of a car's four tires for optimum handling and safety.

Fig. 2-52. This tire is rated at 1201 lbs at its maximum 35 psi inflation. With four of these tires on a car, combined weight of car and occupants can safely total 4804 lbs.

Bias-Ply and Radial

For a very long time, tires were built on carcasses that used strips of fabric laid at angles to each other—or biased—across the tread. The angles of the layers increased the strength of the tire and provided a soft ride. Since the plies wrapped from bead to bead around the tires, the side walls had the same number of plies and strength as the tread area. These tires offer a very comfortable ride and their transition from traction to no-traction is smooth, giving quite adequate warning of impending trouble.

Bias-ply tires were absolutely standard equipment until just after the war, when some sports cars began appearing with radial-ply tires from Pirelli and Michelin. Radials have a much more pliable side wall than the bias-ply tire. Though their side walls usually have only two plies and are more supple, the radial's tread area is reinforced by several beltlike plies that are formed from either fabric (Pirelli) or steel wire (Michelin).

Because of their more flexible side wall, these tires maintain superior contact with the road, while the extra radial belt makes the tread inextensible and the tire does not grow at speed. Radial tires ride a bit harder than bias-ply tires but they offer superior traction. The bad news is that the superior traction lasts only up to a point, when they cut loose rather suddenly.

No one would seriously consider bias-ply tires for a modern sports car. For normal driving, however, a bias-belted tire, which combines some of the best features of both designs, is a fine compromise.

It's critical never to mix bias-ply and radial-ply tires on the same axle of a car. The result of such a mix is a very unstable car, even when driving straight down the street. In fact, mixing ply types is never a good idea.

While there are significant differences in tires from different manufacturers, those differences are apt to be so subtle that different people will perceive the tires differently. After all, if there were a single tire that was so superior to all others that its superiority were easily perceived, it would be the only tire on the market. Which is why, if you ask ten different people what tire is best you'll get ten different answers.

8. Body/Interior

The materials Alfa uses for interior trim range from attractive to garish, and I have found all of them to be somewhat fragile over time. Don't be too quick to conclude that your Alfa has a leather interior, because some Alfa plastics are very leather-like. The best test is smell; after that, looking at the back of the material will verify its composition. The back side of leather is suede; plastic has a fabric weave.

Both plastic and leather seat covers dry out and crack or tear, while the door panels either warp from excessive moisture trapped inside the doors or simply wear out from arms and legs rubbing against them. Alfa has for decades supplied plastic water barriers behind the door panels, but these are frequently discarded by owners, much to the detriment of the door panel's life expectancy.

The goal of interior decoration is to hide the fasteners, and a great volume of cleverness has accumulated over the years. As a result, door panels, especially, may prove to be very frustrating to remove. Alfa uses screws and spring clips of several designs to attach trim. The door panels on Giuliettas have to be slid around a bit to free them from their upper attaching clips. Perhaps the hardest trim item to remove is the inside armrest on the door.

Modern Alfa seats have a foam core which can eventually dry out and crumble. A crumbling foam core leaves a coarse flesh-colored powder on the floor. It's important to keep the foam protected with a covering, especially if dry, so it isn't gouged out or otherwise damaged.

9. Electrical System

A car's electrical system is the most difficult and frustrating component to work on. If you're attracted to a used Alfa "with a wiring problem" take a cold shower and look for another car. Alfa wires themselves are sturdy enough, but Alfa connectors, including the spring clips which hold the fuses, are notorious for corroding and ruining the contact.

Wiring Diagrams

There is an "Alfa way" of wiring a car. Alfa's fuse block is quite logically laid out. As a gross generalization, the #1 fuse at one end of the block is a kind of master fuse which takes care of essentials including, on the Giulietta and Giulia, the ignition circuit. The four fuses next to it take care of all the accessories, while the remaining five fuses handle the lighting. The high and low beams of both headlamps are individually fused. See Fig. 2-53.

Fig. 2-53. *Because of its location, a Giulia Super fuse block without protective cover is as vulnerable to moisture as a covered one. White fuse (#5) is an accessory fuse: to its left are fuses which make the car run. To right are five fuses for parking and high/low beams.*

As labor-saving electrical devices were added to the car, the number of fuses in the fuse block grew, and a two-fuse accessory block is common on later cars.

One of the biggest challenges in troubleshooting an Alfa is to find its proper wiring diagram. Independent repair manuals such as Haynes include representative diagrams, but frequently not for the exact model you're working on. You should pick the wiring diagram which is closest to your model and, rather than trying to follow it slavishly, use it to try to understand how Alfa wires up the circuit you're troubleshooting. That is, use the wiring diagrams as guides, not as gospel.

This approach is especially necessary when it comes to color-coded wires. In the first place, the colors in the diagram are given in Italian, so N(ero) = black, and B(ianco) = white, and so on. Moreover, I have found disparities between wire colors in the diagrams and what's really on the car. If a wire is color-coded with a stripe, both colors are indicated on the diagram. A number, which is occasionally found after the color code on the wiring diagram, indicates the diameter of the wire in millimeters. Frequently, noting the diameter of the wire is more instructive than trying to translate colors.

The first rule of electrical troubleshooting is to verify a good ground. On earlier Alfas, ground is frequently provided by the chassis sheetmetal itself. On later cars that use plastic dashes and light housings, the ground is supplied by a black wire. Voltage in a black wire usually indicates that something in the circuit is leaking battery voltage to ground.

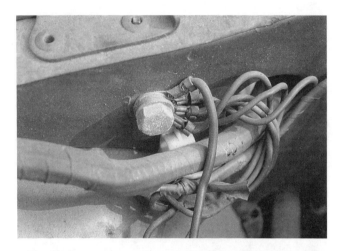

Fig. 2-54. A gathering of grounds. If it's electrical and doesn't work, the odds are it's got a bad ground. Start here, on driver's-side radiator bulkhead of an Alfetta. Be careful, though: not far away, on fender well, there's an insulated nut of very similar appearance with a lot of wires—and all of them are "hot."

Fuses

Most Alfas use the same pointed-end fuses as the Volkswagen Beetle. These fuses are nothing more than fragile strips of metal strung on a ceramic core. The color of the core typically indicates the capacity of the fuse, but a glance at the fuse material itself will tell its relative capacity: the wider the metal, the higher the current carrying capacity.

Relays

When you use the power windows, honk the horn, flash the lights, or operate the turn indicators, you do not yourself actually operate the circuit which controls the device. All the items mentioned require a higher-capacity switch than is convenient to mount on the steering column or the console.

The conveniently small switch you operate controls a larger, higher-capacity switch, or relay. Several relays are placed in the engine compartment along the fender well. See Fig. 2-55. They are little boxes with several wires connected to them. You can't tell by looking which relay controls which item, but you can hear them click if you hold your ear close to them and operate the proper item. (Pray it is not the horn relay! See below.)

Fig. 2-55. A relay looks like this: just a shiny box with wires going to it. Relays operate horn and headlamp flashers.

On Alfetta and later Alfas, the fuse block itself carries relays for the power windows and turn indicators.

The wiring diagram in the owner's manual gives the approximate location of each relay. The black ground wire runs to the horn button.

Horn Relay

If the horn sticks, tap the relay soundly, trying not to smash it in panic. If that doesn't silence the horn, you may want to rip all the wires from it. Don't. Just remove the red one, which carries current from the battery.

Power Windows

Enthusiasts hate Alfa's power windows because they add unnecessary weight to the car. Electrically speaking, they also add unnecessary complication. Alfa's wiring diagram of the power window relay is virtually indecipherable. The weakest link in the power-window chain is the dash-mounted switch itself which corrodes, rendering the windows permanently up or down depending on your luck. The switch is a press-fit into the dash and can be pried free for inspection and cleaning.

You can frequently restore the contacts in the power window switches using an aerosol contact-cleaner, available in electronics stores. The switches really aren't repairable but you can reach the contacts by bathing the back side of the switch with contact cleaner. The cleaner is nothing more than carbon tetrachloride under pressure. The one thing carbon tet dissolves better than corrosion is brain cells. Use the spray only in a well-ventilated area and try not to inhale while spraying.

9.1 Charging System

Alfa has never bothered the driver (until very recently) with information about the charging system by placing an ammeter or voltmeter on the dash. Pity: anything less than a fully charged battery in an Alfa is skating on thin ice, especially if you're in a cold climate. An inexpensive trickle charger is helpful to keep the battery well charged.

You can adjust the charging regulator, but if you don't know the proper procedure (which is given in the Giulietta shop manual) you can overcook the battery. Adjustable regulators are available in the aftermarket.

A simple test of overall charging system efficiency is to turn the headlights on and then start the car. The lights should dim perceptibly when the starter is engaged, then brighten as the engine revs. If the headlights don't get brighter as you rev the engine, it's likely that the charging system needs work.

Fig. 2-56. This is a Bosch (note the logo embossed on top) voltage cutout for the alternator. No adjustment is required.

9.2 Starter

Alfa starters are absolutely generic, and there's nothing at all remarkable about them. See Fig. 2-57. Starters from the Giulietta up to the 1750 are physically interchangeable, though their capacities differ. Starters for the 2-liter have three mounting holes instead of the two holes used for earlier cars.

A starter is designed to run in short bursts, not continuously. If you grind on a starter for minutes at a time, it will burn out.

There are three basic parts to a starter: the motor itself, a gear with a one-way clutch (called the Bendix), and a solenoid which simultaneously moves the gear into mesh with the flywheel ring gear and energizes the starter motor. While each part can be rebuilt (except the

later Bosch solenoids, which are virtually impossible to dismantle because they're crimped together), the modern approach is to trade in the entire unit on a rebuilt.

Fig. 2-57. Starter is a large black cylinder topped by a much smaller one (the starter solenoid). Heavy wire in this photo is main feed from battery. Wire to its right that forms a backward "C" runs from ignition switch to solenoid.

9.3 Windshield Wiper

The wiper motor itself is, like the starter, absolutely generic. Giulietta and Giulia wiper motors are long lived and almost trouble-free.

Fig. 2-58. Windshield wiper motor on a 1979 Sport Sedan is wonderfully accessible. On virtually all other Alfas, it's hidden by bodywork and in some cases requires significant effort to remove.

Beginning with the Duetto, Spider wiper motors are located in a cold-air plenum just forward of the windshield. A drain hole is located at the far end of the plenum on the passenger's side. The drain can clog with

Fig. 2-59. *On Alfetta coupes, windshield wiper is hidden beneath a flying buttress which also carries the hood latch with its large "mousetrap" spring. Note large plastic connector for wires to wiper motor: an improvement in Alfa electricals long overdue. This photo shows hood release cable with a safety pull wrapped around it. Your Alfa should have a safety cable, too.*

an accumulation of leaves and dirt. If it does, rain will cause a little swimming pool to form in the plenum, drowning the wiper motor. When the rain stops and the water evaporates, the inside of the motor rusts solid.

Preventive maintenance is necessary. Use a flashlight to peer past the grillework and locate the hole, then use a piece of strong wire as a roto-rooter. Bend the tip of the wire back into a small loop to keep it from piercing the rubber drain tube.

There are several different models of wiper motors, depending on whether or not a mist-cycle is featured. If you get a used replacement wiper motor, be sure it's for the exact model you own, otherwise the wiring will be wrong.

9.4 Inertia Switch

Alfas from the mid-1970s on are equipped with a "G" switch designed to cut off fuel flow in a crash. This switch is made in England and is also a Jaguar part. The switch looks like a gray firecracker with a short, fat fuse. See Fig. 2-60. A steel ball is held in a spring clip and, during an accident, the ball flies free of the clip, opening the fuel pump circuit. A failed inertia switch is the reason why your Alfa won't start when everything else is in perfect condition. Push the reset button to return the steel ball back into its spring clip. If that doesn't work, you can take the unit apart to clean the contacts. Never disable this switch: it's a serious piece of insurance, especially if you should roll over.

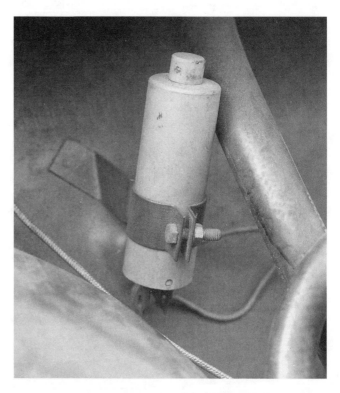

Fig. 2-60. *If your Alfa rolls over, this little "firecracker" shuts off the fuel pump to keep things from burning up. It will also shut things down if you hit a pothole really hard, or crash. Most are disconnected: bad. They're easy to take apart and clean. Disconnect wires and remove unit. Pry outer housing free of little nipple shown here at very bottom of cylinder. Pull base free of housing and clean away after you've figured out how it works (easy).*

Chapter 3

Buying an Alfa

MOST PEOPLE WOULD appreciate the counsel of a knowledgable friend when they set out to buy a car, especially an Alfa. That's what I intend to offer you here.

1. Buying a New Alfa

In the past, many Alfa dealerships/shops operated on a semi-official basis, but were generally run and staffed by Alfa enthusiasts who had learned about Alfa the hard way. From the early 1990s, under Fiat's influence, franchises had to be official with mechanics specifically trained in Alfa repair techniques. The day of the enthusiast-dealer is passing. While there are still Alfa dealers who'll be glad to have coffee with you on Saturday morning in the shop and talk about racing, they are a dying breed. For good reason: Alfa has found that such dealers are often poor businessmen. As a consequence, you're likely to buy your new Alfa from the same kind of steely-eyed professional who sells more mundane marques. If I were buying a new Alfa from a dealership, I'd first want to talk to a few of its service

Fig. 3-1. *GTV6 has the same body as a 4-cylinder Alfetta coupe, but a more robust drivetrain and a V-6 engine. Telltale hood bump and spoiler identifies the model; 4-cylinder coupes had a flat hood line.*

customers to verify their satisfaction. If you don't want to be quite so nosey, just show up when the service lane opens on a Monday morning and listen to the conversations between the service advisors and their customers.

More than the average car, Alfa needs knowledgable maintenance. The 164 comes with a comprehensive three-year warranty that virtually weds you to the dealer's service department for a major portion of the car's life. The warranty coverage is so complete that you may well drive the entire warranty period without spending money for anything but gasoline. As a result, however, if you're buying a new Alfa, first satisfy yourself that the dealer can maintain your car.

You may not feel you have a choice of dealers, of course—in the U.S., for example, there are fewer than a hundred. You can shop the enthusiast magazines or the Times to locate other dealers, but if you buy a car from them (at a discount, perhaps) you're faced with the unhappy prospect of having your local dealer service a car he didn't sell. I strongly advise you to buy any new car from the dealer who will service it (even if it costs more at first). An important corollary of this advice is that if you're not comfortable with your local dealer, then don't buy a new Alfa.

Unlike many cars, new Alfas offer few extra-cost options. Some dealers may try to improve their profit by loading the car up with accessories such as floor mats and "Never-Dull" paint protection. The stock Alfa is well, if not absolutely sumptuously, equipped, and you shouldn't feel you need to buy the "extras."

At some point in your Alfa ownership, you'll consider having your car serviced at an independent garage. It is a fact that most of the Alfas running around are serviced at independent shops. An independent shop does offer the distinct advantage of putting you eyeball-to-eyeball with the guy who actually services your car. Many of the operators of these shops have been trained by the factory when they were working as technicians for Alfa dealerships. As an unhelpful generalization, some of these shops are excellent and some are dismal. Even more confusing, some customers will swear by the same shop others swear at. I recommend talking to the shop's customers, or showing up Monday morning, just as if you were evaluating a dealership service operation. I will venture one bit of advice: avoid the "we work on all foreign cars" shops and try to find a shop that works only on Alfas.

A final word to the new-Alfa buyer: never scrimp on service.

2. Buying a Used Alfa

If you have some mechanical skills, you may be tempted to purchase a less-than-perfect used Alfa and save some money by fixing it up yourself. You will need all of your mechanical abilities to check out the car and identify what needs fixing.

In the example I'm going to use, I actually purchased the car before completing all the recommended checks. Some of the checks, such as pulling the cam cover, may not be appropriate on a used-car lot, but they should certainly be made before deciding to buy an expensive car. If you can't arrange to do them yourself, then have an independent mechanic make them. Generally, if you do what I describe in this chapter, you'll know very well what kind of a car it is you're buying.

I've bought and sold a lot of Alfas in the almost 30 years I've been in love with the marque. I've learned to divide used-car buying into two camps: buying from a private party and buying from a used-car lot.

Fig. 3-2. *Be sure your Alfa has the right engine in it. In a celebrated loss of sanity, the author once installed a 101-series Berlina engine, complete with single-throat Solex, in a 1750 sedan body. Car was so underpowered it was dangerous.*

Fig. 3-3. *There's a 100,000-mile club for Alfas, as this plaque shows. While owner may be proud, prospective buyer might be put off.*

Fig. 3-4. *If you simply must have a classic 6-cylinder Alfa, this is probably the model to buy. Although the 2600 is the most bulletproof Alfa of all time, according to an Alfa engi-* *neer who knows such things, they never had the charm of smaller Alfas and were not successful in U.S.*

Buying From a Private Party

I much prefer to buy from an individual. Over time I've learned that a car's owner is a better guide to the car's condition than a physical inspection of the vehicle. If the owner is compulsive, then his car is likely to be perfect. If he is slovenly, then his car is likely to be poorly maintained. Prolonged ownership of a car usually reflects an individual's lifestyle. If you're buying from a private party, buy from someone you really feel comfortable with. Most private individuals aren't into conning you out of your money.

There is a lot of ignorance involved in buying and selling Alfas from individuals, but not much intentional deceit. I once looked at a GTV coupe owned by a fellow who knew most of its history, though he really didn't know that much about Alfas. Just as well. The car's basic problem escaped me, but not my friend Jim Weber, who was along to notice that the front half of the car was built in 1976 and the rear in 1979. (The '79 models had a gasketed windshield while '76s had glued-in glass.) The car I wanted to buy had been concatenated from two wrecks early in its lifetime. Close inspection showed the weld joints at the rear of the door sill. The owner had no idea his car had been made up of two wrecks.

Fig. 3-5. *If you're shopping for a Milano, be aware that color of four-leaf clover is supposed to identify trim level: Silver, Gold, Platinum, or Green. Silver and Platinum badges are virtually indistinguishable, so plan didn't work well at all. Best buy is probably Gold Milano. Green clover model has 3.0-liter engine and Recaro seats.*

Beware of any seller who is selling a car he's just bought, or professes to know nothing of the car. The latter is an old used-car trick that relieves the nice salesman from telling you what is really wrong with the car.

CHAPTER 3

Buying From a Commercial Lot

If you find an Alfa for sale on a lot, don't bother asking the folks questions. You're on your own. On a used car lot, expect everything to be sparkling clean. They're pros at making silk purses out of sows' ears.

The car you're looking at will probably have a new paint job, but be warned that a good thick coat of paint will effectively cover rust for several months or so. You can usually tell if a car has been repainted by looking closely for paint traces from poor masking on shiny trim, at the rubber gasketing around the windshield and doors, and by looking in the wheel wells where fresh paint overspray can be found. If there are any electrical problems with a car on a used car lot, you can expect that to be just the tip of the iceberg.

First Checks

No matter where you find the car, there are two things you should always look for on any Alfa: a blown head gasket and rust. The other sins you're likely to find are all more easily survived, though a blown head gasket doesn't necessarily mean a trashed engine.

You can tell a head gasket is blown by the chocolate-brown emulsion that attaches itself to the dipstick or the radiator filler cap or excessive oil leakage below the exhaust manifold flanges. A blown head gasket is going to cost around $500. To understand more fully what a blown head gasket means, refer to Chapter 5, where I do a gasket replacement.

Rust is Alfa's fatal enemy. Rust usually reveals itself by bubbling up from under the paint. All rust is survivable — providing you're willing and able to hand-form a new car. Clearly, there is a limit to how much rust you can live with. Structural rust damage to the chassis includes a completely rusted floor pan, insecure motor mounts, or a weakened area around the suspension at-taching points. If the used Alfa has structural rust damage, don't buy it.

There are certain things you can more or less expect from a used Alfa. Second-gear synchronizers never work. You'll always grind going from first to second on a cold gearbox. It's reasonable to expect the carburetors or SPICA fuel injection system to be improperly tuned. Generally, insist that the seller obtain a "smog" inspection (even in states where it's not mandatory) to verify that emissions, especially HC, are within reasonable limits. Excessive oil consumption, like rust, is repairable but probably not worth it.

Wipe your finger around the inside tip of the exhaust pipe. If the residue is oily, then the engine is burning oil (or did, if you can be sure there was a very recent overhaul). Look under the car for leaks: engine oil, transmission oil, brake fluid, fuel, or water. You can tell the difference by appearance (color), smell, and feel— or even tasting (not sanitary, but accurate).

Look at the tires for cupping or uneven wear. Tire wear is an indication of chassis health: unevenly worn tires indicate misalignment or an abused chassis.

To the asking price of a car, you should add the costs of immediate repairs to determine its actual cost. Dollar values are treacherous to quote in a book. These are 1994 values. An engine overhaul costs about $1,500 to

Fig. 3-7. *Alfettas with "Euro" headers suffer crushed front head pipes, shown here. Pipe extends below sump and acts as an early-warning device that protects finned sump from cement parking stops or curbs over which enthusiastic owners drive. One more early warning and this pipe will be completely closed, burning valves and probably destroying the engine. Richly deserved.*

Fig. 3-6. *The Rust Faerie sat here. All Alfas rust, but Alfettas go at it more enthusiastically than others. Note that side of car is still bright and shiny.*

Fig. 3-8. This cracked manifold has been painted white to highlight a fault common to Alfettas. Braze, do not weld, to repair.

$3,500 depending on how sick the engine was to begin with. A good paint job runs about $1,500. A transmission overhaul costs about $500 to $1,000, and a total restoration probably between $10,000 and $30,000.

The value of any car is well known by someone, the trouble is finding the person. For recent cars, price guides available on the newsstand will give average prices. Banks and companies offering credit services will often have up-to-date price guides. If you can't locate one of these guides, look in your newspaper or "Auto Trader" publication to find competitive prices.

In the U.S., the Alfa Romeo Market Letter, published by Keith Martin (see Appendix 2), is a good place to find a specific Alfa you're looking for, though prices are on the high side. You can place an ad for the car you want if you can't find it anywhere else.

In certain areas, the values of Alfas prior to 1970 plummeted after a spectacular rise in the late 1980s. They are now again escalating, so any amount quoted would be very tentative. Generally, you wouldn't buy a '67 or '69 to drive every day except under unusual circumstances. If you are truly innocent mechanically, then stay with an Alfa never more than eight to ten years old.

3. An Example

In order to stay current on asking prices, I usually buy a weekly *Recycler* newspaper, which lists everything from condominiums to kittens for sale. The 1979 coupe picture below was not listed in the used-car section, but rather in the Parts Cars section. The owner, who had been a semi-professional Alfa driver (in a GTA), ran out of patience with this car, which was suffering some kind of fuel system problem. He had replaced the fuel tank and fuel pump without curing what he felt was a problem of clogged injectors. Rather than represent the car as ready-to-run, he offered it for parts.

The car had been stolen for its radio, then hidden in an abandoned garage. It was found about a year and a half later when the garage was about to be razed. The demolition company sold the car to a wrecking yard, where it was purchased by the seller. The seller had driven the car about six months before the fuel problem surfaced.

Fig. 3-9. This 1979 GTV was advertised in the Los Angeles Recycler *weekly as a parts car.*

Initial Inspection

This car was originally top-of-the-line, with factory air conditioning and sunroof. The owner thought the interior was leather (it wasn't). There was no battery, but I was assured the car would run for about 30 miles before "clogging the injectors." There was no evidence of major body damage and all the tires, though of mixed makes, showed reasonable and even wear. Externally, the most significant item was the rust that was developing all over the body.

Sales for this model were slow. Though titled as a '79, the tag on the doorpost showed that this car was manufactured in November 1978. Many Alfa coupes sat outside on the docks for months awaiting dealer orders. During this time, being bathed in salt-air from the ocean, they had ample time to begin rusting. Many Alfas were delivered to dealers already well rusted.

Fig. 3-10. *Factory-issue Campagnolo alloy wheels are a plus for the car. The fact that it has four different tread patterns on its tires suggest that maintenance had taken a low-cost approach.*

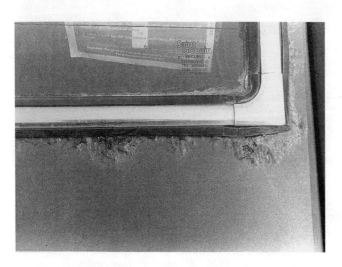

Fig. 3-11. *Bubbles of rust show beneath rear window molding.*

Fig. 3-12. *Rust along the front windshield, another common problem with this model Alfa.*

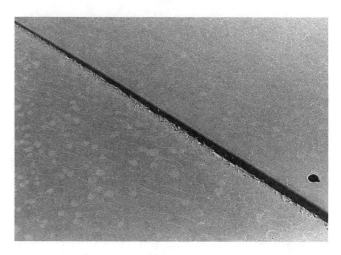

Fig. 3-13. *Not rust — flaking paint. This paint has weathered to point that it has lost its adhesion to primer. A proper repair would go to bare metal for a brand-new start.*

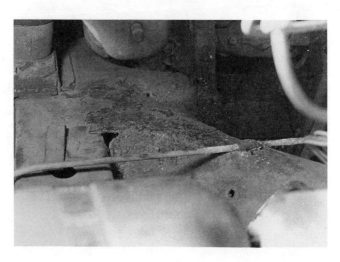

Fig. 3-14. *This rust near driver-side motor mount is common on Alfettas; it's structural and serious. It must be sanded away and metal refinished to stop the damage from getting worse.*

The car's odometer showed only 13,000 miles, but the seller suggested that about 30,000 was more appropriate. The years of idleness certainly contributed to the poor exterior condition, and will no doubt also mean dried-out and leaking oil seals throughout the driveline in a few thousand miles. All things considered, the car more likely had 113,000 miles on it.

The interior of the car showed more wear than would be appropriate for the 13,000 miles recorded on the odometer. The seats were worn as well as ripped, and the floor mats showed wear as well. High-mileage indicators included the condition of the seats and floor mats, the amount of wear on the brake and clutch pedals, and, for truly high-mileage cars, the amount of wear on the gearshift lever.

Fig. 3-15. Torn seat seams could be a result of brittleness caused by the car sitting in a hot place. Torn fabric gives opportunity to verify that material is plastic, not leather. Woven fabric backing indicates man-made material.

Fig. 3-16. Writ large: electrical problems. Dismantled steering wheel (horn) and hanging fuse block say that someone has tried electrical troubleshooting and then given up. Tangle is stock; disconnected or spliced wires aren't. Somewhere in there may lurk a short. It would be much easier to find another car.

My greatest concern looking at the interior of the car was for its electrical system. The steering wheel had been dismantled, the horn wires were loose, and the fuse block was hanging because one of its pivots had been broken. Electrical problems are sometimes baffling and, if you're going to have a mechanic solve them, expensive. Generally, any indication of electrical problems should cool your ardor for an Alfa significantly.

Fig. 3-17. More evidence of electrical problems: inertia switch has been disconnected and its wires (poorly) taped up. Braided line running in front of switch is signal from catalytic converter sensor. Look closely and you can see that there is an electrical burn on line just where it passes over a heater hose. Burn no doubt came from the inertia switch wires.

Fig. 3-18. A dirty engine with several evidences of oil leaks. Some leaking is to be expected: car sat idle long enough for its rubber seals to dry and become brittle. Not expected: much of the oil comes from a leaking cam cover, which is missing nuts for two front attaching bolts.

Fig. 3-19. This car, if it has low mileage, has also had a hard life: broken hood release is not typical and suggests over-use. Makeshift repair argues against buying car. (There was a brand-new cable in trunk.)

Fig. 3-20. Never expect air conditioning to work, anyway. Sight glass (located dead center on drier) is covered with dust, suggesting that no one has checked freon charge for several years. Original decal on fender tells me that engine bay paint is original. Since it matches exterior color in overall condition, I conclude that majority of body has never been repainted.

The Decision

Lots of problems, but the owner was a genuinely nice guy and Alfa enthusiast. Had I seen the car on a used-car lot I would have passed it by. Based more on the individual than the car, I decided to buy it even though I'd never heard it run.

I want to point out that a very low price often makes up for a large number of problems. And even though I bought this car for only $700, I want to emphasize that I took a very big gamble by not hearing the car run or otherwise more closely checking its mechanical condition. Don't do as I do: do as I say.

Always ask a private seller to deliver the car to your home. This assures that it runs and any difficulties with the car will probably be flushed out by your request, especially if the distance is more than a few miles.

In this case, I put the car on my trailer.

Closer Inspection

I had purchased a nonrunning car. The first thing to do was to verify that it will run. To begin the work, I soaked the engine bay in a degreaser, waited for about five minutes, and then used the highest-pressure water source available to wash away all the visible grease.

Then, I dried the coil, distributor, and spark plug wells to give the engine a fair chance when I tried to start it.

I remove d the air cleaner to gain easy access to the distributor and fuel injection pump. I then installed a battery and turned on the ignition to activate the rear fuel pump. This is a closed fuel system, so I removed the fuel filler cap before draining the tank. By running a new fuel line from the main filter into a large container, I drained the tank. Then, I installed a new main fuel filter.

In the process, I noted that the center capsule of the thermostatic actuator was not supported and its attaching bracket was missing. See Fig. 3-21. This is a potentially serious matter because the actuators are very expensive, and an unsupported line will certainly break. Before the engine runs, I bolted up a proper bracket and secured the line. I removed the plugs and foundthem well and truly fouled.

Now that I had a battery installed, I could take a compression check. See Fig. 3-23. I was not interested in a very accurate measurement. My usual rule of thumb is that if all cylinders show over 60 psi compression, the engine will run. In this case, all pumped up to over 125 psi, a good sign for a cold engine with no real lubrication.

Had I encountered a cylinder giving less than 60 psi, I would have squirted some engine oil into the cylinder and rerun the test. If the compression then had come up to over 100 psi, I would have known the rings were bad. Had it stayed low, I would have to tear the engine down for bad valves, or a badly worn piston.

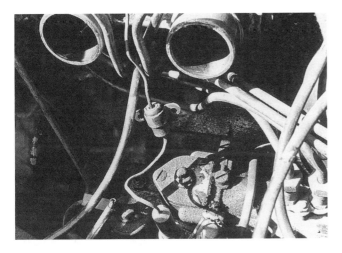

Fig. 3-21. *An unsupported thermostatic actuator line is an invitation to trouble. It's also an indicator of ignorant maintenance.*

Fig. 3-23. *A good compression check helps assure me that I've made a wise choice.*

Fig. 3-22. *Plug removal is one of first steps to evaluating a used car. Correct extended-tip plug, but very fouled. A new set is required.*

Fig. 3-24. *Too loose: a proper tension will deflect between 1/8 and 1/4 in.*

Satisfied that the engine was basically in sound condition (well, I might have been surprised by a rod knock when it first ran), I went on to check the valve clearances. I would never run an unknown engine for long without first checking the valve clearances and the condition of the camshaft drive chain.

The next step was to remove the cam cover. At this point, I discovered that the two nuts for the front cam cover bolts were missing, in part because the front air conditioning bracket makes them almost impossible to reach. The bolts were dropped in place, making them look like they were attached. In fact, they were not.

After removing the cam cover, I checked the timing chain for proper tension. See Fig. 3-24. It needed tightening. In fact, it had needed tightening for some time. The chain was loose enough to belly out and wear itself into

the aluminum head. See Fig. 3-25. This is, unfortunately, a common problem on Alfas that are not carefully maintained. When I tried to loosen the chain tensioner nut on the front of the head, I found that it had been tightened much too tight. Everything about the cam drive encountered so far reflected poor maintenance.

The valve clearances were within limits I'll accept: not perfect, but adequate. The exhaust valves all had between 0.017 and 0.019 in. clearance. One intake valve clearance of 0.012 in. was too tight, but not nearly so critical as an exhaust valve. I wrote down the clearances and checked them again after about 1,000 miles to verify that a valve head was not stretching, closing up the clearance further.

I next located the timing marks on the crankshaft pulley and brought the engine up to top dead center (TDC). I checked the static ignition timing by hooking a test lamp between the terminal on the distributor and

CHAPTER 3

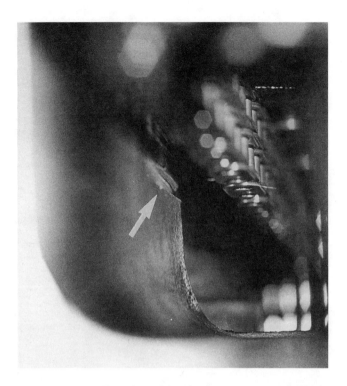

Fig. 3-25. Looking down front timing cover, you can see how camshaft chain has begun to eat into soft aluminum (arrow), feeding a fine shower of metal into engine.

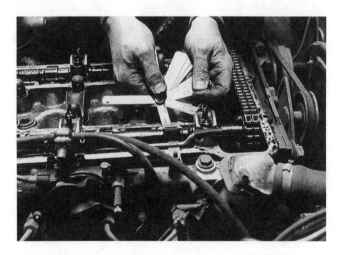

Fig. 3-26. Checking valve clearance. Various thicknesses of feeler gauge are used as a go/no-go gauge.

I re-tensioned the cam chain, put the cover back on, fitted new spark plugs, added some clean fuel to the tank and slipped behind the wheel. The car started. I took it for a brief drive down the driveway, grinning broadly. All the gauges worked. The engine felt very powerful. The Big Question was answered.

I let the engine cool down, then probed around the engine bay some more. I noticed that the oil pressure sending unit was dented. Since the unit is relatively inaccessible, the dent suggests that the engine had probably been removed at some time, or that the intake manifold had been removed (and slipped coming off).

I next put the front of the car on jack stands and crawled underneath. The first thing to look at is the condition of the front driveshaft donut, just behind the engine. This one was in perfect condition. My spirits rose further.

Fig. 3-27. Looking up at front rubber driveshaft donut through rear motor-mount inspection hole. It looks new. A few surface cracks would have been expected.

ground, then switching the ignition on. By turning the rotor slightly against the automatic-advance springs, I can make the light turn on (points open) and off (points closed). The timing was right on.

The cooling system was empty, and the seller told me that the water pump leaks. Undeterred, I filled the radiator with water, then stepped back to watch the engine drain itself from a badly leaking water pump. At this point, the pump was of secondary concern. Getting the engine running was more important to me.

Working along to the rear of the car, I discovered there was no rear fuel filter. There should be. I'm not sure if the pore size of the rear filter is larger or smaller than the main filter in the engine bay, but my experience with the SPICA system is that an absolutely stock setup is the only way to assure proper performance. I installed a new rear filter.

While under the car I inspected the suspension bushings and brake pads and rotors. Everything seemed OK. The bumps and groans I heard during my brief drive were expected and will probably go away with use.

The exhaust system leaked. Somewhere. I heard, but cannot see the leak. Later.

I slipped behind the wheel and tried the wipers and lights. Considering the carnage hanging down from the dash, I was gratified that the wipers worked (both speeds), as well as most of the exterior lights. The turn indicators didn't work, and needed to be repaired before driving the car on the street.

This is really all I need to tell you about checking out a used car. However, since we're so deep into this one, I want to tell you the final outcome.

When I tried to restart the car, it wouldn't start. After repeated attempts, I pulled the plugs to find them fouled. Clearly, there was a fuel-delivery problem, but the likelihood of clogged injectors was fading. I had found no contamination in the fuel I'd drained from the tank and the new filters assured a clean supply. Now, I have had the experience that a SPICA system that is fuel-starved from a clogged filter will foul plugs, so I had not decisively eliminated the possible injector problem.

So, I put in known-good injectors and cleaned the plugs on a wire brush. The car would still not start.

I knew the system was basically good, otherwise the engine would not have run so well for the brief driveway sprint. All my diagnosis pointed to an over-rich, not over-lean, condition. The warning about clogged fuel injectors still haunted me. I pulled the plugs again, backed off the connections to the injectors, and turned the engine over with a remote starter to check for fuel delivery.

There was an even leak around the injector connections, so I had verified that the pump was delivering fuel as it should. To my surprise, a virtual geyser of vapor spurted from each spark plug hole as I turned the engine over. The spark plug wires lying near the openings were covered with liquid.

It could have been water (serious crack or blown head gasket) or fuel (over-rich). A taste revealed a rather uninteresting vintage of gasoline: my engine was running horribly over-rich. Reluctantly, I loosened the lock ring on the SPICA fuel cutoff solenoid.

Well, you'd have to read Chapter 7 to proceed from here on your own. On a system that should be adjusted in increments of 1/8 and 1/16 of a turn, I leaned out the system 9 full turns before achieving an approximately proper mixture. In fact, I was initially guided by the density of the vapor geyser from the spark plug holes. Its intensity began to lessen visibly only after 7turns.

I believe that the initial run was an absolute stroke of luck. The engine was running so rich that it really shouldn't have started. After leaning out the fuel deliv-

Fig. 3-28. Clean and ready for final diagnosis. At this point, fuel injection pump was still running over-rich.

ery, the engine now started reliably and was ready for a smog check, when its final adjustment was made.

A lot remained to be done on this car, but the big hurdle—getting it running—was achieved. Paint will be expensive, for the front and rear glass, as well as the chrome trim around the side windows, will have to be removed to clear away all the rust. Then, the car will be taken to bare metal.

The main mechanical components of the car, however, were very good and I'm convinced this was a wise purchase, especially at $700.

Epilogue:

After putting everything right on this car, it has proven to be one of my favorite Alfas. The amount of work I did on the car—over the months correcting everything except the rear shocks—if charged at mechanics' rates, would have made it a financial black hole. The car has since returned about 30,000 trouble-free miles and a lot of satisfaction. I have now set it aside to refurbish an Alfetta sedan.

4. Buyer's Checklist

The following checklist is a rough guide to the most frequent problem areas likely to be encountered on a used Alfa. For simplicity, both carbureted and mechanically fuel injected cars are grouped in the column headed "1600/2.0." The column headed "Alfetta" refers to 4-cylinder Alfas with the 115 chassis.

How to Use the Checklist

The numbers given in the grid indicate the relative seriousness of the problem. If you add up all of the appropriate numbers for the car you're considering buying, then the higher the total, the less enamored you should be of the car. Generally, if you get a total of more than 50, you are probably looking at a car that requires excessive repairs.

Please consider the fact that this chart is not a substitute for vision or reason; we have all had our heart overrule our head at one time or another.

Item	Giulietta	1600/2.0	Alfetta	V6
Engine				
Mechanical noises	5	5	5	5
Low compression	5	5	5	5
Exhaust smoke (white or blue)	4	4	4	4
Oil (emulsification)	5	5	5	5
Coolant (emulsification)	3	3	3	3
Oil leak (head gasket)	2	2	2	2
Oil leak (timing belt detensioner)	-	-	-	3
Radiator fan removed	2	3	3	4
Water pump leak	3	3	3	3
V-belts broken/missing	1	1	1	1
Hoses broken/missing	1	1	1	1
Fuel/Ignition				
Air cleaner housing missing	2	2	2	5
Plastic seal on fuel injection pump missing	-	-	5	-
Hard starting	2	3	4	5
Rough idle	3	3	3	5
Erratic power under acceleration	3	3	4	4
Mixture richness (sooty exhaust smoke)	2	2	4	5
Ignition timing spark knock	2	2	2	3
Fuel pump pressure low (light on)	-	-	5	5
Thermostatic actuator tube broken	-	4	4	-
Emission controls missing/disabled	-	2	4	5
Carburetor conversion	-	3	4	5
Generator mounted securely	4	2	-	-
Erratic starter switch on ignition	2	1	3	3
Electrical				
Battery charge light inoperative	4	4	4	4
Some gauges inoperative	2	3	4	4
Fuse block visibly damaged	1	2	3	3
Lights/accessories inoperative	2	3	3	3
Power window switches don't operate	-	3	-	3
Subtotal				

Item	Giulietta	1600/2.0	Alfetta	V6
Transmission/Driveline				
Oil leaks	2	2	3	3
Continuous mechanical noises	5	5	5	5
Hard shifting (ignore 2nd gear)	4	4	2	2
Clunks or erratic noises while driving	4	4	4	4
Vibration	1	1	4	4
Torn/broken rubber driveshaft joints (donuts)	-	1	4	4
Abnormal clutch operation	4	4	4	4
Brake booster inoperative	-	2	3	3
Chassis				
Uneven tire wear	2	2	2	2
Body bounces after a bump	1	1	1	1
Clunks over bumps	1	3	2	2
Rust				
Body pan	4	4	3	3
Rocker panels	1	1	2	3
Sub-chassis	5	5	5	5
Wheel arches	1	2	3	4
Upper front suspension mounts	-	-	5	5
Trunk/spare tire well	1	2	3	4
Bottoms of doors/door sills	2	1	2	2
Areas surrounding windshield/rear window	-	-	4	4
Unequal tracking (front/rear wheels don't align)	5	5	5	5
Steering pulls to one side	2	2	2	2
Power steering rack fluid leak	-	-	3	3
Bodywork/Interior				
Obvious body damage (sheet metal wrinkles)	5	5	5	5
Loose seat mountings	4	4	1	1
Missing seat belts	1	2	2	2
Subtotal				
Subtotal from other column				
Total				

Chapter 4

Maintaining Your Alfa

SOME PEOPLE FEEL that Alfa is a complex, unreliable vehicle, the workings of which can only be fully known to those who speak dialectical Italian. There are, I will confess, those whose experiences confirm the belief. These people will forever be at the mercy of repair shops, ignorant even of whether or not they're getting the kind of competent service they're paying for.

Alfa is not a complex car. Quite the opposite: its racing heritage has contributed significantly to leaving Alfa free of complex work-arounds or exotic design compromises. The reason Alfa is considered exotic is that it is not built like the Fords, Chevrolets, or Toyotas most service shops know so well. And many mechanics, rather than admit their own ignorance, blame the car. The mechanic who tells you that Alfa is a disreputable mistake is probably describing the quality of his own service skills.

There are several compelling reasons why you should learn something about your car. At the least, you can defend yourself from incompetent service. Ideally, you can save some money by doing simple maintenance tasks. Altruistically, you can learn about one of the best-designed vehicles on the road. Hopefully, you can also have some fun and gain a sense of satisfaction.

I encourage you to learn enough about your Alfa so you can perform all the routine maintenance procedures given in this chapter. My words alone won't be enough to give you the confidence to open the hood. You need to cultivate one or two friends who are knowledgable about Alfas. The best way to do this is to join an Alfa club. If you just can't bring yourself to be a joiner, then find a specialist parts store that will give advice and sell you parts at the same time. These stores advertise in enthusiast magazines, especially in *Autoweek*, which is available at your local newsstand.

The Tune-Up Myth

"Points, plugs and condenser. Fiddle with the timing and carb, change the oil and give 'er a grease job." We're talking history, here. (Remember how often you needed a new coil?) The classic tune-up began to disappear at least 30 years ago. First to go was chassis lubrication, which was eliminated by rubber bushings and permanently-sealed joints. Electronic ignition systems

Fig. 4-1. *Proper way to examine a restored—and valuable—car is with your hands grasped behind you. This immaculate Giulietta 750 Veloce was photographed in 1982. In interim, another zero has been added to car's price tag. Veloce, in this model, means dual Weber carburetors and other mods. Standard engine had a single Solex carburetor, but was never referred to as a "Normale."*

did away with points and condensers and significantly lengthened the service life of the spark plugs. And, owner-adjustable fuel-delivery systems haven't been legal since 1971. In other words, there is no longer any such thing as a tune-up. What the mass-market service shops actually do for their "$59.95 tune-up," Lord only knows.

The Italian Tune-Up Is No Myth

You've sunk your next five years' paychecks into this sweet Italian beauty and you're not going to abuse it, by golly. You shift at 2500 rpm and never go over 70 mph. To prove your point, the car needs tuning so often now that if you really put your foot into it you're sure something would explode. That's Italian cars for you.

Somewhere, right now in Italy, someone's driving exactly the same car as yours. His right foot is pressed firmly to the floor, where it has been for the last three hours. The rubber on the brake pedal shows no perceptible wear. The speedometer and tachometer are both trying to bury themselves offscale. His arms are weary

CHAPTER 4

from cranking the wheel back and forth as he negotiates the mountain roads. His eyes gleam and his heart is glad for he has never once had cause to do more than regular maintenance to this thoroughbred of a car that has uncounted kilometers on its engine.

He's right and you're wrong.

Alfas don't just like to be driven hard. They *need* it. When your car limps in for service, the plugs are fouled and the oil is so diluted with volatile hydrocarbons that the engine is grinding itself to death. To correct this sorry condition, the mechanic waits until you are just out of earshot and then takes off in your jewel, driving it with an abandon that would give you seizures. After a few minutes at full-throttle and something nearing 100 mph the spark plugs clear and all the junk in the crankcase begins to evaporate. He gets a big grin, you get a bill for $59.95, and the car never ran better.

Driving Tips

There is a definite technique for driving an Alfa. The guidelines are both straightforward and important. If you ignore them you will miss much of Alfa's appeal. In worst cases, you will damage the car.

Each model Alfa has a unique starting procedure. It's important to determine the best starting procedure for your particular car and then follow it each time you start the car. If your car won't start easily using your optimum technique, it needs a tune-up. The details of handling a nonstarting Alfa are given in Chapter 6.

I always start my Alfas with the clutch pedal pushed in to avoid the extra drag of the transmission gears. This can cause accelerated wear of the clutch release bearing, but it's the way I prefer. As a safety precaution, always start a car with the parking brake engaged.

If your Alfa has carburetors, starting depends on the right balance between choke, throttle opening, and the number of pumps you give the accelerator pedal. If it's near or below freezing, pull the choke out all the way; otherwise leave it in. Pump the accelerator pedal once or twice. This adds a little extra raw fuel to help the initial charge light up properly.

Hold the accelerator pedal down about an inch and operate the starter in a series of short bursts: don't keep it engaged for more than five or ten seconds at a time. Grinding away on the starter runs down the battery quickly and can also overheat the starter.

Starting a Giulietta Veloce in freezing weather requires a minimum of 10 to 20 strokes of the accelerator pedal before engaging the starter. Once the car is started, ease the choke in slowly, using it to maintain the idle speed.

The cold-start enrichment mechanism on Alfas with SPICA injection is built-in. Getting started depends almost entirely on finding just the right initial throttle opening, and I've found that varies from car to car. Be-

cause the fuel injection system does its own thing, grinding away on the starter will only flood the engine.

Turn the ignition on and wait until the fuel pressure light goes off. Hold the accelerator pedal down about an inch and operate the starter in a series of short bursts. If the car won't start after about ten tries, stop, and try again in about 15 minutes.

Some SPICA systems seem to start best if you slowly depress the accelerator pedal while cranking. Using this technique, you continue to open the throttle progressively until the engine catches.

Bosch fuel-injected Alfas require no starting technique at all. In fact, if you try to embellish start-up on a Bosch-injected car, you'll only make the car less likely to start. With your feet off the pedals, just turn the key.

Once started, let the engine warm up at idle briefly. Leave the transmission in neutral so that the idling engine spins the transmission gears. If you're late to work, make haste slowly: never use full throttle on a cold engine. And shift slowly.

As you shift gears, pause in neutral momentarily. Don't think in terms of first-to-second, rather first-to-neutral-to-second. This technique is the same whether you're going up or down through the gears. Work at achieving smooth gearshifts which are lurch-free. Some Alfa drivelines have rubber donuts which are intended to help absorb shocks, but if you needlessly overstress them they'll not last long.

Alfa clutches are strong but not idiot-proof. Never slip the clutch unnecessarily. Above all, don't slip the clutch to hold yourself on a hill.

Alfa engines make up for their diminutive displacement with lots of high-end horsepower. Because they are tuned to be efficient at wide throttle openings, they shouldn't be driven constantly at low engine speeds. Stay between 2000 rpm and 5000 rpm for most driving. When thoroughly warmed, there's no harm at all to take the engine up to its redline. Indeed, it's fun.

1. Engine

1.1 Cooling System Maintenance

Never remove the radiator filler cap if the engine has reached normal operating temperature. In fact, there is little need ever to remove the cap on the radiator of a recent Alfa. They have a plastic bottle located in the engine compartment that is connected to the radiator by a rubber tube. You check and add coolant using this bottle. See Fig. 4-2.

To change all the coolant in the car, you need to drain both the radiator and the engine block. Place a large, shallow pan (it should hold at least a gallon) beneath the front of the car. With an absolutely cold engine, remove the radiator cap and loosen the circlip that

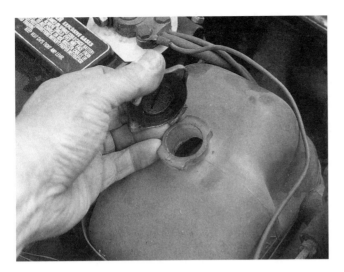

Fig. 4-2. *Radiator overflow bottle with its cap removed. This is how you check coolant level, not by removing pressurized radiator cap.*

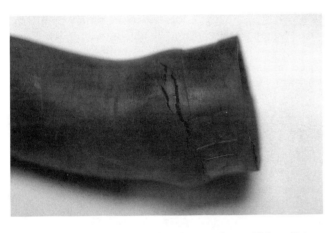

Fig. 4-3. *Hoses typically crack near clamp. This radiator hose hasn't failed yet, but the end is near. Test a hose for brittleness and cracking by bending or crushing it slightly with your thumb.*

holds the lower radiator hose to the radiator. This is typically an awkward, frustrating task. Coolant will begin to seep out of the connection and drip into the pan. Position the pan so the drip falls in its approximate center, then pull the hose free from the radiator. You'll get a big gush of coolant that will spray all over everything except the pan you've carefully positioned.

The 4-cylinder Alfas drain the block from the rear, beneath the exhaust pipes. Early Alfas have a petcock while later ones have only a brass plug. The petcock is usually plugged with debris and should be cleaned out with a piece of wire. The petcock's handle is parallel to the petcock when open. The V-6 has a brass drain plug on both banks, located toward the back of the engine.

Take some time to inspect all the rubber coolant hoses for cracks. Replace any that are questionable. It's cheap insurance. Let things drain for a minute or so, then button everything back up for the refill.

Regardless of climate, use equal parts of distilled water and a reputable antifreeze to get a 50% solution. I have seen some electrolytic damage from using tap water, so fastidious types will insist on using distilled water as well as a sacrificial metallic slug in the radiator to lessen electrolysis.

1.2 Engine Oil Change

Check the oil regularly. How often depends on whether or not your Alfa is consuming (burning or leaking) oil. Normal oil consumption ranges from a quart every 1,000 miles to no perceptible consumption between oil changes.

The owner's manual gives specific oil recommendations. If you don't have your manual, you'll find that a new Alfa will run just fine on 10W-30 detergent oil. Or, you can use a straight 30-weight oil in the summer and 10-weight when temperatures get around zero. For older Alfas, 20W-50 or straight 40-weight oil will help reduce oil consumption if the rings are getting weak.

The choice between a straight or multiple-viscosity oil is personal. A multiviscosity oil gives a slightly lower oil pressure reading but probably lubricates better. Don't stint on the quality of the oil or the frequency with which it's changed. Every 3,000 miles is the proper interval to change oil, but don't let it get much older than that. The larger percentage of its life the engine runs cold or just warming up, the more frequent the oil change should be. Clean oil is the best insurance for your engine's longevity.

An oil change is unavoidably dangerous to the amateur who doesn't have a proper hoist. In the first place, the car could fall on you. Always use at least two jackstands to support the car and never crawl under a car that is supported only by a jack, no matter how sturdy. In the second place, you could fail to snug up the oil drain plug when replacing it, risking the chance of a catastrophic oil loss. Don't be ham-handed, just careful, when tightening the plug. Finally, oil should be changed when the engine is hot, so there is some danger of burning your hand. Work carefully.

A large brass oil drain plug is located on the bottom of the engine. Place a drain pan with a capacity of at least seven quarts under the plug and then unscrew the plug. Try to keep the plug from falling into the drain pan and at the same time avoid having hot oil course down your arm toward your armpit.

While the oil is draining, wipe off the magnet on the drain plug so it's absolutely clean. Let the oil drip for at least five minutes before refitting the plug. While it's dripping, change the oil filter.

CHAPTER 4

Oil Filter Change

Change the filter each time you change the oil. The cost of a filter/oil change is a very small price to pay for engine longevity. You change cartridge-type filters by removing the entire filter assembly where it bolts to the block and then dismantling it.

Spin-on filters are most easily changed from beneath the car. You'll need an adjustable filter tool to get them off. Never pierce the filter or try to use a chisel to free it from the engine. Be sure to lubricate the sealing ring on the new filter and never tighten it more than 3/4 of a turn after it seats against the block.

Fig. 4-4. Spin-on oil filter is tucked under alternator. Space is further limited with an air conditioner compressor fitted. On Spiders, it's hard to get to because of hammer-headed sump. On Alfettas, it's easy to spin off from underneath.

1.3 Timing Chain Tensioner

If you don't regularly tension it, the chain will loosen, hit the timing cover, and feed the engine a constant supply of aluminum shavings. The factory procedure: Every five or six thousand miles, put the car in fifth gear and loosen the tensioner bolt about one turn. See Fig. 4-5. Push the car forward through one complete engine revolution, being careful not to roll backward, then tighten the tensioner bolt. Recheck the cam timing after retensioning.

My experience is that, with older engines, the tensioner bolt has been tightened so much that the adjusting idler is wedged in its bore. One turn may not free the adjuster so that the full spring tension is obtained. I feel that the tensioning spring is actually marginally strong for the job. Because of this, I prefer to set the ten-

Fig. 4-5. This is lock bolt for timing chain tensioner. Overtightening it will render tensioner useless.

sion by hand, with the cam cover removed. I use a large screwdriver to wedge the tensioner pulley to the desired tension, then tighten the tensioner bolt. See Fig. 4-7. I tighten the chain so that I can obtain about 1/8 to 1/16 in. deflection of the chain measured between the cam sprockets.

Fig. 4-6. Chain tensioning. This is author's approach, not factory's. Screwdriver can be used to exert excessive force very easily, so be careful.

Fig. 4-7. *Looking down on chain drive to head. Spur-shaped assembly, seen from top, is chain tensioner pulley (arrow). Base of assembly fits into a mounting boss in head that contains tensioning spring and lock bolt.*

1.4 Timing Belt Tensioner

As a general rule, the belt should be replaced at regular 50,000-mile intervals (the newest belts have a 100,000-mile service life) whether it appears to need replacement or not. Always replace a belt soaked by leaking engine oil. The damage wrought by a broken timing belt is sorrowful to behold and may include anything from a bent valve to a completely trashed engine, depending upon how fast the engine was turning when the belt let go. Regular replacement is cheap insurance.

If you have a modified fuel-injection control unit (ECU) which allows engine speeds over 7000 rpm (and you use such speeds occasionally), belt replacement may be indicated at anything from 30,000 down to only 3,000 miles: "Let's see, that'll be the usual oil change and a new timing belt, please."

There is a device on the V-6 to detension the camshaft drive belt at start-up, idle, and high speed. This is a maintenance-free item with a propensity to leak oil on V-6s prior to 1994. (Beginning in '94, a thermostatic unit replaced the problematic hydraulic version.) If the front of your V-6 engine is smeared with oil, you may need to have the detensioner renewed. Checking regularly for leaks is a maintenance item in this case. Renewing the detensioner is a procedure that can be performed by an experienced owner as described in Chapter 5.

1.5 Cylinder Head Nuts

All modern Alfa engines run some risk of leaking head gaskets. I'm not alleging a basic product fault in observing this. What I am saying is that the odds are, if you have any problem at all with your engine, it's most likely to be a leaking head gasket. The only defense you have against this failure is to verify that the cylinder head nuts are tight. Check their torque every six months, even if you have to buy your own torque wrench and learn to use it properly. (45 to 70 ft. lbs. is typical; verify the exact value from a shop manual.) The sequence of tightening the nuts is simple: it follows an outward spiral path starting with either of the nuts between cylinders 2 and 3. See Fig. 4-8.

Some owners feel that overtorquing the nuts by 10 ft. lb. provides additional insurance against head gasket leakage. I have a basic objection to recommending overtorquing, but must observe that the typical inaccuracy of some torque wrenches may exceed 10 ft. lb. anyway.

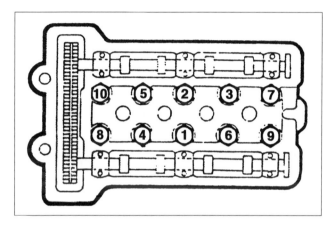

Fig. 4-8. *Cylinder head nut torque sequence for 4-cylinder engine. 6- and 8-cylinder are similar: work outward in spiral pattern from either bolt 1 or 2.*

1.6 Valve Adjustment and Camshaft Timing

These two procedures should always be performed together. Valve adjustment should be checked about every six months of regular driving. This is more a preventive routine than anything else. It is unlikely that the valves will require adjustment. While the valves are being checked, it's easy to line up the TDC (top dead center) mark on the crank and check to see that the cam marks are also lining up. There is never any reason for them to change except for gross timing chain/belt wear.

If you find an exhaust valve beginning to tighten up (that is, the cam-to-follower clearance is decreasing), you should check the valve clearance every 1,000 miles or so. An exhaust valve that progressively loses its clearance is stretching and will eventually break, probably trashing the engine in the process. If you do identify a stretching exhaust valve, remove the head and replace the valve.

CHAPTER 4

4-Cylinder Engine Valve Adjustment

To check the valves, make sure the engine is cold and then remove the camshaft cover by removing the six decorative nuts on the top of the cover and the two 10 mm bolts holding the front of the cover to the timing chain cover. Try not to tear the gasket.

Have two feeler gauges handy, one 0.019 in. for measuring the intake gap and a 0.021 in. for the exhaust side. Measure the clearance between the base of the camshaft lobe and the follower when the tip of the camshaft lobe is pointing at you (directly away from the follower). See Fig. 4-9.

Fig. 4-10. **Top:** *Camshaft follower (or bucket) fits in bore over top of valve stem.* **Bottom:** *Adjusting shim (or cup) sits on top of valve stem, under follower. (Camshaft removed for clarity. Removal procedure is described below.)*

Fig. 4-9. Checking valve clearances using a feeler gauge. Note how camshaft lobe points away from camshaft follower. This is the only position for checking a valve. Write down the clearance after you measure it.

If you find that the feeler gauge either won't fit, or clearly fits too loosely in the gap, then you'll need to measure the gap using other feeler gauges. Don't be alarmed if the measurements are off by up to 0.005 in. However, if the exhaust clearance gets around 0.010 in. and the intake clearance around 0.005 in., then the valve absolutely has to be adjusted or you'll risk damage to the engine.

I'm fairly casual about my valve clearances, so a difference of 0.005 in. doesn't cause me to readjust a valve. If you race, or are even slightly compulsive, you'll want your valves all set to the exact specification because camshaft timing and valve lift are both affected by improperly adjusted valves. One of my good (compulsive) friends insists on no more than a 0.0005 in. deviation from factory spec.

Alfa valves are adjusted using small replaceable cups that sit directly on the valve stem. These cups fit between the valve and its follower, a much larger cup that fits in cylinders machined into the head. See Fig. 4-10. The follower bears directly on the cam lobe and is subject to wear. Always check the bearing surfaces of

the follower carefully to verify a mirrorlike smoothness. The surfaces are hardened, but the hardening goes only a few thousandths of an inch into the metal. If the follower is worn past the hardened surface, or shows hairline cracking, it will be destroyed in a very few miles and can possibly damage the engine.

Write down the clearances of all the valves on the cold engine. I usually draw a diagram of eight circles corresponding to the followers and then draw an arrow indicating the front of the engine. See Fig. 4-11. The measured value for each follower is written under the circle. I keep these measurements in a file to establish a history of each valve's health.

Turn the engine over by hand. Many people prefer not to remove the spark plugs for this operation out of fear that removing the spark plug from a cylinder with an open valve could drop a bit of carbon from the plug threads on the valve seat and cause a bad valve-clearance reading. To avoid this problem, with the plug wires removed, loosen the plugs several turns and turn the engine over several times using the starter. This will blow any deposits away.

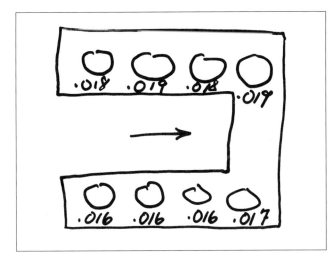

Fig. 4-11. An example of a valve adjustment sheet used to keep track of valve clearance measurements and to calculate dimension of new shims.

WARNING —
Do not loosen the plugs too far. Stay away from the engine area. A spark plug launched from its bore under engine compression could cause severe bodily injury.

After all the clearances are noted, calculate how much change is required to bring the clearance to the desired specification. Write this value next to the circle representing the follower, preceding the value with a "+" or "−" to show whether the gap must be widened or narrowed.

Now, turn the engine to TDC so that the master link of the cam timing chain is accessible. Verify that the camshaft timing marks line up with the marks on the backs of front camshaft bearing halves. See Fig. 4-12. If they don't match exactly, you'll be able to correct the timing on reassembly of the cams. Don't move the crank from this point on. Note whether the frontmost cam lobes are pointing at themselves or away from each other. They can be in either position, just remember which.

Loosen the timing chain tensioner bolt, push the tensioner all the way into its boss, and tighten the bolt to hold it there. See Fig. 4-13.

In order to get at the replaceable cups, the camshafts have to be swung out of the way so the followers can be taken out. The traditional procedure is to remove the cams completely, but this requires a careful retiming procedure. Instead, you can simply remove the cam bearing caps (don't mix them!) and, without disengaging the cams from the chain, swing one or the other toward the spark plugs so it's out of the way. The retracted detensioner allows the chain to "bend" enough so the cams can be swung out of the way.

Fig. 4-12. A quick check puts engine on Top Dead Center. Surprise, in this car, cam timing marks don't line up.

Fig. 4-13. Loosening camshaft idler sprocket (note wrench in left hand) to pull up on chain so that it bellies out as shown.

To remove the camshafts (and avoid bending them), turn all the bolts holding the camshaft top bearing halves in equal increments so the cams are released simultaneously along their entire length. See Fig. 4-14. At some point, the valve springs will probably pop the cam free. If they don't, tap on the top bearing halves lightly to free them and the cams from the head.

If you elect to remove the camshafts entirely, the next step is to remove the timing chain master link and

Fig. 4-14. Once camshaft chain is loosened, you can remove camshaft bearing halves. Begin with center half, loosening incrementally in a widening circle so that stress is relieved evenly on camshaft.

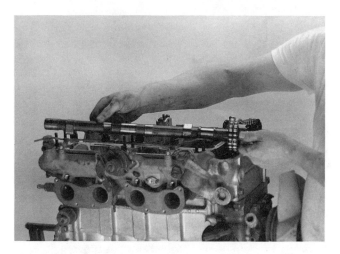

Fig. 4-15. With three bearing halves removed, you can either remove camshaft completely, or leave the chain on the cam (making sure it doesn't jump any teeth) and position cam out of the way.

Fig. 4-16. Use a micrometer to measure adjusting cup thickness. Make sure micrometer tip fits inside adjusting cup.

tie up the ends of the chain so it doesn't fall into the front case. See the procedure on head removal in Chapter 5.

Remove each follower and set it aside so you can return it to the same bore on reassembly. Never casually interchange followers.

The adjusting cups will probably be stuck to the inside of the follower. Measure the thickness of the cup using a micrometer. See Fig. 4-16. You want to measure the effective shim thickness, not the overall outside dimension, so the spindle of the micrometer must fit inside the cup when you make the measurement. Write down the measurement next to the corresponding valve values already noted on your sheet of paper.

You now have three values noted: original clearance, amount to change, and cup thickness. Basically,

you just reverse the pluses and minuses and apply the value to the cup thickness. If the original intake valve value was 0.015 in. and you noted +5 (to bring it to 0.020 in.), you'd need to subtract 5 from the thickness of the cup to finish up with the right clearance.

And where do you get these adjusting cups? At your local Alfa dealer. Now, it may be possible to juggle cups from one valve to another to cut down on outright purchases of new cups. You'll find, however, that most Alfa owners who do their own work build up a selection of used adjusting cups. The cups are the same for all Alfas with 9 mm valve stems from 1959 on. They won't work on the 750-series engines because those cars had 8 mm stems.

It may also be possible to grind the bearing surface of the cup slightly to obtain the needed clearance. You should do this by rubbing the flat bearing surface of the cup on a piece of 240 sandpaper. You'll remove one or two thou every five minutes or so. It's slow work, and you tend to burn your fingers because the cups get hot.

I have found Alfa's quality control on the large followers to be just short of unbelievable. Over the years, almost all measure to exactly the same thickness, so it usually doesn't do any good to try to interchange followers to help get the needed valve clearance. Under no circumstance should you try to grind the followers down. They get enough pounding on the job.

With the new adjusting cups in place and the followers placed over them, bolt the camshafts back on and check the valve clearance on those valves that can be checked.

> **NOTE —**
> You won't be able to rotate the cams 360° because some valves will foul against the pistons. So just check those that are at the right position.

The purpose of this check is to avoid, if possible, getting everything put back together only to discover that several clearances are wrong. The odds are that if the valves you can check are OK, then you've probably got the technique down so that the others will be OK too.

When you're finished checking the clearances, put the cams in their original position, reconnect the timing chain (if necessary), and then tension the chain. (See earlier in this chapter for the procedure.) The chain must be properly tensioned before the cam timing can be checked. Very slowly rotate the engine two full revolutions, being careful to return it to top dead center with the master link exposed. Then, recheck chain tension. If you need to retension the chain, turn the engine through one complete revolution again after tightening the tensioner.

With the engine on TDC and the chain properly tensioned, verify that the cam marks match the timing marks on the top half of the front cam bearings. If the marks are off (or if you want to increase overlap) you will want to retime the cams.

Camshaft Timing

Sometimes, the cams simply won't go back into proper time without some adjustment, so here is the procedure for that.

The cam sprocket is attached to the camshaft itself through a vernier plate. The plate is drilled so the cam can be adjusted in 1.5° increments. A small vernier bolt is used to establish the setting and the entire assembly is secured with a large bolt that threads into the camshaft itself. You may need to rotate the engine 180° to get at the small bolts. Of course, cylinder #one is on TDC.

Bend back the lock tab on the camshaft bolt and then loosen the bolt. See Fig. 4-17. Always replace this lock-tab with a new one. To do this, you will have to remove the camshaft.

Very carefully remove the cotter (if fitted) from each small vernier bolt and remove the bolt, being careful not to drop the nut or bolt into the engine. See Fig. 4-19. Verify that the engine is still on TDC and then position the cams so the timing marks line up exactly. See Fig. 4-20. Be careful not to rotate the crank. This is a baseline

Fig. 4-17. *Cam timing begins with carefully prying back lock tab for big bolt on cam end.*

Fig. 4-18. *With lock tab bent clear, 22 mm wrench loosens bolt.*

Fig. 4-19. *Small cotter used on older cars to lock the castellated nut in position. Remove it, being careful not to drop it into engine. Then, remove nut and pull bolt free.*

Fig. 4-20. *This photo gives purists nightmares, but I still use Vise-grips to rotate cam so marks line up. This can create aluminum chips that fall into the engine and wreak havoc. Careful attention is essential. A stiff strip of leather wrapped around the cam will help protect it from the jaws and give additional purchase. After rotating cam, I recheck to verify crank is still on top dead center.*

setting. If you're only re-setting the cams to stock, then just poke the vernier bolts into adjacent holes until you find the position where the holes line up and you can insert the bolts without rotating the cams. See Fig. 4-21.

Fig. 4-21. *Test locating bolt in different cam sprocket holes until you find one where bolt goes all the way through. It's usually only one or two holes away, since cam hasn't rotated very much. Be careful not to move camshaft.*

If you're increasing overlap, you'll count the number of vernier holes to move and then rotate the cam slightly so the proper vernier hole will accept the bolt.

Slip the bolts home and then rotate the engine 720° to verify the setting. See Fig. 4-22. If the cams are positioned to your satisfaction, fasten everything up and you're through.

Turn the engine over very carefully by hand before trying to start it after a cam adjustment. Even the most skilled mechanic can make mistakes, and an out-of-time cam can trash a running engine in one revolution.

Fig. 4-22. *Carefully replace castellated nut and cotter, then tighten 22 mm nut and rebend lock tab. Turn engine over 720° to verify cam timing is correct.*

V-6 Engine Valve Adjustment

Intake valve clearance on the V-6 is adjusted exactly like the 4-cylinder engine and requires removal of the cam in order to replace the adjusting cups. The exhaust valves are adjusted by a screw with locknut as on a conventional pushrod engine. On the V-6, exhaust valve adjustment is eased over the typical cam-in-block pushrod engine because you can see that the pushrod follower is properly free of the camshaft.

The camshaft sprocket does not adjust on the V-6. Timing marks on the camshafts, distributor, and crankshaft are simply aligned for proper timing.

Fig. 4-23. One bank of V-6 with cam cover removed. Exhaust valves are adjusted easily by a screw. Intake valves are adjusted as in 4-cylinder, using shims.

Fig. 4-24. Camshaft timing marks for V-6 are on middle cam bearing, not front one as on 4-cylinder cars. Timing pointers on plastic timing cover are not accurate enough for anything but a gross check.

Fig. 4-25. Top dead center mark on V-6 crankshaft pulley.

2. Ignition

2.1 Spark Plugs

Lodge HLE spark plugs are recommended by the factory. However, in my experience, Lodge plugs seem more prone to fouling, and so I recommend Champion N5 spark plugs on everything up to the fuel injected cars; N6Y for everything after. Clean the plugs with a full-throttle run in fourth or fifth gear every time you get a high-rpm miss.

Changing the spark plugs is dead simple. First, unsnap the spark plug wire momentarily and place a vacuum cleaner hose around the plug to suck up any debris. Then, just screw the old one out and screw a new one in.

Ha. The number of misthreaded spark plug holes I've personally experienced suggests that the skill of getting a plug started properly in the head ranks right up there with brain surgery. Spark plug threads are steel, while the head into which they screw is aluminum, a clearly uneven match. Plus, the head, even on the 4-cylinder, is offset from vertical, so the plug never goes straight in.

Never use anything but your bare fingers to start a spark plug into its threads, and screw it in as far as you can before fitting a wrench to it. Spark plug threads in the head can be renewed, but not without removing the head from the block. For this reason, use antiseize compound and a lot of care when fitting spark plugs.

Fig. 4-26. Use a wrench to remove spark plugs and for final tightening, but use your fingers to start plugs in their holes.

When you tighten down a new spark plug you'll be able to feel added resistance as the new gasket compresses against the plug. Stop turning just as soon as the gasket won't compress any more.

CHAPTER 4

Oh, yes, returning the spark plug wires to the proper cylinders seems to be another exotic skill. If this is the first time you've done it, use some method of associating the wire with its cylinder before pulling it free of the spark plug.

Wipe the insulating ceramic of each plug with a cloth every once in awhile. If the plug insulator gets oily and dirty, a little bit of rain could create a spark leak and leave you stranded.

If you find that your spark plugs last only briefly before fouling, something is wrong with (1) the fuel mixture, (2) the engine itself, or (3) the selection of plugs other than Champion.

I want to note here that one of the most challenging problems in diagnosing an engine is to distinguish between a misfire caused by the ignition system and one caused by the fuel system. Generally, an ignition miss is sudden and dramatic while a fuel-related miss comes on gradually.

2.2 Distributor

The simplest maintenance to the bakelite distributor cap is also its most effective: keep it clean. Some dark night, open the hood with the engine running. Look closely at the distributor cap and the spark plug wires. You may be able to see flashes of electricity leaking to ground. Current will leak across oily dirt with surprising ease. When you wipe the cap, be sure to get between the posts with a dry rag or Kleenex. If you have any hesitation about removing the cap, leave the rest of the maintenance to your mechanic.

The cap removes with a screwdriver. See Fig. 4-27. Early caps had spring clips that are just a bit too strong to pry back with your finger; later caps are attached using captive screws. Inside the cap is the same number of contacts as you have cylinders. The contacts are typically corroded, and you scrape off the corrosion with a sharp-edge screwdriver. Be careful to remove any debris.

Just under the distributor cap is a Bakelite rotor that directs the spark from the coil to the proper spark plug wire. Early rotors are a single small piece of Bakelite that slips over the center rotating piece of the distributor. On the Giulietta, the rotor is held on lightly with a spring and can be pulled off easily. See Fig. 4-28. On Giulias, the rotor is held securely with a screw. See Fig. 4-29. On both the Giulietta and Giulia, you should be able to turn the rotor slightly in one direction, working against spring pressure. If you cannot, the advance mechanism has rusted solid and the distributor needs to be dismantled and cleaned.

Later Alfas put the spark advance mechanism up in the rotor, and the rotor on these units is an assembly of Bakelite and metal that can be removed when servicing

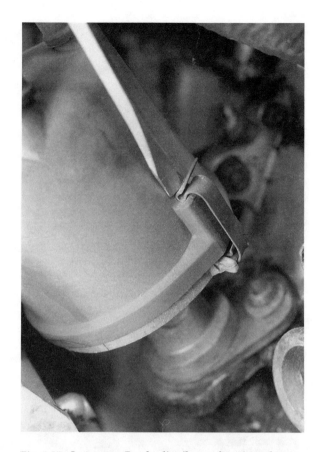

Fig. 4-27. Large-cap Bosch distributor has its advance mechanism as part of rotor assembly. Here, one of two spring clips is being released with a screwdriver.

Fig. 4-28. On small-cap Bosch distributor advance mechanism is located in base, below contact points. Note small rotor.

Fig. 4-29. Large cap has been removed to reveal large rotor with advance mechanism. Points are located below rotor and are somewhat hard to see for regular maintenance. Note rotor mounting screw (arrow).

Fig. 4-30. Points are adjusted by loosening lock screw slightly (shown just to the left of screwdriver blade) and then wedging stationary point in or out with a screwdriver. Be sure to retighten lock screw. Be sure to adjust both point sets on newer dual-point distributors.

the points. The large size of the rotor makes it hard to see the points.

Beneath the rotor, for all Alfas except those with electronic ignition, is a set of points that open and close against a cam on the distributor's center rotating shaft. In the old days, many owners simply filed and cleaned these points using a special ignition file (which is available at most parts stores) with the ignition switched off. Considering that the stock ignition system is often simply adequate, point replacement at routine intervals is probably a better bet.

If you rotate the engine (use care, please) you can watch the points open and close. Rotate the engine so the points are at their widest opening and then use a feeler gauge to measure the gap between them. Look up the proper gap setting in a service manual, being careful not to confuse metric and English dimensions.

If the specified gap doesn't match what you measure, you can adjust the points by moving the stationary member of the pair. See Fig. 4-30. It's held in place by one or two screws. Loosen the screws slightly and then use the blade of a screwdriver to move the point in or out to achieve the desired gap. Tighten the screws and recheck the gap. Frequently, tightening the screws changes the gap, so always recheck. A standard dwell meter gives the most accurate check of point gap.

The cam on the distributor shaft that opens the points should be very lightly lubricated with heavy grease. Don't be at all generous here. If you get too much lubricant on the cam, it'll be thrown off at high engine speeds and possibly short out the points, shutting you down.

Electronic-ignition distributors are virtually maintenance-free, and an occasional cleaning and routine replacement of the cap is all that is required.

When you replace the distributor cap, check, then check again to see that it is properly seated before trying to start the engine. If you get it wrong you'll smash the rotor and possibly the cap itself.

A single nut holds the distributor to the engine. See Fig. 4-31. By loosening this nut slightly with a 10-mm wrench, you can rotate the distributor through a small arc to adjust ignition timing. The actual procedure, which involves the use of some kind of timing light, is explained in all service manuals.

There is also a clamping bolt located just under the distributor body. If you loosen this 10 mm bolt, which lies in a roughly horizontal plane, you'll be able to rotate the distributor for as many complete turns as your heart desires, thus completely confusing the engine timing for yourself and any mechanic you retain to bail you out of the difficulty. Further, if you lift up on the distributor with the clamp loosened, the entire unit will come free of the engine in your hand. If you're a novice, keep distributor movement limited to the small arc provided by loosening the holding nut. If more movement is required, have someone more experienced set up the ignition properly.

CHAPTER 4

Fig. 4-31. After points are adjusted, distributor may be rotated to achieve proper ignition timing. Ten mm locknut is very hard to get to. This photo makes it look simple only because some coolant hoses have been removed. Use this nut for final adjustment, not 10 mm bolt further up on distributor body.

3. Carburetor(s)

Everybody loves to fiddle with carburetors. They are complex enough to be mysterious. If you think you have a fuel problem, for heaven's sake replace the fuel filters first. All of them. You may find several spliced into the fuel line.

With Alfa, you get Solex or Weber (dell'Orto carburetors are Europe-only). The Solex is a durable never-need-to-touch, moderate-performance item fitted to any Giulietta or Giulia that doesn't say "Veloce." Webers (you get two) are only a little fussier but have infinitely more mystique attached, in part because a pair appears to take up at least as much of the engine bay as the little 4-banger itself. U.S. Alfas beginning in 1969 had fuel injection, not carburetors.

The best routine maintenance you can give a carburetor is to keep your hands off it. Indeed, most carburetors fitted to post-1971 cars are not adjustable. While that observation may not seem to apply to U.S. Alfas, it does in one specific way: once set up properly, a carburetor needs no adjustment.

On both Solexes and Webers, you can adjust idle speed and idle mixture richness with a screwdriver. The two adjustments are actually interrelated: adjusting the idle mixture richness changes idle speed. The two screws are adjusted to opposite goals in alternate adjustments. The idle speed screw is adjusted so the engine runs as slow as possible, then the idle mixture screw is adjusted so the engine idles as fast as possible. See Chapter 6 if you are compelled to adjust the carburetor.

4. Fuel Injection

4.1 SPICA

Short of changing the filters (two elements in the air cleaner, two fuel, and one oil), there is no routine maintenance to a SPICA-equipped Alfa. If you must do something, verify that the throttle cable clamp next to the firewall is free to rotate in the arm and then smear a bit of grease on its other end where it wraps around the bellcrank pulley.

Change the air filter elements every 50,000 miles. More frequently in the desert.

On some Alfas, the large fuel filter in the engine compartment is located so that, when you try to change it, it shorts out against the starter because you forget to remove the big negative wire from the battery like you're supposed to. You could blow yourself up. Otherwise, it almost never needs attention. The in-line filter near the gas tank should be replaced regularly (depending on how dirty the gas is in your area) using only the genuine Alfa part. As a rough guide, I'd replace the in-line filter every 30,000 miles, and whenever the low-pressure fuel warning light comes on.

Fuel Injection Pump

Terra incognita for all but the cognoscenti. There's a three-position climate switch on some SPICA units that never needs adjustment unless you're in the Arctic.

It's critical to change the fuel injection pump oil filter about every third time you change the main oil filter. See Fig. 4-33. Be careful or you'll snap off all three of the teeny little studs that hold its cover on.

Fred DiMatteo recommends changing the oil in the fuel injection pump every 30,000 miles or seasonally. Sounds reasonable. He suggests that, with the engine at idle, you remove the three convex-headed screws that hold the barometric capsule carrier, pull the capsule out, and squirt a pint of engine oil into the hole. Replace the barometric capsule and its attaching screws. Excess oil will be expelled back into the engine through a drain hole when you start up the engine.

If you live in a cold, humid climate, it's better to remove the brass drain plug on the bottom of the pump to drain any water that has become trapped. If there is water inside the pump and it freezes, the plunger assembly for the enrichment device will be disabled. In addition, you can be sure that rust will quickly follow any water, causing the rack and other free-moving devices to stop working properly.

Fred has also devised an external oiling system for the fuel injection pump that draws clean oil from the filter cover plate and injects a stream through the side inspection plate to keep the logic section well-lubricated. A Clippard fitting controls the oil flow through a small Teflon tube.

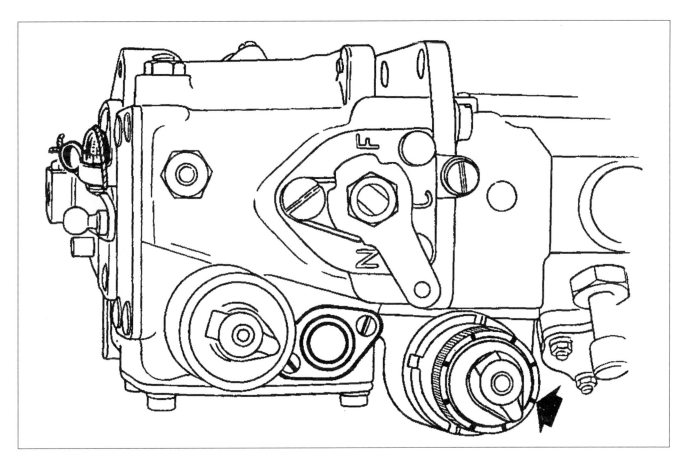

Fig. 4-32. *Looking down on typical SPICA fuel injection pump. Arrow points to notches on top of fuel cut-off solenoid. They're used to keep track of fine-tuning fuel mixture adjustments. Just to left of cut-off solenoid is thermostatic actuator and to left of it, cold-start solenoid. Above thermostatic actuator is barometric sensor (letters N and F are easily visible on its roughly triangular outline).*

Fig. 4-33. *Oil filter on SPICA pump is frequently hard to get to and usually undermaintained. Its regular maintenance is important.*

Regular maintenance of the fuel filter is truly necessary to help prevent water contamination. The in-line fuel filter located near the fuel tank should only be replaced with a genuine Alfa part. The smaller and easily obtained mass-market filters don't have the capacity to deliver adequate fuel to the pump.

Drive Belt

The toothed drive belt for the fuel injection pump should be checked occasionally for signs of excessive wear. With a mirror, you can inspect enough of the belt from the backside of the belt cover. See Fig. 4-34. If you ever have occasion to remove the crankshaft pulley or fuel injection pump, always put on a new belt as cheap insurance.

Fig. 4-34. *Incipient failure. This SPICA belt has a notch torn in its body, indicating ham-handed maintenance. Profile of belt cleats should be sharp. These are rounded, another indication of wear.*

4.2 Bosch EFI

With the exceptions of regular air and fuel filter replacement, the Bosch system is designed to be maintenance-free. The oxygen sensor is a fragile part (drop it and it's broken) that usually lasts at least 50,000 miles. When a sensor goes bad or gets old, it gives the engine control unit a signal indicating that the engine is lean even if it isn't really. So increased fuel consumption may be a sign of a bad sensor.

5. Emission Controls

By U.S. federal law, all emission controls are warranted to have a service life of 50,000 miles. That means they are practically service-free. If you're conscientious about such things, the rubber hoses in the systems should be inspected occasionally for cracking or tearing.

The life of the catalytic converter is variable. Some last less than 60,000 miles, due to the engine running rich. It should be replaced if there is evidence of exhaust gas restriction (excessive heat from the converter, reduced power), or if there is no other reason that your car fails an emissions test (i.e. fuel/ignition fault).

Vapor Separator

A small amount of oil needs to remain in the line that runs to the dipstick. In cold climates especially, keep the separator itself clean. Water trapped in this tube can freeze, raising crankcase pressure and possibly blowing oil out of the engine to bathe your engine bay.

To clean the tube, remove it completely from the car and blow it out with compressed air. You can also clean the tube by beating it against the ground like a whip. If the tube is brittle enough to break, you'd want to replace it anyway. Cleaning the separator is a regular maintenance item.

Fig. 4-35. *When emission controls on an Alfa get messed up by devious owners, one symptom is a gas tank that wants to "oil can." You'll notice a rush of air when you remove gas cap. One makeshift solution—short of a complete system redo—is to drill a small hole in gas cap as shown. Hole maintains gas tank at atmospheric pressure and helps eliminate danger of tank collapse or rupture. Technically, this is an emissions-control violation, so go the extra mile and replace the cap.*

6. Transmission and Drivetrain

The transmission oil level/filler is located about one-third of the way up on the transmission housing. Many look like a yellow-metal disc with a hex depression cast into it. Others are a hex-headed plug. You unscrew the plug using a large allen wrench and stick your finger inside, feeling for the top of the oil level. You should be able to touch oil. If you can't, you'll have to use a pump to add enough 90-weight gear oil so it just flows out the opening. Pre-1966 Alfas should use only Shell Dentax oil in the transmission.

The differential oil filler is located on the rear of the differential housing near the parking-brake linkage. It is maintained in essentially the same way as the transmission.

While you're under there, check the rubber fitments (donuts and driveshaft support) that are located along the length of the driveshaft. If you can see any major tears (not just cracks), the part should be replaced.

6.1 Chassis

As noted above, the Alfa chassis has been maintenance free for several decades. There is a single grease fitting located in the middle of the drive shaft on pre-

Fig. 4-36. Transmission oil filler plug on all Alfas is located about halfway up on casing side. This one happens to be on an Alfetta transaxle.

Fig. 4-37. Separate transmission/transaxle drain plug is located low on the side of the box. Remove using 14 mm hex wrench. Again, this is on an Alfetta.

Alfetta Alfas and it should receive some grease perhaps biannually.

My personal experience suggests that the joints of the suspension and steering require regular inspection and occasional replacement. That may be because I'm attracted to run-down Alfas that need a home.

If your Alfa seems to weave a bit on the road, or goes "clunk" when you drive over a speed-bump, it's time to get under the car and inspect the joints in the suspension and steering. Steering joints can be tested by hand. Just see if you can make the joint move by pushing and pulling on it. Suspension joints have to be worked in the same way, but using a pry-bar because hand pressure won't budge even a bad one.

If you think you have found a bad joint, have it inspected by a professional. If he replaces anything in the steering or suspension, ask him to realign the front suspension.

6.2 Troubleshooting Tires

Federal law requires that tires have built-in wear indicators. Most tires have cast-in rubber webs between the lands of the tread to define minimum tread depth. It is OK to run a tire that has its tread worn down to the wear indicators, but you should not use a tire that is so worn that it is smooth at any part of its tread.

Uneven Wear

You will notice that tires occasionally wear unevenly between the inside and outside tread. On some cars, the outside tread can be almost completely worn away while the inner tread is virtually untouched. That wear pattern can be reduced by rotating the tires on their rims. On high-performance cars using radial plies, front-to-back rotation used to be the only acceptable practice (radial plies take a directional set and reversing the tires reverses the direction as well), but some manufacturers may have changed their rules. If in doubt, check with your tire manufacturer. Careful alignment will help the situation, but bear in mind that some cars are simply set up so that the tires wear unevenly across the tread. Alfas should wear evenly, however, so uneven wear across the tread indicates an out-of-alignment condition.

A tire that wears unevenly around its circumference was laid up improperly during manufacture and is probably so out of balance as to be unusable at freeway speeds.

Cupping (as you run your hand over the tread, you may notice that edges of the tread pattern flip upward slightly, forming a shallow cup or scallop) is caused by small vibrations that are normally controlled by the shock absorbers. Some shallow cupping is tolerable, but severe cupping suggests you should have the front suspension checked for wear.

You can extend the life of cupped tires by changing them from one side of the car to the other, causing them to run in the opposite direction. This switch used to be considered improper when using radials, because of the directional set of the plies, but some manufacturers have changed their recommendations. Check with them to be certain. If you don't ask a lot of your tires you may be able to extend their useful life by judicious rotation.

Tires are expensive. A really good tire will cost over $100. If you're accustomed to driving hard and at high speeds (otherwise, why have an Alfa?) then any money invested in a good tire is well worth it. You wouldn't drive around with your suspension held together with paper clips. Driving on cheap tires invites problems.

7. Body and Interior

Leather should be treated as a living thing. It needs regular feeding of a leather preserving cream. My personal recommendation is Lexol. Lexol is available at any tack shop, where equipment for the horsey set is sold.

Plastic exhales hydrocarbons when it gets hot. The greasy mist on the inside of your windshield comes from the plastic on the dash. You can restore some of the hydrocarbons and prolong the life of plastic by feeding it Armor-All, 303 Protectant, or similar.

7.1 Air Conditioning

In the dead of winter, run your air conditioning about five minutes a month to keep the system water-free and to circulate oil to the seals.

From time to time, inspect the sight glass on the receiver-dryer. That may be a hard item to locate. Begin by locating the air conditioning compressor in the engine compartment. It's the thing that has a rubber V-belt drive up front and two black rubber hoses at the rear. Follow both rubber hoses. One will go to a black cylinder that is about 2 to 3 in. in diameter, sits upright and has a small concave glass window on its top. This is the sight glass. It may be covered with dirt, so wipe it off with your finger. See Fig. 4-38.

Have a friend start the engine and turn on the air conditioning, full blast. Let the engine idle until the frontmost part of the pulley on the air conditioner begins to spin with the rest of the pulley. You will hear a solid "click" when this happens. If it doesn't, either the system is totally discharged (a pressure safety switch prevents it from turning on) or the electrical circuit is faulty.

Now, look at the sight glass and ask your friend to rev the engine several times. You should see just a few bubbles in the sight glass as the engine revs. If you don't, the air conditioning system needs to be charged.

You can no longer buy R-12 refrigerant recharging kits at auto stores, and I'd suggest you leave the operation to a professional.

8. Electrical System

On an Alfa with an electric problem, you always suspect a fuse first, and then a bad ground. Alfa electrics corrode with wonderful subtlety. Just looking at the fuse will not verify that it is making proper contact. A voltmeter or test lamp has to be used to verify that a fuse is carrying current from the top contact on the fuseholder to the bottom contact. Similarly, just looking at the black wire bolted to the chassis ground won't verify that a good ground exists. An ohmmeter is necessary.

Fig. 4-38. Air conditioning sight glass is a clear window inside hex-headed plug on top of receiver/dryer. With air conditioning on and engine running, a few bubbles might be seen as engine is revved. No bubbles means either an over-charged or empty system. Froth in sight glass indicates it's time for recharge.

8.1 Alternator/Generator

Both are maintenance free. If the red charging light gets brighter as you rev the engine, the ground wire to the generator is probably bad.

The V-belt that drives the unit should be checked for proper tension. About 1/2 in., midway between pulleys, is the maximum deflection allowed using thumb pressure.

If you want to adjust the tension, loosen the top clamping bolt that runs through the slotted adjusting strap and the bottom mounting bolt. Use a large screwdriver or hammer handle to lever the unit out if you need to increase drive-belt tension. Some leverage will be required if the bottom attaching bolts are as securely tightened as they should be. Be sure to retighten the top clamping bolt. As a final check, locate the nut at the other end of the slotted adjusting strap and verify that it is tight.

8.2 Starter

The starter, like the alternator or generator, is maintenance free. If something goes wrong with it, it's a lot cheaper to get a rebuilt unit.

To troubleshoot the starter, turn on the headlamps and twist the ignition key. If the lights go away, you've got a dead battery. If there's a click before the starter fails to work, the starter's brushes are probably bad. If there's no click and the lights stay bright, the starter solenoid is probably bad.

Sometimes, however, the ignition switch itself gives up, especially in original-equipment Giuliettas and Alfettas. It's easy to check. Just remove the skinny black wire from the starter solenoid and connect a test light between the connector on the black wire and ground. Then, turn the ignition switch to "start." If the light doesn't come on, the switch is bad.

8.3 Fuse Block

Alfa is the only car I know of that requires fuse-block maintenance. About once every six months, take each of the fuses out and inspect its tips for corrosion, then use an old toothbrush to burnish the spring contacts on the contact side. If you don't do this and corrosion builds up you will find you're plagued by unexpected power failures. Fortunately, the newer Alfas use less-troublesome "plug-in" fuses.

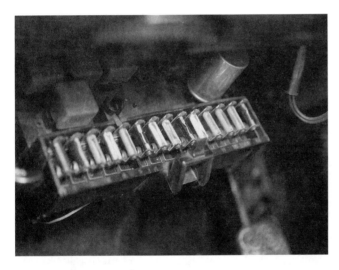

Fig. 4-39. Progress brings complexity: Alfetta fuse block adds circuits but keeps them safely (relatively) inside car tucked up under dash. Regular brushing with a toothbrush (no kidding) will keep contacts on both block and fuse ends functioning. Otherwise, be prepared for electrical gremlins when you least need them. Shiny cannister on fuse block is turn indicator relay.

9. An Introduction to Troubleshooting

If you're comfortable changing the oil or adjusting the ignition timing, you will probably want to tackle some nonroutine problems with your Alfa. It's important to approach troubleshooting with a clear understanding of just what it involves. To that end, I want to take a bit of space to explain just what you do when you troubleshoot a car.

A True Story

One weekend, I was checking over a 1979 Sport Sedan that I intended to use as a daily commuter. One of the tires was badly out of balance and so I simply mounted the car's spare tire in its place, telling my wife to have the local tire shop spin balance the offending tire.

Monday night on the way home from work the clutch hydraulic system began to fail. Only about the last inch of pedal travel did anything and I limped home, barely able to engage the gears.

The next night, my wife came out to check on my progress in troubleshooting the system. I was looking at the master cylinder under the hood when she asked: "Would changing that tire have anything to do with the clutch failing?"

Experienced mechanics might howl at the *non* sequitur, but in point of fact, my wife had asked an absolutely reasonable question. The last thing I did was change a tire, then the clutch failed. It was reasonable to suspect a cause and effect relationship.

Reasonable to expect, but not appropriate to conclude. I might have damaged a clutch hydraulic line, either in changing the tire or just jacking up the car. But my wife did not know that the hydraulic lines are not near where I was working.

Another True Story

Let me give an instance when such a simplistic cause-and-effect conclusion did work. The next night (remember, this is a true story) I was troubleshooting the rear turn indicator lamps. The right rear lamp didn't work. On the Alfetta, the lamps are mounted on a large plastic snap-on base and they all come out as a unit. With the plastic units in place, I was probing the connections with a voltmeter to determine exactly where I was getting power and where it was failing. The testing was not going well, so to improve my access to the connectors I unsnapped both plastic light holders and laid them on the floor of the trunk.

Suddenly, everything worked. Now, at first blush, the location of the panels should have no effect on their electrical performance. But there was (pun intended) a connection. The challenge was to find out just what it was.

I tested each of the wires with a voltmeter, verifying that it was getting current when it should. This one works; this one works; this one doesn't. One of the wires in the connector was loose. In its proper location, it failed to complete the circuit. When the panel was placed on the floor of the trunk, the wire bent just enough that it completed the connection. Now, there seemed an essentially unreasonable cause-and-effect relationship here: placing the panel on the trunk floor made the lights work. That situation was exactly analogous to the tire/clutch query of the previous night.

The moral of all this is to test the most obvious cause-and-effect relationships first when you're trying to diagnose a problem with your car. Might I have damaged a clutch line while changing the tire? It's worth checking. If the car won't start just after you've changed the spark plugs, then something you undid while changing the plugs was probably put back together wrong.

A Two-Phase Approach

Troubleshooting is not so much a matter of using the right tools, having volumes of reference manuals, or even years of experience. It is rather a process of sorting things out using logic. I also need to emphasize, and this comes from over 20 years' experience working on Alfas, that troubleshooting is conducted while sitting quietly and thinking calmly. It is not done while probing nervously with a screwdriver or tapping things randomly with a hammer. The brain is the basic troubleshooting tool.

The way top-notch mechanics approach troubleshooting is instructive. Studies have shown that they approach any job in two phases. In the first phase, and in order to save time, they use their vast experience to jump to conclusions: the last ten jobs I had that acted like this were cured by doing that.

You can't do that. It's no great loss, for good mechanics try only the most likely two or three fixes before going on to the next phase. After all, chasing hunches could take all day and still not fix the problem.

The second phase is an exercise in diagnostic logic. An automobile is made up of systems, most of which work in an easily identifiable way. There are not many systems in a car, and the organization of this book reflects them, engine, driveline, chassis, electrical, and so on. Each system is made up of subsystems: the engine has ignition, carburetion, lubrication, and so on. If you verify that a whole system is good, you remove it and all its subsystems from further consideration. This process allows you to eliminate large portions of the car from consideration very quickly. You discover the bad system using a process of elimination.

Logical Steps

You will probably start with the system that most obviously seems bad. That's fair. No need to check the shock absorbers when the problem is a car that won't start (not right away, anyway). Again based on (bitter) personal experience, the most frequent error is to jump to the conclusion that a whole system—or some part of it—is good when in fact it isn't. The most challenging distinction is to determine whether carburetion or ignition is the cause of a problem. I have spent hours trying to correct a "carburetor" problem that turned out to be electrical.

Be meticulous when you verify something as good. But if a system does exactly what it's supposed to do, it's good and you go on to other systems, verifying them until you find the bad one. Once you have identified the bad system, you begin by verifying the good subsystems (or parts) within the system until you find the bad part.

If the clutch doesn't work, I should verify the individual parts of its system: the master cylinder, slave cylinder, and connecting tubing. I might find that the tubing had been damaged jacking the car up, but that discovery would have been made in the process of working through one system, and not jumping randomly to another.

The thing that separates the good mechanic from the great mechanic is that the good mechanic tends to interrupt his diagnostic search to chase more hunches. The great mechanic meticulously follows the logic tree to its proper conclusion.

Engine

ALFA ENGINES ARE the center of the marque's mystique. They are not, however, mysterious. A properly maintained Alfa engine is dead reliable. Still, there are a few important things you should know about your Alfa engine.

1. Failure Modes

Writing a section on engine failure modes is very much like using the terms "dealership" and "legal ac-

tion" in the same paragraph. No matter how well intentioned the author may be, someone is going to jump to conclusions.

There is a pervasive uncertainly among Alfa owners about the reliability of the car, primarily I think, because those who do have problems are so vocal about them. Make no mistake, there are those whose beloved Alfa has languished in the shop for six months waiting on an essential part from Italy. The fact that this also happens to other marques, may in fact be unnecessary,

Fig. 5-1. *The 3-liter 164 engine is set transversely in body so manifolds for intake and exhaust are quite different from Milano or GTV6 installation.*

Fig. 5-2. *One of the hottest—and fussiest—of all Alfa engines was Giulietta Veloce powerplant. Twin Weber carburetors demanded a separate air cleaner, hung from the bulkhead.*

or is just a random case of bad luck, does not minimize the owner's grief. Before beginning a section about the bad news, then, I want to give some good news. The Alfa engine is designed along the very well established principles of a racing engine. As such, it is quintessentially rugged. All its components are designed to withstand full-throttle loads virtually indefinitely. Be assured, your Alfa is basically reliable. In fact, most owners err in never using anywhere near the potential of their Alfa. It is not kind to the engine to drive it around town in fifth gear. An Alfa needs exercise, and this means full-throttle runs in top gear.

Most of what goes wrong with an Alfa is not the Alfa's fault. The important corollary to this is that if your car is running well, leave it alone. The reasons Alfas are messed up are twofold: ignorance and ham-handedness. Not everyone takes the time to learn how to work on engines, and occasionally those who do make grossly stupid errors. What I intend to do now is to identify the things that are most likely to go wrong and to suggest how these problems can be avoided.

Fig. 5-3. *Twin-cam head design is clear from this shot of a head's backside. Gas flows in a virtually straight line from right to left (interrupted only by the—considerable—combustion process. Central position of spark plug is also clear.*

Won't Start

Don't feel like the Lone Ranger, it's happened to us all. The causes are many, but the most frequent cause is an almost-discharged battery. Try starting the engine with the lights on. If they seem to go off when the starter is engaged, the battery is too low to start the car.

The starter draws a lot of current from the battery. In marginal cases, there is not enough current left to saturate the coil and the resulting spark isn't hot enough to light off the engine. In such a case, push-starting the car will get you going. If you try a jump start, take a look at the precautions noted in Chapter 10.

If you're not sure of your battery's condition, remember that extremes of temperature can degrade battery performance. I've also noted several times in this book that Alfa electrical connections tend to be mischievous in that they look good but don't work. Don't presume your electrical system is ship-shape until you've verified a good battery charge and cleaned the battery terminals and fuse connections.

If you have run down the battery trying to start a SPICA-injected car, the odds are that you've fouled the spark plugs in the process.

Giuliettas can be made to start more easily if a heavy wire is run directly from the negative (ground) battery terminal to one of the two starter mounting bolts. For Giulietta Veloces, you may wish to follow the old rule that, to determine the number of times to pump the accelerator pedal before starting, you subtract the ambient temperature from 100.

Faulty Head Gasket

We'll play a little game. Your Alfa's in the shop, sick, and I have to guess its problem. If I say it has a leaking head gasket, the odds are about ten to one that I'll be right.

As discussed in Chapter 3, the wet-sleeve design puts quite a responsibility on the head gasket. Alfa has chosen to exacerbate the problem by running oil lines through the head gasket to supply lubricant to the camshafts. These lines are sealed by small rubber O-rings fitted into the head gasket. The pressures involved and the eventual deterioration of the gaskets contribute to the high incidence of leakage.

This is not the typical head-gasket failure most mechanics will be prepared to notice. Only rarely do Alfas have failed head gaskets that leak pressure from the combustion chambers. Most failures are oil, not compression, leaks.

You'll see ample evidence that the O-rings are leaking because oil will usually bathe the back half of the engine block and drip on your parking spot.

The big danger of the leaking oil is that it eventually erodes a path across the head gasket and opens a way for water to leak back into the oil. If this happens, the lu-

Fig. 5-4. *Emulsified oil (caused by a leaking head gasket) is a warning sign to a situation that can damage your crankshaft. This particular head gasket leak was caught early, yet I can still scoop a finger-full of emulsified oil from coolant.*

Fig. 5-5. *One-piece Alfa head gasket for both 4-cylinder and 6-cylinder cars looks pretty similar (one has an extra hole). Obvious danger is erosion of gasket material between cylinders, marked by an hourglass shaped shiny section on this gasket. This is, in fact, a fairly unusual failure.*

bricity of the oil goes away and the main and rod bearings can be ruined.

There is a warning sign that water has mixed with the oil: your oil dipstick will have the malted-milk foam on it and your radiator cap may have a foamy, malted-milk looking deposit on its sealing side. You may notice the same foam in the radiator overflow bottle. The light brown foam is a signal that it's time to replace the head gasket.

Fig. 5-6. Look here for problems: Oil-line sealing rings have given Alfa owners more problems than everything else on car combined.

The technique for lowering the odds of a leaking head gasket is to keep the head torqued properly. Since the engine is composed almost exclusively of large alloy castings, there's a lot of expansion and contraction going on as it heats up and cools down. The result of all this motion is to loosen the head bolts. It's worth getting a good quality torque wrench and learning to use it to torque your head bolts to the correct specifications listed in your shop manual. The torque should be checked about every six months.

If you do repair a leaking head gasket, you can help reduce the danger of future failure in several ways. First, make sure that the head's mating surface is absolutely flat. Take it to a machine shop and have them make a finishing cut on the head. You don't want to remove any more material than is absolutely necessary to assure its flatness. If the head has been warped, it may be necessary to line-bore the cam bearings to regain a straight bearing surface. Some shops now offer aluminum head-straightening services, where the head is clamped to a heavy jig and heated to straighten it, bringing both the mating surface and the cam journals into parallel. I wouldn't recommend this approach, however, unless there is no other alternative.

Second, check to see that each cylinder liner protrudes above the block's deck by 0.002 in. See Fig. 5-7. If it does not, fit new O-rings under the liners to obtain the proper extension of the liner. If new O-rings still don't stand the liners 0.002 in. proud, the block should be removed, dismantled, and then its deck milled to provide the proper cylinder height.

Third, be scrupulously clean in preparing the mating surfaces. Use a strong solvent (acetone or lacquer thinner) to remove all traces of gasket material and dirt from the surface.

Finally, use Viton O-rings instead of the O-rings supplied with the head gasket. Viton is less susceptible to deformation than the original ring material. As an in-

Fig. 5-7. It's a little hard to see, but then 0.002 in. or so isn't much, anyway. Lay a straight edge across top of cylinder and slip a feeler beneath it and block, as shown. Block surface must be absolutely clean. If there isn't any clearance, try seating another used cylinder with a new sealing ring. If that doesn't give needed clearance, you must fit a new cylinder. If there's still no clearance, the top of the block must be machined.

surance, especially if you don't use Viton rings, insert roll pins in the oil passages at the head joint. The pins should extend above the face of the block to help locate the O-rings in the head gasket and keep them from collapsing. This procedure is shown in detail in the section on head gasket replacement.

Crankshaft

Most owners will never see their Alfa's crankshaft. The bottom end of an Alfa is one of the most sturdy structures running down the road.

Having put the crank's essential reliability into perspective, I can now list the two most likely reasons for being concerned with an Alfa crank: missing oil plugs and scored bearings that result from coolant mixing with the oil as a consequence of a blown head gasket.

Crankshaft Oil Plugs

Pressurized oil from the main bearings is delivered to the rod bearings through passages drilled into the crankshaft. The ends of these passages are plugged with aluminum dowels that are pressed in place. Occasionally, one or two of these plugs will work free and fall out, especially in engines that are revved hard when cold.

The failure is not catastrophic, but can be identified by a sudden drop in oil pressure. Every plug that falls free will cause an additional loss of about 10 psi pres-

Fig. 5-8. Surest way to secure a crank plug is to tap threads into crank and screw an in-hex plug into place. Loctite will keep it from vibrating out.

sure. The plugs will be found in the pan and can be pounded back in without removing the engine from the car (providing the pan can be removed in situ).

I need to remind you here that late-model Alfa oil pressure gauge sending units are notoriously fickle and can indicate low oil pressure when, in fact, oil pressure is fine. Thus, if there's any question about your engine's oil pressure, fit a mechanical gauge before concluding something is wrong with the bottom end of your engine.

Note also that older-model Alfas show measurable oil pressure at idle while newer Alfas may indicate no pressure (don't worry, it's usually there; it just doesn't show).

Some crank plugs can be replaced just by removing the oil pan. This may sound like a simple procedure, but in the case of most Alfas, it's easier to remove the engine and transmission than to remove the pan.

You may wish to freeze the plugs, then coat them with Loctite as they are pushed in place to form a chemical bond. Under any circumstance, the plugs should be staked in place by upsetting the crankshaft metal around the circumference of the plugs. This is not an easy task, as the crank is surface-hardened. Hard-core racers may wish to try to thread the crank and screw in newly-fabricated plugs.

Scored Crankshaft Journals

It's a head-bone-connected-to-the-shoulder-bone sequence: the head gasket gives up, oil and water mix, and the emulsified mixture causes the rod to rub on its bearing journal, scoring the bearing surface, causing a rod knock and, eventually, bearing failure.

A knowledgeable Alfa owner will always check for emulsified oil each time he looks at the dipstick. Those who find the phenomenon interesting but not worthy of action will end up gazing at their crankshaft and wondering if it can be saved.

The answer, fortunately, is almost invariably yes. An Alfa crank can be ground to three undersizes before it has to be scrapped. If the crank is severely scored, the third-undersize dimension may be necessary even if the crank had never been ground before. Some machinists like to jump to third undersize because it eliminates the possibility that a first or second undersize grind won't clean the crank up and they'll have to do third undersize anyway. Insist on grinding conservatively, even if it means paying a little more to find out that first undersize isn't enough. The less you take from the crank the better, and if the first undersize cleans it up, you've still got a regrindable crank.

Under normal wear, a crankshaft's bearing journals will become slightly scored. This normal scoring can be cleaned up by polishing the journal with a length of crocus cloth, using the same technique as you do polishing your shoes. The rule is that, if the scoring is deep enough to catch your fingernail as you draw its edge across the bearing journal, the journal must be reground undersize rather than polished.

For a truly long-life rebuild, have the surface of the crankshaft hardened after it has been ground undersize.

The rod and main journal diameters of the 1600, 1750, and 2000 crankshafts are all the same even though their strokes are different. While this opens the possibility of a "stroker" kit, I have never heard of anyone actually doing the job.

Spun/Loose Rod/Main Bearing

This is a failure everyone fears because it's so expensive to correct. Fortunately, Alfa almost never loses its bearings. The symptom of a failing bearing is a drop in hot oil pressure. When cold, the oil may be viscous enough to mask a loose bearing.

A rod bearing that is failing will cause a hard, metallic knocking sound on acceleration. This sound shouldn't be confused with the higher pitched and much more staccato sound caused by the pre-ignition (commonly called pinging or knocking) of low-grade fuel.

The most terrifying sound you will ever hear is a failed bearing just before the engine seizes. If you're suddenly aware of a sound that resembles a hammer beating on the side of your engine with swift, increasingly harder blows, immediately shut the ignition off (don't lock the steering wheel!), quickly put the transmission in neutral, and coast to the side of the road. If you react quickly enough, you may save yourself the cost of a new crankshaft or block.

If you do spin a bearing, you have to remove the crank plugs and clean out all the oil passages.

Oil Pump

If you have low oil pressure and the crankshaft bearing journal oil clearances are correct, it is quite possible that the oil pressure relief valve seat is leaking. You can lap in the seat to the valve using valve grinding compound. Polish the valve's outer diameter to remove scoring using the same crocus cloth you use to polish the crank bearings. Be sure to clean the pieces thoroughly after using an abrasive. Every time the engine is torn down, this valve should be disassembled and cleaned.

The shop manual gives instructions on measuring the end-play and radial clearances of the oil pump. While these checks may be diverting I have never en-

countered a bad Alfa oil pump. You can buy a new oil pump driven gear but I'd recommend buying an entire replacement pump rather than trying to rebuild a bad one.

Timing Chains

The 4-cylinder Alfa engine uses two timing chains. A short primary chain drives an idler gear that halves the speed of the longer secondary chain. The secondary chain drives the camshafts and has a master link.

You can check the secondary chain for wear by laying it out flat on a table and compressing it to its shortest measure. Then, stretch it out by pulling on its ends. The two measures should not be greater than about one inch. If they are, the individual rollers of the chain are wearing the links and the chain should be replaced.

The primary chain is checked by rotating the crank back and forth to discover how much slack there is in the chain. You'll need to remove the cam cover and peer down the front of the engine with a flashlight. If you get

Fig. 5-9. A modular Alfa engine being tested on a dynamometer. This engine was designed to run on fewer cylinders when power demands were low. Though very successful, it has yet to see widespread application, possibly as a result of Cadillac's disastrous experimentation along same lines.

more than about 5° of slack, you can begin worrying about replacing the chain during the next rebuild.

It's important to reassemble the secondary chain so the master link falls halfway between the two camshafts when the crankshaft is on top dead center and both camshaft timing marks line up with the marks on the camshaft top bearing halves, and the lobes on cylinder one point outwards.

Toothed Drive Belts

The SPICA fuel injection pump introduced toothed-timing-belt technology to Alfa owners. The same kind of belt—only larger—is used to drive the camshafts of the AlfaSud and V-6 engines. These inextensible belts are virtually maintenance free and long lived. As virtuous as they are, there are two things they cannot stand: being bathed in oil or being bent sharply.

The first problem is not likely to bite you, even though your V-6 detensioner leaks oil. After all, the oil bathes more than just the belt, so you're going to do something about it, probably, before things get serious with the belt.

The second problem, the fact that the belt should not be bent over a small radius, is more insidious. Someone can wad a toothed belt up to stuff in a box, pocket, or parts bag and you'd never know just looking at it. The sharp bend, however, fractures the fiberglass strands that make the belt inextensible and significantly shortens its life. The belt fails by breaking, which can also mean bent valves and possibly a broken piston or two.

In practical terms, there's nothing you can do to protect the belt before it lands in your hands. But, you can be careful once it's in your possession. A bend over a radius of less than one inch should be reason enough not to use the belt.

Typically, the belts have a service life of 50,000 to 60,000 miles. Considering the damage a broken or slipping belt can do, replacement at 30,000 miles is cheap insurance, and highly recommended.

Cooling Systems

There is probably no engine more inherently cold-running than the Alfa. I used to make a 250-mile weekend commute in a Giulia Super in near-zero weather. In order to get the engine up to operating temperature, it was necessary to slip a piece of cardboard in front of the radiator to block it completely from air flow. In the 4- and 6-cylinder engines, coolant washes the entire swept volume of the cylinder. That is one virtue of wet liners. Early Alfas with oil temperature gauges revealed that five to ten minute's driving was required to get the engine's oil up to proper operating temperature.

The emission-controlled engines, especially those in cars built between 1980 and 1986, run hotter by design. An engine is healthier if it runs somewhere around 180°

to 195°. Don't worry if the temperature gauge indicates something near 212°. All modern cooling systems are pressurized, and boil only at temperatures significantly higher than normal. Remember, too, that your temperature gauge is probably not dead-accurate.

If your Alfa overheats, something is definitely wrong. The first thing to check is the coolant level in the radiator! Check the radiator with the engine running (*do not* remove the cap if the system has pressure in it!). If revving the engine causes a lot of large bubbles to burp out of the radiator filler neck, you may have a blown head gasket.

Another possible cause of overheating is a sticking thermostat, which is easily checked by removing it and heating it up on the stove in a pot of water. You may also find that the radiator fins are clogged with dirt, or the inside of the radiator may be clogged with rust.

I have a 1980 Spider that, box-stock, overheated not only in heavy California traffic, but also at high speeds. Spiders of this era have a large black bumper that almost completely fills the air intake and tends to deflect airflow away from the opening at higher speeds. A cross member, located behind the bumper and just in front of the radiator, also blocks cooling air. I found that I could lower engine operating temperatures by making a larger opening below the bumper, where the surgery is not very evident, and cutting the cross member out with a hacksaw.

One caution: if coolant level drops so low that the engine temperature sensor is no longer in contact with fluid, you'll see a drop in temperature, even though the coolant itself is critically hot. If you're watching the temp needle climb and then notice suddenly that it's

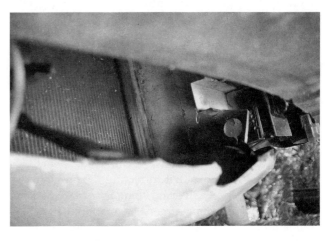

Fig. 5-10. *Keep cool: we're looking beneath bumper, up toward right side of front radiator. Underside of bumper runs across top of photo. Bright C-section brace has been cut away, and bottom sheetmetal droops dramatically from its original position, the result of a very heavy foot. This modification was made to keep author's 1981 Spider from overheating, and works wonderfully well.*

dropped, that's just the time to shut things down and start walking.

Removing the plastic radiator fan from the water pump on Alfas so equipped can gain a few horsepower. The car won't overheat as long as it's moving, but you can only idle it one or two minutes in hot weather.

A final note: if your car is overheating because of a small leak, you may be able to limp home with the radiator cap loosened. This eliminates coolant system pressure and lowers the rate of water leakage. Be careful. It also lowers the temperature at which coolant boils.

Oil Leaks

Not really a failure, more of a characteristic of the very elastic aluminum parts used throughout the engine, transmission, and differential. There are no doubt Alfa engines that are oil-tight. I've never owned one, and I don't know of any owners who have. Oil tends to seep from around the camshaft cover and the front timing chain cover. The tendency of oil to find its way through the head gasket has already been noted above.

In theory, two smooth, flat aluminum surfaces with a gasket between them should be oil-tight. You should honor this theory by assuring that all mating surfaces are parallel and scratch-free. Attach the cam cover gasket to the cover with nonhardening Permatex, and then carefully fit the assembly to the engine (never Permatex both sides of a gasket, ever). If oil leaks from the rear crankshaft seal, you have to remove the engine to replace the seal. There's no easy fix.

Most knowledgeable Alfa owners regularly wash their engine down using an aerosol solvent and garden hose. Really. Don't bother covering the ignition system or taking any other special measures. It's hard to get the spark plug wells dry (though a wet/dry shop vac works well), and you should avoid squirting water directly into the intake manifold if you're running a non-stock set-up. But, generally, anything that keeps your car from starting just because it got wet needed your attention anyway.

Oil Burning

There's a durable and well-known legend that all Ferraris burn oil, especially when cold. The likely source of the legend is that correcting the problem in a Ferrari is very expensive and some Ferrari owners are surprisingly stingy in maintenance matters. There may be a tendency to extrapolate the Ferrari legend to Alfas, or to all Italian cars in general. No engine, including an Alfa, "should" burn oil excessively, hot or cold.

All engines do burn oil, however. They're designed so a small amount of oil is drawn to the walls of the cylinders and valve guides to provide lubrication for the moving parts. Engines may consume oil at a rate somewhere around a quart every 1000 miles. If you burn a quart in 500 or fewer miles, you're heading toward a rebuild. However, don't panic if your Alfa seems to burn no oil; that's OK, too. The worst thing oil burning indicates is a complete engine rebuild. If you do have to have the engine torn down, replace the rod and main bearings whether they need it or not. That's very cheap insurance.

Oil burning is caused by several failures. The most common cause in an Alfa is oil working its way past the valve stems. If your Alfa smokes on deceleration when the throttle is closed, then your engine is sucking oil past the valve stems because the intake valve guides have worn out. To test for this condition, warm up your Alfa by driving it at cruising speeds. Suddenly, floor it, then let the throttle snap close. Repeat this two more times in quick succession. On the third sequence, check the rear view mirror for a cloud of blue smoke. If you see it, then oil is getting past the intake guides.

Modern Alfas have rubber seals around the intake guides to help keep oil flow to the valve stems under control, but when a guide begins to wear oval, the seals can't cope with the added movement and simply leak.

Sometimes, the rubber seals harden and crack. In that case, oil consumption can be reduced simply by replacing the seals. This is a job that can be done without removing the head (thread rope into the combustion chamber and then bring the piston near TDC to hold the valve in place), though the work goes much more slowly with the head in place. The latest valve guides have Teflon seals and should be used during a rebuild.

If your Alfa burns oil only on hard acceleration, that's an indication that the piston rings have worn and need replacing.

My personal experience has been that the top ring groove of the piston is frequently worn so that the ring-to-groove clearance becomes excessive. Never replace just the rings unless you're absolutely certain that the ring-to-groove clearance is less than 0.003 in. on all compression rings on all pistons. If the clearance exceeds this amount significantly, you'll need new pistons. In that case, you might as well get the piston/cylinder pair. To put this more succinctly, if ring-to-groove clearance is excessive, buy new pistons and cylinders.

There is an easy and accurate test to determine the source of oil burning called a "wet-dry" compression check. You begin by warming up the engine and then measuring the compression of each cylinder with the throttle held wide open. After this "dry" test, squirt a little oil into a cylinder and repeat the compression test. Test every cylinder in turn. If the compression reading rises significantly, then the rings are at fault; if there is no increase in compression, you know the rings are good so the valve guides must be the source of the oil burning.

Fig. 5-11. Permissible clearance between a ring and its groove is very small: 0.002 is as wide as I'm comfortable with. If in doubt, get a new set of pistons.

Not all smoke out the tail pipe is caused by oil burning. Oil burns with a blue-tinted smoke. If the smoke out the tailpipe is black, that's an indication of an excessively-rich air-fuel mixture; if the smoke is white, you have a coolant leak into the combustion chamber. The distinction between white and blue smoke is fairly subtle, and some experience is needed to diagnose smoke properly.

Alfas can burn oil for a very long time if you're willing to keep pouring oil through them. As noted above, if oil consumption climbs above a quart in 500 miles, it's time to rebuild something. You can stave off the inevitable by going to a heavier-grade oil. Most people use 10W-30 multiple viscosity oil simply because it's the most readily available at the gas station. Except for temperatures near freezing, I feel that 10W-30 is too thin. A 10W-40 is preferred and, in warm weather, I suggest 20W-50.

A straight-weight oil will be consumed more slowly in an engine with an appetite for lubricant. If you've reached a quart in 500 miles, try a change of oil to 40- or 50-weight, or pour a can of STP oil treatment in the engine. STP increases the viscosity of the oil temporarily and usually slows oil consumption for about 500 miles. Don't keep feeding the engine STP as a substitute for oil, or for fixing the problem.

Low Performance

If you think your car is performing poorly, the first thing to do is take it through a good car wash. It's amazing how much better a clean car seems to run. This is not a joke. Performance tends to be very subjectively measured. Indeed, without benefit of a stop watch, fifth wheel and accelerometer, performance is impossible to measure. So, beware the "hunch" that your car is slower

Breaking-In Rings

A modern engine's break-in period is designed primarily to get the rings properly seated in the cylinder bores. When poured-metal bearings were used, the break-in period was also a time when the bearing surfaces got used to working together. Modern precision insert bearings have eliminated this need in the break-in cycle.

My personal experience with Alfa engine rebuilds is that getting the rings to seat is the toughest part of the job. I've done rebuilds where the rings seated in perfectly with no special care. For other rebuilds, I succeeded in glazing the rings so they would never properly seal.

In theory, you should progressively increase pressure on the rings by accelerating slowly to full-throttle performance over perhaps 30 miles or so. In stark contrast to this theoretical approach, one of my good friends felt the best way to seat rings was to drive the engine hard, at full throttle, within the first mile or so. He would take a freshly rebuilt engine and flog it almost unmercifully to get the rings to seat. His approach worked for him every time.

Another friend advises using nondetergent oil for the first 500 miles of light-throttle driving. His approach works for him.

I can tell you one thing for sure, if you let a new engine idle for very long without putting it under some load (that is, driving it), you will almost certainly glaze its rings and it will always burn oil. There is no practical recovery from glazed rings short of replacing them.

There is one classic last-ditch approach to breaking the glaze on rings, and that is to sprinkle a small amount of powdered, non-abrasive cleanser (such as Bon Ami) into the intake of a running engine. After about five minutes, shut the engine off and replace the oil and oil filter. The problem with this approach is that you're introducing an abrasive not only to the rings but to the rest of the engine, including the very fragile surfaces of the precision insert bearings. The technique is something akin to chemotherapy. You want to administer a dose that is just short of fatal. If you were going to tear the engine down anyway, it's worth a try. But I cannot recommend the approach except as an absolutely last-resort effort that may in fact destroy an engine instead of fixing it.

or faster than the next. Or, that because of feared low-performance, something must be wrong with your Alfa.

Some Alfas do seem to perform better than others, even when both are box-stock. Well, some Alfas sit differently, too, so I suppose these wonderful cars have their little idiosyncrasies.

If you fear your Alfa isn't performing up to par, try a 0–60 mph run against a stop watch and compare your results to a road test published in any reputable car magazine. Don't panic if you're slower by a second or so. The guys who do these tests are good at it, know ex-

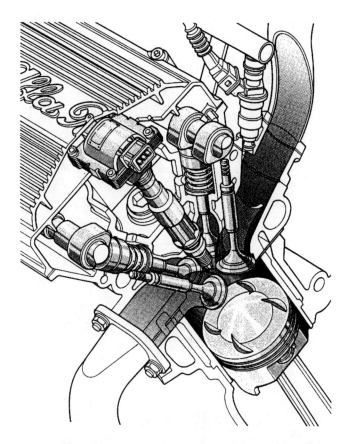

Fig. 5-12. Alfa has been a bit of a Johnny-come-lately to multi-valve heads, preferring to use instead twin-spark and variable-valve technologies to keep output high and emissions low. This is 4-valves-per-cylinder set-up on 164 Super. Note that each spark plug has its own coil.

actly how to get under way most efficiently and shift at exactly the right time.

The first thing to check is proper ignition timing.

In the same league as low performance is the suspected flat spot in the acceleration curve. If your Alfa is equipped with Webers, the most likely cause of a flat spot is poor jetting. Uneven performance is much less a problem with fuel-injected cars, and a flat spot could indicate several possible problems associated with the fuel injection system or timing.

Poor ignition timing can cause low performance, of course. Generally, as you advance ignition timing you improve the engine's top-end responsiveness.

Restrictions in the intake and exhaust systems also cause poor performance. Be especially careful that an exhaust pipe doesn't get crushed going over something. The excessive restriction can burn valves as well as deteriorate performance.

Wear-related poor performance is undetectable by the owner because it happens so gradually. Marginal compression, poor ignition timing, and a dirty air filter will all reduce performance in subtle increments. The realization of deteriorated performance usually comes when the owner drives a friend's Alfa. The flip side of this whole discussion is how to get better performance without fitting competition parts such as high-compression pistons, wild camshafts, or a turbocharger. The easiest answer is to make sure that everything is properly set to exactly stock specifications. Many Alfas run around mis-tuned either out of neglect or the mistaken notion that the owner knows better than the factory what the engine needs to run.

Finally, it's possible to increase performance slightly by careful reassembly during rebuild. We're talking porting and polishing, cc'ing the chambers, indexing the plugs, blueprinting—whatever. The cost of all these operations is high and, if you really want more power, either fit a turbo or go out and buy a car with a larger engine.

2. Head Gasket Replacement

To the novice, replacing the head gasket may seem more akin to brain surgery, but it is well within the abilities of most owners who have a weekend or two and a basic set of hand tools. As with any first-time procedure, proceed slowly and don't force anything. The most frustration on 4-cylinder engines will come from trying to get the exhaust manifold bolts off cylinders three and four; on V-6 engines, removing the brake booster to get off one of the heads is the biggest pain. The biggest physical strain will come trying to lift the head once everything is freed. The greatest challenge is to get the mating surfaces of the block and head scrupulously clean so the new gasket will have a chance to seat properly. Don't worry about magical cures for head gasket leaks.

There is a great deal of myth about the six small O-rings that fit inside the 4-cylinder head gasket to seal the oil passages. Many head gasket kits come with roll pins and Viton O-rings. Use them. Tap the roll pins into place in the block, letting them sit just high enough to hold the O-rings. Just before you lower the head onto the gasket, check that the O-rings are properly in place.

Every time you remove the head, have it checked for flatness, and resurface it if necessary.

2.1 In-line (4- and 6-Cylinder) Engines

It is best to let an Alfa sit overnight before beginning a head gasket replacement. This assures that the engine is cold. Removing the head from a hot Alfa engine will almost certainly warp the head, quite possibly beyond the limits of repair.

Every shop manual gives a step-by-step procedure for removing and replacing the head. The following information supplements the "official" procedure.

Fig. 5-13. Bad news. See foam in radiator? Goo on cap is evidence of a leaking head gasket.

Fig. 5-15. Clean, but water in the spark plug wells has to be removed before proceeding to dismantle engine. Use a turkey baster or similar item to suck the water out, blow it away with compressed air, or use a wet/dry shop vac.

Fig. 5-14. Cleanness is essential in engine repair. Author begins head gasket replacement by spraying solvent over entire engine. Front surface of engine won't come clean unless you use a wire brush and more solvent to dislodge dirt. Note that hood has been removed completely to ease access. A rag under hood (relocated to top of car) saves roof of this 1969 GTV from scratches.

Fig. 5-16. Removing hoses. Generally, inch your way completely around engine, removing and labelling everything attached to head. This is crankcase ventilation hose that goes to separator canister. Don't forget rear water hose that goes to heater, and hose that runs between thermostat housing and water pump.

Drain the block by removing the radiator cap and then opening the petcock (Giulia and Giulietta) or brass plug (1750 and after) at the driver's side rear of the engine. If coolant doesn't flow freely from the petcock, you may need to run a sturdy piece of wire through the petcock to clear it of rust. Removing the lower radiator hose will not drain the block, and the block must be drained to avoid the danger of coolant leaking past the lower cylinder seals and into the sump when the head is removed.

Fig. 5-17. *Loosen four hose clamps that attach air cleaner spacer hoses to intake manifold and pull. Large hose to separator canister and one or two other small hoses will have to be detached once air cleaner is pulled free.*

Fig. 5-18. *Use a large screwdriver to snap free upper connector of long link. Short link behind it stays in place. You can see another sin: cooked accelerator cable housing is evidence of a poor electrical ground between engine and body. A large braided ground wire runs to starter mounting area and should be checked for damage.*

Fig. 5-19. *Disconnect end of accelerator cable from bell crank. Attaching nut must be free to rotate in its housing. If it isn't, fix it now. Then remove accelerator cable from bell crank pulley.*

Fig. 5-20. *Disconnecting front fuel line at fuel injection pump. Line just to left must come off, too.*

A carbureted head is removed with carburetors in place. A fuel injected engine requires complete removal of the fuel injection lines and thermostatic actuator line. The fuel injection lines should be held to the underside of the manifold by relatively inaccessible clamps that are secured by 10 mm bolts.

On some engines, these clamps will not have been connected from the last time the head was off. The clamps provide an important function and should always be used to reduce the vibration of the coiled fuel injection lines. If they are not used, the fuel lines typically break inside the tube nuts. The lines go back only

Fig. 5-21. *Thermostatic actuator tube is held midline by a bracket. It's held in place with two small bolts.*

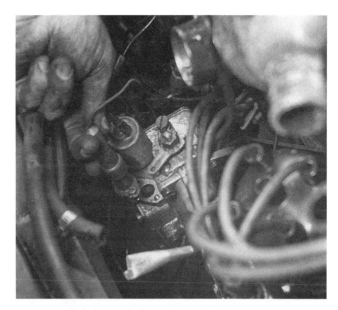

Fig. 5-23. *Remove bottom of actuator from pump with screws still in place, as shown. That way, they won't fall into pump itself.*

Fig. 5-22. *Thermostatic actuator tube is removed both from fuel injection pump housing and head. Be careful, tube is irreplaceable. Also, studs attaching tube to head break easily.*

Fig. 5-24. *Metal fuel lines must be removed very carefully, without any twisting force applied to lines themselves. Two 17 mm wrenches are used to loosen flange nuts. Hold rear one steady.*

one way and should be clearly labeled before you remove them so you know which cylinder each serves, and which end goes to the pump and which to the manifold.

Remove the eight 13 mm brass exhaust manifold nuts. See Fig. 5-25. All are easy to remove using a 1/4 in. drive 13 mm socket, except the two lower nuts of cylinders three and four. Remove the lower nut for cylinder four by swinging a box-end wrench below the manifold

Fig. 5-25. *Exhaust manifold bolts are made of soft metal so they don't freeze in place. Front ones come off easily; bottom ones on cylinders three and four are difficult. For number three, work from top. Number four comes off most easily if box-end 13 mm wrench is swung from beneath.*

and roughly parallel to the block. The lower nut for the third cylinder is easiest to remove with a 13-mm box-end S-wrench , working from the top of the engine. Hopefully, this nut can be removed with your fingers after one or two flats of turning with a wrench.

The exhaust manifold is then slipped carefully off its mounting studs. If you have to pry, use a screwdriver judiciously, without gouging any precious aluminum mounting surfaces.

Next, remove the cam cover by removing the six cam cover decorative nuts and, on Giulia and later engines, two bolts at the front of the cover. Try not to destroy the substantial gasket that forms the seal between the cam cover and the head. The gasket will probably stick to one or the other mating surfaces. If it doesn't come free easily, use a thin knife blade to try to separate it without tearing. My experience is that this gasket can be reused indefinitely without leaks. There are two semicircle "half-moon" rubber seals beneath this gasket at the rear of the head. For a leak-free engine, they should be replaced every time the cover is removed. Be careful that the seals don't drop into the large oil return passage.

Fig. 5-26. *Removing front camshaft cover bolts.*

Fig. 5-27. *Early cam cover bolts require a 14 mm Allen wrench for removal. Tool shown is part of original Alfa Giulietta tool kit, but bent to author's preference. A 14 mm bolt head also fits.*

Bring the engine to TDC with the distributor rotor pointing toward the #1 plug wire. Verify that the camshafts are on their marks before disconnecting the timing chain. Then look for the master link on the timing chain. If you can't see it, rotate the engine 360° and you should be able to find it about midway between the cams. (If you couldn't find it, the chain was improperly installed. On reassembly, make the master link fall between the two cams when the distributor is pointing to the #1 spark plug wire.)

Leave the engine at TDC and then loosen the 14 mm nut that locks the tensioner. See Fig. 5-28. Retighten the

Fig. 5-28. *After you've found master link in cam timing chain, loosen chain tensioner bolt and belly out chain as shown, then snug up tensioner bolt.*

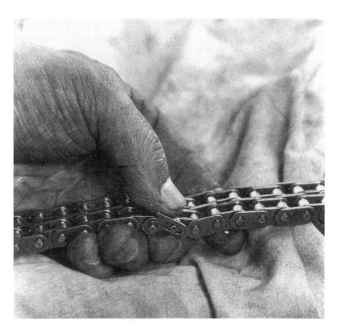

Fig. 5-29. *Snap master link clip free with a screwdriver. A large rag is placed under chain to catch clip when it falls—as it surely will. In position shown, clip can be lifted free.*

nut once the tensioner has been pushed all the way into its boss. You'll find that the cams will move slightly when the chain is detensioned, but you'll be close to the proper marks for reassembly.

Drape a towel under the master link so that its pieces cannot fall into the engine as they are removed. Only remove the master link after the chain has been detensioned, and immediately reassemble it to one end of the cam timing chain. Remove the master link clip using a broad screwdriver blade. See Fig. 5-29. Wedge the blade between one "tail" of the open end and an adjacent chain outer link. Don't let the link pop off. Hold it against the chain with your finger as you wedge it along the chain axis. After the open end has popped free, slide the clip forward a bit and lift it off the other end using a needle-nose pliers.

At this point, you can simply let the chain fall to the bottom of the engine where it won't catch the crank. This is a good procedure if you plan to rotate the crank at all. Getting the chain back up requires a bit of fishing with mechanic's wire or a magnet on a long stalk, but it's not all that tough to do.

Alternately, you can wire the chain now to keep it always accessible. The cam sprockets will hold the chain in place until you tie up its ends. Use about 12 in. of mechanic's wire for each end of the chain. You can let the chain fall down into the engine with the ends of the wire sticking up where they're accessible.

If you wish, you can also wire the ends of the chain in a loop, using about 24 in. of mechanic's wire. This is

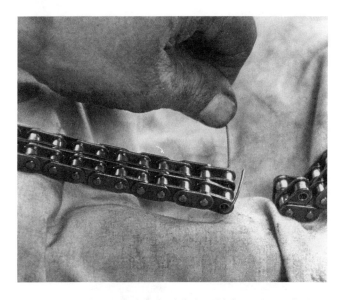

Fig. 5-30. *Secure both ends of chain with long pieces of wire.*

my preferred method simply because that's the way I learned to do it. As the head is removed, you'll rotate it slightly, set it on the top of the head bolt studs and then hold the chain securely below the head with one hand. With your other hand, undo the twisted wire ends and let the chain ends fall through the head passage, taking the wires with them. Reattach the wires from under the head to some part of the engine bay to keep the chain from dropping completely into the block.

The head bolts should be loosened incrementally following a spiral from the center out. Work slowly around the engine double-checking that everything is removed before trying to lift the head free.

Typically, the head won't come free. First, double check that you removed the two upside-down bolts at the front of the timing case cover. See Fig. 5-31. Resist the temptation to wedge a large screwdriver between the head and block. With the head bolts and spark plug wires removed, but the spark plugs screwed firmly in place, crank the engine over briefly so the compression will lift the head free.

If that doesn't work, try to jar the head free by hitting it with a plastic or rubber mallet. Strike the sides of the head, being careful not to bend or break anything in the process. Whale away for a while, then try to lift the head using engine compression as described above. Be aware that even this approach may not work.

Fig. 5-31. Two upside-down head bolts are easily forgotten. Second one is similarly placed on other side at front of head.

Fig. 5-32. When head is ready to come out, get a good grip on things.

Fig. 5-33. Rest head on studs and get another grip on it before trying to lift it over the fender. You could save some body work.

There is an Alfa tool specifically designed for head removal. If you think you'll be removing a lot of heads, then you should consider buying it from your local Alfa parts department.

You can also make your own version of the tool using the following directions given by Fred DiMatteo. Essentially, you fabricate a puller from a 6-in. square piece of 1/2 in. steel stock. The puller is large enough to cover four of the cylinder head studs. In the center of the plate, drill a hole about 9/16 in. diameter. Break out the porcelain from an old spark plug and weld an approximately 6-in. length of 1/2 in. threaded rod to the plug. To use the puller, thread the plug into cylinder two and slip the puller over the threaded rod so that the corners of the puller rest on four head studs. Run a nut down the threaded rod and then continue tightening it with a wrench to pull the head free. Cut the head gasket with a knife where it does not come free.

If this is quite likely the only Alfa head you'll ever remove, there's another simple procedure that uses coiled rope. Free the upper camshaft chain from the crank sprocket and rotate the crank no more than 45° backward (BTDC) to lower the pistons. Thread coils of rope down spark plug holes one and four. Put the car in first gear and push forward just enough for the rope to lift the head free.

CAUTION —
If you raise the head significantly this way, there's a danger of pulling a cylinder off its seal, so work carefully.

Once the head is removed, the cylinders are free to move up and down in the block. If the crank is rotated for any reason, the friction of the piston and rings may be great enough to lift the cylinder off its bottom seal and allow any residual water to leak from the block into the sump area. Some cylinders will be very easy to remove from the block and others will resist your most heroic attempts. The only way to get a truly stuck cylinder free is to remove the block from the car, dismantle the lower half of the engine, and drive the cylinder free from the crankshaft side using a block of hard wood or a brass dowel.

There is an official Alfa tool for clamping the cylinders. You can achieve the same end with a length of plumber's lead pipe and a large washer. Slip a large washer over a cylinder head stud so it captures the lips of two adjacent cylinders, then slip a length of lead pipe onto the stud and clamp the washer in place by tightening a cylinder head bolt against the pipe. See Fig. 5-34.

It's theoretically possible to do a ring job at this stage by slipping each cylinder free of the block, honing it and fitting new rings on the pistons while they are still in place. Since the bottom internal diameters of the cylinders are relieved, you can use your fingers to compress the rings enough to slip the cylinders back over the pistons. Just be sure the end gaps of the rings are properly staggered. This is not an operation that a novice should attempt, however.

I always fill the cam chain drive well with a large rag to catch the several things I always manage to drop. The most fatal error is to drop the little "fish" that acts as the outermost clamp for the master link.

NOTE —
The little master link "fish" always swims upstream. Point its head toward the exhaust side of the engine.

Examine the old head gasket carefully. It carries more clues than you might expect. Each cylinder liner should be uniformly embossed into the gasket. There should be no signs of erosion between the coolant passages and the combustion chamber. If the rubber O-rings are deformed you may wish to fit roll pins to help locate the new ones on reassembly. See Fig. 5-35 and Fig. 5-36. Under any circumstance, always have a machine shop check the head for flatness every time it's removed. Now is also a good time to have a shop replace any broken exhaust flange studs or renew marginal spark plug threads.

While the head is off, refit the spark plugs and turn it so the combustion chambers are up and level. Fill each chamber with gasoline and watch carefully for leaks around the valve heads. If you notice a wet trail of gasoline in either the intake or exhaust ports, you've found a leaking valve. With good valves, the chamber should hold fuel overnight. Some fuel will evaporate, lowering the levels, but the levels in each chamber should be roughly equal at all times. You can make this an instant test if you have an air compressor. Direct a stream of compressed air through the ports toward the valve heads. A leak will show a steady stream of bubbles in the gasoline.

If you dismantle the heads, you should check for valve stem to guide clearance after the valve springs are removed. Lift the valve off its seat by about 10 mm and try to move it in an orbital motion. If you detect any angular motion at all, the guide is suspect and the stem-to-guide clearance should be measured using microme-

Fig. 5-34. A homemade cylinder retainer is nothing more than a short piece of electrical conduit with a big washer to hold cylinders in place. This is a way of assuring that cylinders don't unseat themselves. In this particular engine, the cylinders were so stuck that they required a mandrel to get them out.

Fig. 5-35. Pounding roll-pins in place. This is a modern modification designed to help reduce danger of oil leaks at head joint.

Fig. 5-36. *A roll pin tapped to proper depth: just enough to keep O-ring in place.*

Fig. 5-37. *With roll pins in place, fit head gasket. Note that roll pin extends slightly above O-ring. When head is torqued down, pin will be pressed further in place.*

ters. If you remove the valve springs, always fit new intake oil seals.

Clean the mating surfaces of the head and block so that they are almost surgically clean. When you're ready to refit the head, rest it first on the studs and thread the cam chain through the front of the head. You can use the wires to hold the chain in place. Then, sight down one of the stud holes and move the head around until you locate the proper stud. Lower the head slightly onto the stud and then rotate it just enough to pick up the rest of the studs so they go into their holes. Slip the head slowly in place then check that nothing has been trapped between the head and the gasket.

Use a file to smooth the mating face of the exhaust header before refitting it. The exhaust gaskets are soft copper and can be reused if they are first annealed in a fire. However, if they show any sign of leaking, then you'll have to replace them. They are a one-way fit. If you get them on upside down or backward you'll create

Fig. 5-38. *Heavy carbon deposits on an exhaust valve. A wire brush will (eventually) remove these, but careful scraping with a screwdriver is faster.*

Fig. 5-39. *If using a wire brush, work slowly and carefully to avoid gouging the combustion chamber walls.*

an unwanted blockage in the exhaust system. See Fig. 5-41. Never use a silicone caulk on these gaskets.

The final steps include torquing the head down, reconnecting the timing chain, and checking cam timing. See Fig. 5-42 and Fig. 5-43. The cam timing procedure is described in the maintenance chapter.

Fig. 5-40. Clean cylinder block surface very carefully. Here, some head gasket material has remained toward back of block. Note how timing chain ends are laid out.

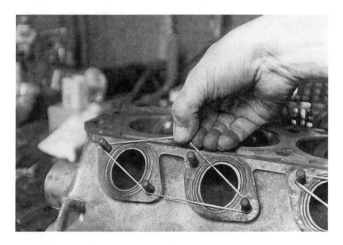

Fig. 5-41. Exhaust gaskets can be fitted wrong. This is right way, so that no part of gasket intrudes into exhaust port. Hold the exhaust gaskets in place with rubber bands as shown. When refitting exhaust manifold, thread one or two nuts onto studs to hold manifold in place, then pull hard on rubber bands to break them free.

Fig. 5-42. Head bolts are tightened incrementally using same center-out spiral pattern as they were loosened. Use a torque wrench to obtain factory specifications. Here cam chain has fallen down into the casting, but is being held safely by wire.

Fig. 5-43. We meet again: chain master link needs to be centered between two cam sprockets.

2.2 V-6 Engines

The main difference between doing a head gasket on a V-6 versus a 4-cylinder engine is that the V-6 has a toothed camshaft timing belt, which easily slips off its drive sprockets. It's easier to take down a six than a four. If you've jumped to this section without reading the procedure for the 4-cylinder engine above, take a moment now to read it. Most of the techniques and problems associated with removing the head on a V-6 are identical to those of the 4-cylinder. After all, they're both Alfas.

Fig. 5-44. *A V-6 powerplant undergoing tests at a factory facility. Note serpentine drive belts and detensioner (arrow) just to right of circular oil filter.*

Begin by disconnecting the battery, removing the alternator, air conditioning pump (without disconnecting the hoses), air intake system, and fuel rails. Drain the block by removing the two brass drain plugs, one on either side of the engine just under the exhaust manifolds. Set the engine on TDC with the rotor pointing at the #1 cylinder mark. See Fig. 5-45.

Remove the distributor. Unbolt the distributor drive idler wheel (a holding fixture is required). Don't try to undo the bolt depending on the toothed belt to hold it. Compress the hydraulic detensioner and hold it in place with a dowel (about the size of a 3 in. nail) inserted through the hole in the stamped metal movable piece and a mating hole in its cast body, then slip the timing belt off its sprockets. See Fig. 5-46.

Remove the water pump/thermostat assembly, both cam covers, spark plug tube seals, and cylinder head nuts. Finally, carefully soak the brake line bolts with brake fluid to act as a penetrant, then unscrew them and remove the brake booster from the bulkhead. It is attached from the inside of the car by four Allen-head bolts, and the pedal is held to the operating shaft

Fig. 5-45. *A properly timed V-6 engine on TDC has distributor rotor pointing exactly at mark notched in upper surface of distributor housing. Note that inductive rotor fingers, below, don't line up exactly opposite pick-up posts.*

by a clevis pin. The cylinder heads should now lift off easily.

Fig. 5-46. Detensioner removed from engine shows pin used to lock detensioner in compressed position (arrow).

Fig. 5-47. V-6 engine with one head removed, showing cylinders and pistons. At very front of left block, hydraulic timing belt detensioner is clearly shown. Note water pump outlet bolted along front of head.

3. V-6 Detensioner

The V-6 camshaft belt detensioner has a propensity to leak. Replacing it may help, but there are plenty of reports of new-part leakage. There was no reliable factory solution until early 1994, when a new detensioner was introduced. Some folks in New England are re-machining old detensioner housings to accept a larger, more robust seal (a lot of work for a little drip). Replacement hydraulic de-tensioners are warranted of course, so you may have to go back to the shop an extra time if you get another leaker.

Hydraulic Detensioner Rebuild

The detensioner assembly is normally held against the camshaft drive belt under spring tension. This is fine at low speeds, but the belt wants to belly out a bit at higher revs, so oil engine pressure acting in the detensioner is used to back off some of the spring pressure. Since the cylinder is fed from the engine oil pump, pressure tends to rise with engine speed (up to the point that the blow-off valve opens in the oil pump to stabilize maximum pressure).

Extensive steps have been taken to make the detensioner leak-free. If anything, there are too many seals involved. The top mounting stud for the detensioner is drilled so it is the source of pressurized oil to the detensioner cylinder. This stud requires a total of three seals by itself. Surprisingly, a common problem is that the detensioner tends to seize against this stud (a little oil leak here would help!).

The actual hydraulic detensioner is nothing more than a piston working in a cylinder. A single large O-ring seals the piston to the cylinder walls. Another seal is used between the actuating rod attached to the piston and the end of the detensioner housing. This is the seal that is usually the cause of the leakage, and the one the New Englanders are replacing with an oversize. Finally, a bellows boot covers the end of the detensioner where the actuating rod exits. To further discourage leaks, some enthusiasts pack the detensioner with

Fig. 5-48. Before Alfa produced its own nonhydraulic detensioner for the V-6 engine, Tom Zat's device was an alternative to a perpetually leaking unit. Pieces from the original idler, including idle wheel, are attached to this fabricated replacement piece. Since this unit makes no provision for thermal expansion, check the belt more frequently for wear.

heavy grease. This may work, but I need to point out that engine oil will dilute a grease rather quickly.

A detensioner rebuild consists of simply replacing all the seals. Replacing the hydraulic unit with a newer thermostatic detensioner is preferable, but I'll give the hydraulic rebuild procedure here since there are so many of the old ones around. You should inspect the cylinder bore for wear, but it is very unlikely that the bore will be damaged. You should be very careful if you hone the bore, since any abrasion on the cylinder walls will eat the rubber O-ring and certainly cause a leak.

The real difficulty in doing a rebuild is getting to the detensioner. It cannot be removed without first removing the right half of the black plastic cover for the cam drive belt, then removing the distributor driver idler. Removing the plastic cover is straightforward enough, but the distributor drive idler requires a special tool to keep it from turning as its mounting bolt is unscrewed. The tool is similar to the one used to hold the camshaft while retiming it and is a V-shaped adjustable spanner with two short studs at the ends of the V that engage the two holes in the idler. If this sounds at all mysterious, it will become very evident what kind of tool is needed as soon as you try to remove the idler bolt.

The idler will jump a tooth or two on the rubber belt if you don't use a tool to protect it from turning, and you can suddenly find yourself with an out-of-time engine. If you cannot borrow the proper tool, you can make one using two lengths of strap iron and some aircraft-quality bolts to make the V-shaped device. Under any circumstance, it's wise to bring the engine up on top dead center just so you'll have a proper starting point if something does slip. You'll need to recheck distributor timing anyway after everything is put together again.

Once the distributor idler is removed, you'll need to lock the detensioner in its compressed position. To do this, slip one end of the detensioning spring off its mount and collapse the detensioning mechanism by hand. Then, insert a dowel (another special tool, Alfa part number is A20363) through the hole in the spring-loaded stamped metal lever until it seats in a mating hole cast in the main body of the detensioner. You can use a nail or bolt to do this, providing its diameter is a snug fit inside the mating hole.

Approximately below the pin you've just fitted you'll be able to feel the lower mounting nut. Use a mirror to locate it first, so you'll know where to grope with the 13 mm socket. The top nut, which is threaded to the oil-supply stud, is easily seen. Remove both nuts and pull the detensioner off the engine. A short, strong spring is captured between the detensioner and a lip on the steel spacer plate located between the detensioner and the engine. This spring will fall off, so be ready to catch it. Note also that a rubber oil seal is fitted into the plate. Replace that seal now.

Mount the detensioner in a vise, then use an Allen wrench to remove the two screws that hold the roller bearing. See Fig. 5-49. Remove the stamped steel piece and the pin you inserted. Unscrew the 13 mm bolt that holds the bearing assembly together, then remove the roller and the two O-rings. See Fig. 5-50.

Use an Allen wrench to remove the two screws holding the cylinder top to the main casting. There is a fairly strong spring just under the top, so remove the screws incrementally and evenly. See Fig. 5-51.

With the cylinder top removed, release the C-clip on the end of the piston rod and remove the bellows assembly that covers the piston rod. Push up on the rod at the bottom of the cylinder and pull the piston free of the casting. See Fig. 5-52.

Fig. 5-49. *Removing two Allen screws that hold roller bearing.*

Fig. 5-50. *Using a 13 mm wrench to remove roller bearing assembly.*

Fig. 5-51. *Removing two Allen screws from cylinder top.*

Fig. 5-53. *Piston rod oil seal.*

Fig. 5-52. *Pushing out detensioner piston, which is being held by my right hand.*

Fig. 5-54. *Replacing circlip. Note how bellows is retracted with my left hand.*

There is an oil seal for the piston rod that is a snug fit inside the main casting. Use a screwdriver to press or pry the old seal free, noting which way the seal fits. Replace it with a new one, then replace the large O-ring around the piston itself. See Fig. 5-53.

To secure the bellows in place over the piston rod, hold the bellows back with one hand and insert the C-clip in its notch, then snap it in place with a needle-nose pliers. See Fig. 5-54.

Reassembly of the unit is straightforward. Replace the two O-rings on the top mounting stud. The only real

challenge when refitting the unit to the engine is to get the short spring wedged between the detensioner and its mounting lip. With the detensioner just started on the mounting studs, put one end of the spring on its mounting lip and the other end on the detensioner. You'll be able to wedge the spring so it's approximately in place. Sliding the detensioner home will properly compress the spring. Check to see that the spring sits squarely on the sheet-metal mounting lip and hasn't caught up on one of the coils instead.

Reassemble everything but the black plastic cover and then check the ignition timing. When the timing checks OK, replace the cover.

Chapter 6

Fuel and Ignition

1. Carburetors

Several times throughout this book I comment that the best maintenance you can give a carburetor is to keep your hands off it. There is something about a carburetor that demands adjustment. Because a carburetor is somewhat complex, people may wish to adjust their idle circuits just to assert their knowledge of things automotive. Or, perhaps it is just that those knurled screws are irresistibly inviting. Major adjustments to the main fuel circuits require swapping jets and are much less inviting.

The siren-call of carburetor adjustment creates a vicious circle: it's safe to assume that (1) the carburetor has been adjusted from its original settings, (2) the adjustments are wrong, and (3) it is up to you to Set Things Right.

1.1 Carburetor Idle Mixture Adjustment

Remember, you're only fiddling with a mixture adjustment that your engine hardly ever uses, once underway. If your car starts easily, idles well and runs smoothly, leave the carburetors alone. Of course, you could probably make it run even better just by turning that knurled screw.... OK, I'll tell you how. While performing this procedure, especially for Webers, repeat continuously the mantra: "I shouldn't be doing this, I shouldn't be doing this...."

Fig. 6-1. *Solex carburetor installation on a Giulia Spider shows characteristic forked intake air hose.*

CHAPTER 6

Solex Idle Mixture

We'll start with the 2-barrel, downdraft Solex, which is less complicated and so offers a virtually generic procedure. To do any good, the valve and ignition timing has to be verified as correct before tackling the carburetor. Make sure the engine is fully warmed up by taking it for a drive first. When you get back home, use the idle *speed* screw to reduce the engine's idle speed to as slow as it will go short of stalling. Next, note what happens to engine speed as you slowly turn the idle *mixture* screw about 1/2 turn clockwise, then back to where you started and another 1/2 turn counter-clockwise. Return the mixture screw to its original position.

Turning the idle mixture screw in one direction will cause the engine to speed up. Turning it in the other di-

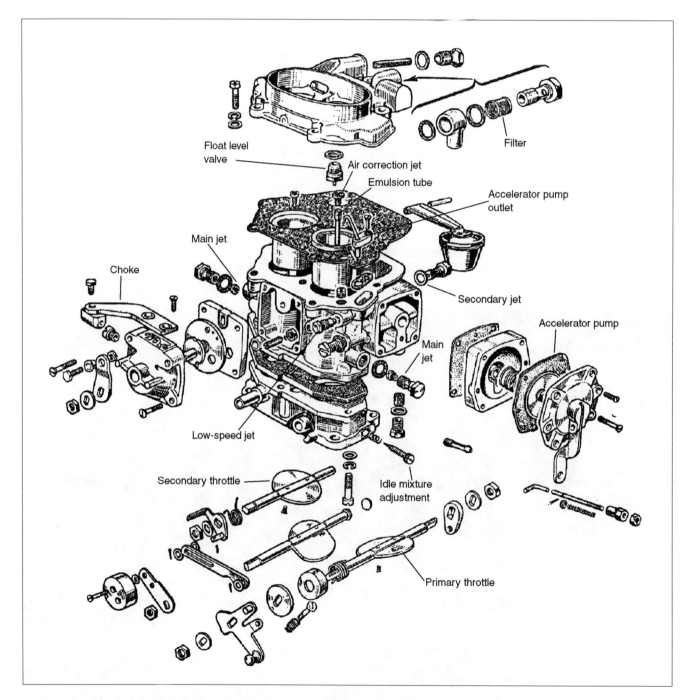

Float level valve

Air correction jet

Emulsion tube

Filter

Accelerator pump outlet

Main jet

Choke

Secondary jet

Accelerator pump

Main jet

Low-speed jet

Secondary throttle

Idle mixture adjustment

Primary throttle

Fig. 6-2. *Classic Solex 35APAIG carburetor that was standard on all non-Veloce Alfas. Later versions added extra dashpot to control opening of secondary throttle, and some were fitted with an electric choke heater to improve cold-start emissions.*

rection will probably stall the engine. If there's no change in idle speed, the initial engine speed has been set too high or the car needs a basic tune-up. Other more serious possibilities are that the idle circuit(s) could be clogged with dirt or there is an air leak in the intake manifold.

Restart the engine and slowly keep turning the idle mixture screw in the direction that causes engine speed to increase. You may need to back off the idle speed screw from time to time to keep the engine idling as slowly as possible. At some point, additional rotation of the idle mixture screw in the selected direction will cause the engine to begin to slow down. The idle mixture screw should be set so that the engine idles as fast as possible. Then, use the idle speed screw to set the idle to factory specifications.

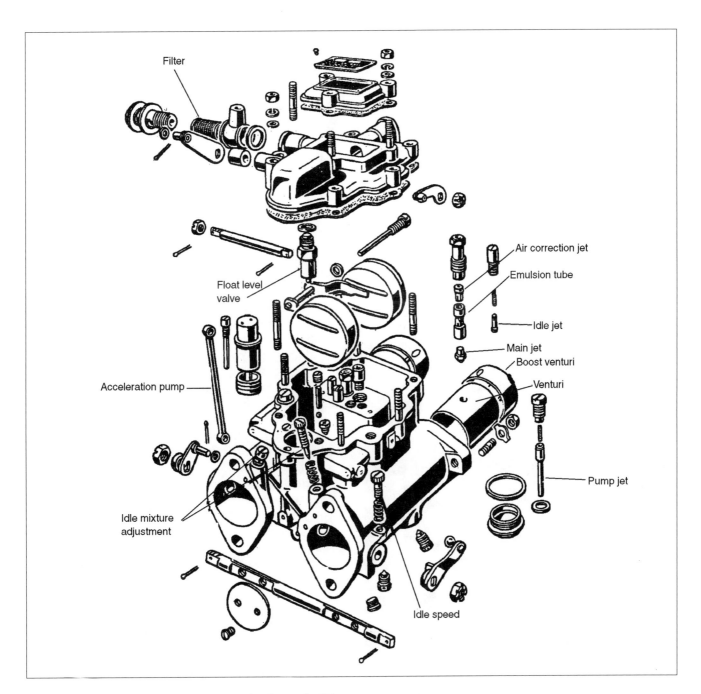

Fig. 6-3. *Weber 40DCO3 carburetor as fitted to early 750-series Veloces. Later Veloces used 40DCOE units, which are essentially identical to this carburetor.*

Weber Carburetor Setup

Contrary to what you may have heard, Webers don't go "out of tune." They do get dirty. They do get fiddled with. But they don't go out of tune. The only incrementally adjustable fuel circuit on a Weber is idle mixture richness, and that circuit is used—you guessed it—only at idle. Therefore, if a Weber has been properly set up there is never any need to fiddle with its jets, its venturi sizes, the stroke of its accelerator pump, or the operation of its starting circuit (choke).

We'll presume you have a set of Webers installed and now need to verify that all the possible adjustments are put right. Basic to this discussion is the assurance that you have a properly tuned engine with no manifold air leaks and that the carburetors have been properly jetted for your engine. The proper venturi and jet sizes for almost any engine are available from Weber and dellOrto so there's no need to recapitulate that information here. Similarly, there's no need not to use the proper sizes either out of ignorance or a mistaken belief that your experience makes you more qualified than Weber to determine engine requirements.

Synchronize Weber Throttle Plates

The first step is to assure that the throttle plates of the front and rear carburetors are properly synchronized. This is basically a mechanical adjustment. While there are several ways of doing this, the easiest is to remove the cold air box and use a UniSyn (or similar device) to verify consistent air flow between the two carburetors as the engine runs. The balance adjusting screw is located on the spring-loaded coupling between the two carburetors.

If you don't have a UniSyn, you can make the adjustment with the engine off, but it will be less precise.

Fig. 6-4. Carburetor balance adjustment screw is hard to get at, but here it is. Giulia Super pictured.

The balance adjusting screw should be turned so that, with the idle stop backed completely so the throttle plates are free to seat themselves, the throttle plates of the front and rear carbs close at exactly the same time. There are several ways of verifying this.

On many Webers, there is a brass inspection bolt placed just over the intermediate circuit jets. By removing the bolt and using a bright flashlight, you can see the edge of the throttle plates through the jets. The balance adjusting screw should be turned until both sets of throttle plates bottom equally as seen through the jets in both front and rear carburetors.

If you are not hurried, you can watch the motion of the front and rear throttle shafts and, by turning the balance adjusting screw in and out in progressively shorter arcs, zero in on the correct setting without seeing the throttle plates at all.

Weber Idle Mixture

The goal of this process is to get all the throats feeding just the optimum mixture of fuel at idle. You'll know when you get the optimum idle richness setting because the engine will idle faster at that setting than at any other. To obtain the optimum mixture, you keep enriching (turning out) the idle mixture richness screws while at the same time reducing the idle speed set by the idle adjusting screw. The purpose of progressively lowering the idle speed adjustment is to assure that it is the idle mixture, and not the idle speed screw itself, that controls idle speed.

With the Webers, adjusting a single idle mixture screw gives an imperceptible change. To overcome this difficulty, you'll be adjusting all four screws an equal amount each time: four equal turns of the screws equals one adjustment step.

With the engine off, turn each idle mixture richness adjusting screw in until it just seats, then turn it out exactly one-half turn, or 180°. Note carefully the exact orientation of each slot on the idle mixture richness adjusting screws. The slots act as pointers that will help you keep the mixture screws equally adjusted. For the next step you will always rotate each screw exactly the same amount.

Start the engine, bring it up to operating temperature, and adjust the idle speed screw so you get the lowest possible idle just short of stalling the engine. The first two or three passes at the following routine are apt to be rough, indeed. (If the engine will not idle at all, you'll have to turn each of the idle mixture richness screws out equally one turn to obtain a mixture rich enough for the engine to idle.) Turn each of the idle mixture richness adjusting screws out 90°. The engine will probably speed up. If it slows down instead, turn all of the idle mixture richness screws in 90° to their original position and then turn them in further another 45°. Readjust the idle speed so the engine idles as slowly as possible without stalling. Again, turn each of the idle mixture richness screws out equally in increments of 90°, then readjust the idle speed screw to obtain the lowest possible idle speed without stalling the engine.

Continue to turn the idle mixture richness adjusting screws out in 90° increments and then readjust the idle speed. Repeat this process until you have obtained the fastest possible idle speed. At some point, turning the idle mixture richness screws further out will cause the engine to begin to slow down. It is at this point that the idle mixture transitions from its ideal mixture to over-rich. You can fine-tune the idle mixture richness by repeating the process using 45° increments. Once you're satisfied you've obtained the highest possible idle, the mixture adjustment job is done.

Finally, use the idle speed adjusting screw to bring the idle to about 850–900 rpm.

It's clear that this procedure for Webers gives only an approximation of the optimum setting for each cylinder. Resist the temptation to find the absolute optimum setting for each cylinder by adjusting each idle mixture screw individually. Trying to do so compromises the basic balance you've given the system.

Color Tune, a product imported by Gunson, is an easy way to verify mixture richness. It is essentially a transparent spark plug that allows you to see the color of the burning mixture. Color Tune is useful both on carbureted and fuel injected engines.

1.2 Fuel Jets

This applies both to Solex and Weber carburetors.

Fuel mixture values for anything other than idle are controlled by replaceable jets. For the Solex, these jets are located in brass carriers on the side of the carburetor casting. The 14-mm carrier holds the main jet; the 10-mm carrier holds the idle jet. Weber carburetors have the main and idle jet carriers located under an access plate that is held to the carburetor by a thumbscrew. The jets themselves are at the bottom of the carrier.

The main jets control the overall volume of fuel delivered. Unfortunately, because fuel is more dense than air, an uncorrected flow would draw much more fuel at high speeds than at low speeds; that is, an uncorrected system goes rich as air velocity increases.

To reduce fuel flow at higher speeds, air is bubbled through the gasoline to emulsify it. The amount of air drawn in is controlled by the air correction jet and the place where emulsification occurs is called the well. The air correction jet on a Solex carburetor sticks into the well. On a Weber, the main and idle jets are threaded into cavities that form the well, and the air correction jet is located at the top of the jet carrier.

Basically, fitting a larger main jet increases fuel richness at the low end of the jet's operating range, and fitting a larger air correction jet causes a leaner mixture at

Fig. 6-5. *Solex main jet screws into this holder. There's another one, on other side, for the second venturi.*

Fig. 6-6. *Solex intermediate jet hardly ever needs changing, but this is where it's located.*

the top of the jet's operating range. That is, if you want a richer top-speed mixture, fit a smaller air correction jet. If you want a richer mid-speed mixture, fit a larger main jet.

The solution is not quite so straightforward as that, however. If you fit a larger main jet, you'll need a larger air correction jet to keep from going over-rich at maximum speed. And, if you fit a smaller main jet, you'll need a smaller air correction jet to keep from going lean at maximum speed. That is, you have to change both the main and air correction jets to maintain a mixture of optimum richness over the entire operating range. Finding the ideal combination of main and air correction jets can be very time consuming. If your Alfa is box-stock, there is never any need to change the jetting from stock. If it isn't, well, running with the big dogs isn't always easy.

CHAPTER 6

Fig. 6-7. *The idle jet sometimes clogs. If engine won't idle but runs great above 2000 rpm, chances are idle jet is the culprit. Here's where it's located.*

Fig. 6-8. *Air correction jet for main circuit is located here on Solex. Be careful when removing this jet. Tube into which it screws can become loose if you twist too hard. You can prevent this by holding the tube with a pair of needlenose pliers.*

2. SPICA Fuel Injection

Alfa is the only car in the world to use SPICA fuel injection. As a result, it is an unknown system to mechanics who are not Alfa specialists. Alfa has wanted to keep the SPICA system as unknown as possible, too, to guard against tampering. It has a policy never to repair, but always to exchange the entire injection pump. Over the 13 years that the SPICA system was offered Alfa exchanged over 4,200 SPICA pumps, quite a significant number taken against the fact that annual new-car sales in those years ranged around 3,000 units. But before you conclude that one of ten SPICA pumps failed, I need to report that many of the SPICA units returned for service were found to be working properly.

Because of the lack of general information about the SPICA system, any inexplicable problem is usually laid at its doorstep. This leap of logic is not really necessary. Since the fuel injection system still manages to do exactly the same thing as a carburetor (although more accurately), the symptoms of misadjusted carburetion also apply to the SPICA system: poor fuel mileage, black smoke from the exhaust pipe, failure to idle, and no top-end power are identifiable fuel-delivery faults. The only trick is to eliminate all the other possible causes of the same symptoms before concluding the SPICA unit is at fault. That is, verify that the ignition is properly timed, there is good cylinder compression, the valves are properly adjusted, and there are no blockages in the air or fuel filters.

Some Background

When the SPICA system was introduced on U.S. Alfas in 1969, Alfa was one step ahead of the industry in providing a low-emissions vehicle that did not require either a catalytic converter or an air pump. As a result, in 1971, when government regulations mandated strict emission controls, Alfa was one of the best-running cars you could buy off a showroom floor.

Unlike Alfa, the emission controls on domestic marques in the early 1970s were rather hastily grafted on. Because they were so poorly engineered, they did serious damage to power output and fuel economy. Even the average owner quickly realized that he could improve overall performance simply by having the emission control devices disconnected and a cottage industry of illegal desmog shops grew up almost overnight. The fine for a franchised dealer caught disabling emission controls was $10,000 but that did not deter the practice.

In this era, manufacturers were very frustrated. They did not like power-robbing and costly add-ons any more than the consumer. Yet, most manufacturers feared that unless they protected the emission controls from tampering, more strict government regulations would surely follow.

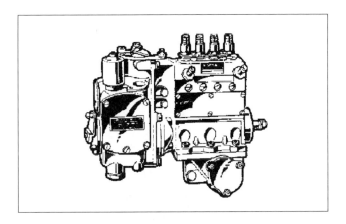

Fig. 6-9. Early 1750 SPICA unit. Note absence of fuel cut-off solenoid. Mixture richness is adjusted using a screw and locknut on the top of pump body.

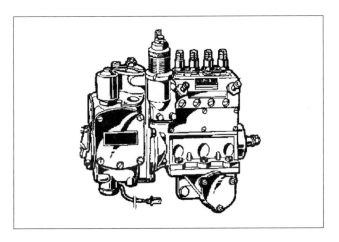

Fig. 6-10. SPICA fuel injection pump as fitted to all 2-liter 4-cylinder engines. On these pumps, mixture richness is adjusted by fuel cutoff solenoid.

The SPICA injection pump was regarded as an emission control device by enthusiasts (probably inappropriately) and the general consensus of the time was that it robbed the engine of power. Putting Weber carbs on fuel-injected Alfas became the goal of many owners. In the meantime, they were quite willing to fiddle with the SPICA system in an attempt to discover some way of disabling its presumed power-robbing characteristics. In retrospect, we have come to realize that the SPICA system is every bit as efficient as Webers and there is no real performance increase to be gained by changing systems (providing Venturi sizes remain the same). That fact was not known in the early 1970s.

It is in this context of frustration and fear that Alfa's attitude toward the SPICA system must be judged. Alfa did publish a technical manual on the SPICA system, but there were no repair parts and Alfa would only ex-

change entire units. No field repairs were permitted because a calibration bench was required for successful set-up. Owners' questions directed to Alfa regarding SPICA went largely unanswered. The Service Manager for Alfa during this era was Bob Francioni, a dedicated Alfa enthusiast who was totally committed to following the company line of keeping everything box-stock. Bob once vehemently complained to me that owners were using other than the standard Lodge spark plugs in their cars—that will give you some idea of his reverence for original equipment.

The technical editor for the Alfa club magazine *Alfa Owner* in those days was Joe Benson, a trained automotive engineer. In 1975, Joe began publishing a series of articles in the club magazine that must stand as the single most popular and most significant series ever to appear in that publication. He told all about the SPICA system, including how to take it apart—and, in a later series, how to modify it. The EPA calls this "tampering," of course.

Though most readers didn't know it, Alfa was contributing several thousand dollars a year to the magazine, and here that magazine was efficiently subverting Alfa's strict, no-tamper policy. So, there were some hard feelings.

The troubleshooting and repair portions of Joe's articles follow. I've edited them somewhat for brevity.

2.1 SPICA Pump Operation

The SPICA unit is a diesel injection pump modified during ten years of development by Dario Radaelli for a gasoline engine. The basic concept of using the SPICA diesel pump for a gasoline engine came from the Alfa engineer Garcea. Any number of in-line diesel pumps are similar to it, but most passenger-car technicians are simply not exposed to that technology. See Fig. 6-11.

It helps to think of the injection pump as consisting of two distinct sections. The front pumping section, which contains a small crankshaft, piston-like plungers and output check valves, is responsible for the actual delivery of the fuel to the cylinders. The rear logic section is best described as a mechanical computer that controls the output of the front pumping section.

Pump Section

The basic fuel injection is controlled by the up-and-down motion of the plungers. This motion compresses the fuel for injection at the injectors. The plunger stroke is constant, that is, there is the same amount of lift and fall every time the crankshaft rotates. So, all things being equal, the pump would inject the same amount of fuel for every rotation of the pump crankshaft.

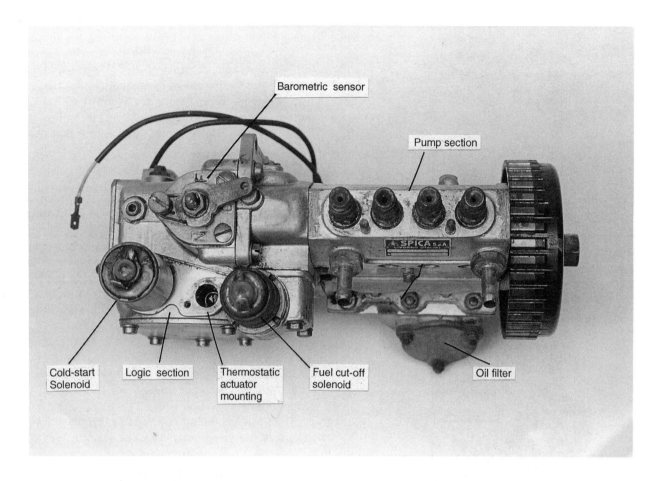

Fig. 6-11. *From top, two sections of SPICA pump are clear. To right half, pump itself; to left, logic section.*

The *amount* of fuel injected is changed by a rotary valve. The position of the rotary valve is in turn controlled by the gear rack. All those levers you see in the logic section affect the position of the gear rack, and therefore the amount of fuel injected for each rise and fall of the plungers.

Logic Section

The movement of the rack link in the logic section determines movement of the gear rack in the pump section. Various other links and levers are connected to the rack link, all intended to change the amount of fuel injected. See Fig. 6-12. To try and sort it out, let's take it from the point where you step on the accelerator:

Accelerator pedal movement moves the control lever, which then moves the 3-dimensional cam in a circular fashion around the crankshaft. This causes the cam follower to move (not much, but enough), which moves the notched lever, then the connecting link, then the rack link.

Here you should know that the 3-dimensional cam is actually composed of two parts: the first part just described, on which the cam rides, and whose movement is determined by the control lever; the second part is splined and fits onto the pump crankshaft, and therefore turns at crankshaft speed. See Fig. 6-13. This second part traps a number of balls that spin with the cam in a cup. Centrifugal force pushes the balls up the cup as the cam rotates, so the cam is pushed outwards at the same time, toward the control lever. This also changes the position of the follower on the cam, and therefore injection quantity.

From here you can take it yourself to figure out the relationship of the other parts (the thermostatic actuator, the cold-start solenoid, etc.) and how they affect the relationship of the cam and lever. It should be obvious that if any one setting of the pump is wrong, all the relationships become distorted

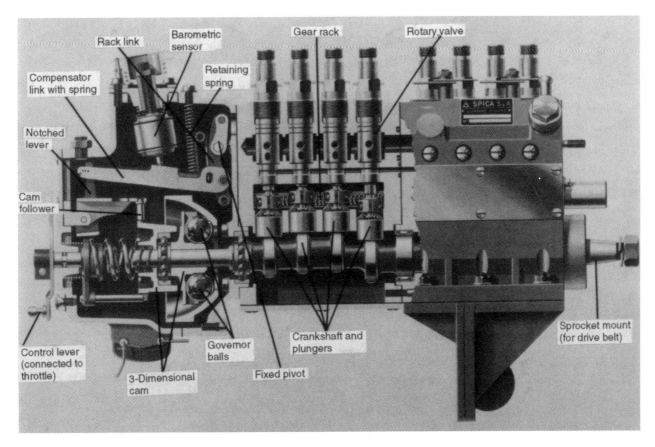

Fig. 6-12. *Phantom view of Montreal SPICA pump shows levers in logic section, as well as crankshaft, plungers, and gear rack.*

Fig. 6-13. *Three-dimensional cam controls fuel delivery volume according to engine speed and throttle position. There are 3,200 reference points used to plot cam surface! At bottom are guides for heavy steel balls that act in cupped surface to move cam backward and forward according to engine speed. Depressed groove in surface is idle position of cam follower. Fuel delivery volume increases with elevation of cam surface. Widest throttle openings fall to the right of the depression while highest engine speeds fall toward the bottom of the cam surface in this photo (nearest the ball races). Notch to left of cam is connected by control lever to accelerator pedal linkage.*

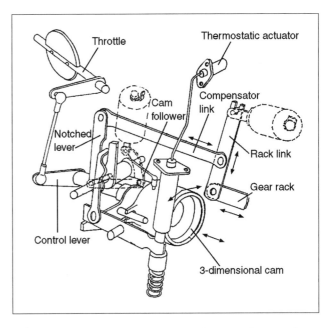

Fig. 6-14. *Compare this functional drawing of SPICA logic section to cutaway drawing (above). Deciphering what this drawing really shows is best way to understand SPICA logic section.*

Fig. 6-15. *Fuel cutoff switch, located on bottom of SPICA pump. Switch is operated from 3-dimensional cam, to cut off fuel flow on deceleration.*

2.2 SPICA Tune-up

The following tools are essential:

1. Mercury manometer (Alfa part C.2.0011). An inexpensive manometer is available for tuning 4-cylinder motorcycle engines.

2. Manometer connecting hoses and restrictors (Alfa part C.2.0012).

3. Throttle stop setting fixture (Alfa part A.4.0121).

4. Wrench to loosen nut that locks fuel cutoff solenoid (1971 and later models only—this is a castellated wrench but a workable tool can be made by slightly bending in the loop of a 2 in. muffler clamp and grinding its ends to fit the slots).

All these items are shown in the injection manual, and they can be ordered through your local dealer. The purpose of the restrictors of item two is to adjust the measured vacuum so it can be measured by the manometer. If you get the motorcycle-type manometer, you can make your own restrictors by drilling plugs with number drills until you get a restriction that gives easily readable values on the manometer.

Before diving in, a few words of advice.

First, the Dos and Don'ts:

Do buy Alfa's injection manual and read and re-read it until you're well oriented with the system.

Do be sure the engine is freshly tuned before trying to adjust the injection. It must have good compression, proper valve adjustment, and be in proper time.

Do have all the necessary special tools before starting any injection work or you'll be sorry.

Don't ever tamper with the reference screw that acts as a stop for the pump throttle bellcrank. All adjustments are made by measuring the gap between the bellcrank and this screw, and if this reference is lost the pump must be sent back to Alfa for recalibration.

Now for some encouragement. The SPICA system is most forgiving, very logically designed, and, once you understand it, very simple to adjust. In fact, it will tolerate gross misadjustments and still run quite acceptably. For any given misadjustment, there always seem to be at least several compensating tweaks that will get you safely home. The correct adjustments, however, are what you really want in order to obtain the optimum overall performance and best emissions, which, happily, seem to occur simultaneously.

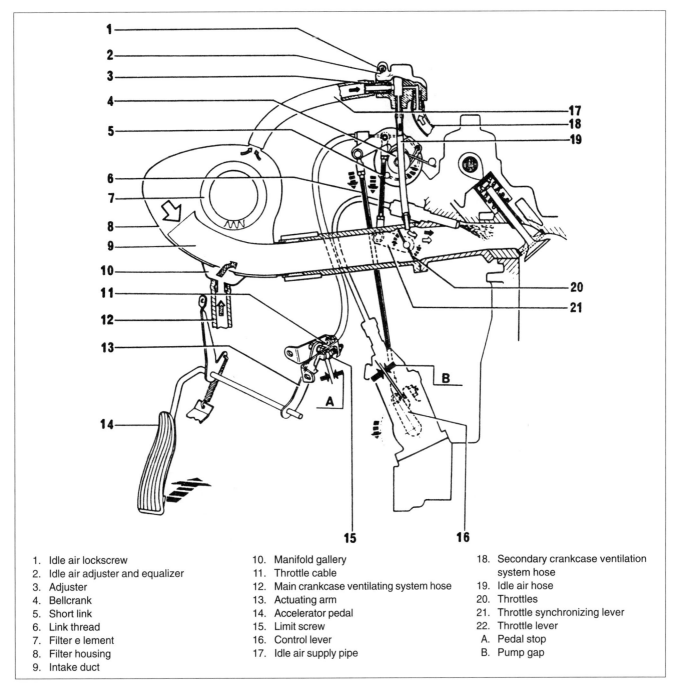

1. Idle air lockscrew
2. Idle air adjuster and equalizer
3. Adjuster
4. Bellcrank
5. Short link
6. Link thread
7. Filter e lement
8. Filter housing
9. Intake duct
10. Manifold gallery
11. Throttle cable
12. Main crankcase ventilating system hose
13. Actuating arm
14. Accelerator pedal
15. Limit screw
16. Control lever
17. Idle air supply pipe
18. Secondary crankcase ventilation system hose
19. Idle air hose
20. Throttles
21. Throttle synchronizing lever
22. Throttle lever
A. Pedal stop
B. Pump gap

Fig. 6-16. *Mechanical details of SPICA fuel injection system and its complicated accelerator linkages. Note especially connection of bellcrank, long link, and short link.*

Cold Adjustments

The cold settings are somewhat misnamed since they may all be done on a hot engine with the exception of the thermostatic actuator adjustment. However, since a completely failed actuator can still be adjusted to the correct hot setting and may escape detection until your next (attempted) cold start, I recommend that the procedure be made on a cold engine.

Bellcrank Adjustment. A special Alfa tool (P/N A.4.0121) is used to adjust those two inviting stop screws that sit highest on the Alfa engine on the same casting that carries the throttle-cable relay crank. Using the tool, it's a simple matter to set the screws to their proper values. Once set, they need never be changed. Repeat: never changed. Wes Ingram has discovered that, from the factory, both screws were set so they were

Fig. 6-17. Intake manifold of a SPICA fuel injected engine. Throttle bellcrank is top center and four venturi are aligned across bottom. Injectors are located directly above venturi. One adjustment screw is just visible behind bellcrank.

of even height. A quick check with a caliper will verify this. If the screws measure the same, then leave well enough alone.

Thus, if you are the original owner, you'll never have to worry about setting the stop screws. If you are not the original owner, you will always suspect that the last owner changed them. Or a mechanic. You will certainly want to have the screws checked, but the cost of the tool is very high considering its one-time use. You may be able to borrow the tool through a chapter of the Alfa Romeo Owners Club (see Appendix 2) or a sympathetic dealer or independent repair shop. The idle stop screws can also be set using a protractor and some care. The idle setting of the throttle cable relay crank is 10° open from the fully closed throttle position and the full-open setting is 86.5°. See Fig. 6-18.

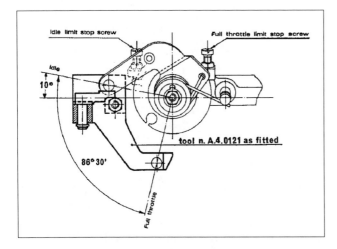

Fig. 6-18. From full-closed, bellcrank is rotated 10° for its "closed" setting, then a total of 86.5° for full-open. A special tool is available, or you can achieve same results with some careful persistence and a protractor taped to bellcrank.

The tool bolts in place on the cable clamp studs using special shoulder-bolts for exact centering and offers two stops to the throttle cable relay crank. The first stop, the idle setting, is obtained by swinging the arm on the tool and adjusting the rearmost screw so the relay crank just touches the arm. The arm is then swung out of the way and the other stop screw is adjusted so the relay crank just contacts the fixed tip of the tool. With the stop screws properly set, the throttle stop at the firewall end of the throttle cable should have a clearance of 0.40 to 0.60 in. With the accelerator pedal fully depressed, there should be a gap of 0.080 in. between the relay crank lug and the full-throttle stop screw. Adjust the bolt on the floor under the accelerator pedal to obtain this clearance.

Pump Gap. Measure the gap between the pump control lever (bellcrank arm) and its reference set screw with a feeler gauge (this distance is called the pump gap). See Fig. 6-19. It should be 0.146 to 0.165 in. If it is greatly different there is either a maladjustment or a failed thermostatic actuator.

Fig. 6-19. Pump gap is distance between pump control lever and its set screw. Note that stop screw has an anti-tamper plastic cap wired in place. If cap is removed, unit should be returned to Alfa to verify setting. Don't ever remove cap!

To check the thermostatic actuator, remove it from the pump (two screws), being careful not to kink its tube. If the probe at the end of the actuator is almost flush, the actuator is bad and must be replaced.

If the actuator is OK, turn the adjusting screw inside the pump that the actuator probe bears against until you can obtain a pump gap clearance of 0.146 to 0.165 in. See Fig. 6-20. Turning the screw clockwise increases the pump gap. This is a rough adjustment. The hot setting later in this procedure is more critical. Re-fit the thermostatic actuator.

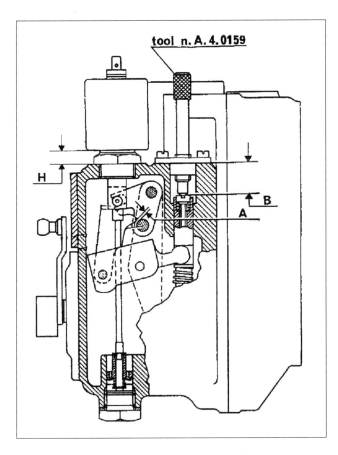

Fig. 6-20. *Special tool A.4.0159 in place to adjust pump's mechanism for thermostatic actuator. Just below tool (at line B) is adjusting screw to obtain proper pump gap setting on a hot engine and is very critical. Distance B is nominally 27–29 mm. Distance A can only be adjusted with pump removed.*

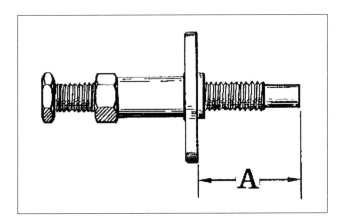

Fig. 6-21. *John Shankle has designed a universal dummy thermostatic actuator for SPICA-injected cars which simplifies a tune-up. Tool is available from many Alfa enthusiast mail-order stores.*

Throttle Plates. Reinstall and adjust the short link between the throttle bellcrank and the throttle plates so the throttle plates act as a stop when the bellcrank is released gently and the throttle stop screw acts as the stop for a slightly more forceful release. After the throttles are synchronized at a later step, it may be necessary to repeat this adjustment. If the throttle plates do not close completely the engine will backfire on deceleration and it won't idle properly.

Throttle Cable. Reconnect the throttle cable to the bellcrank and refit the cable housing locating clips. Make sure there is a small amount of slack in the cable when the bellcrank is against the idle stop screw. The cable must not be allowed to hold the bellcrank away from the stop. Adjust, if necessary, at the other end of the cable where it attaches to the pedal linkage.

Linkage. Adjust the linkage to obtain full throttle opening. Have a friend sit in the car and hold the gas pedal all the way down. Check the bellcrank to ensure it is opening fully against or (preferably) almost against the throttle stop screw. Adjust the stop under the gas pedal to obtain a fully opened throttle bellcrank.

Hot Adjustments

Refit the air box and any other miscellaneous hoses and pieces you may have removed. Start the engine and warm it up by driving it at least ten minutes at full indicated temperature.

Pump Gap. Quickly remove the air box and disconnect the long link. Immediately measure the pump gap (the thermostatic actuator starts changing within minutes of shutting off the engine). The gap should be 0.019 in. If it isn't, remove the actuator (refer to **Cold Adjustments** above) and turn the adjusting screw a bit, then check the gap quickly by manually pushing the actuator back down into the pump until fully seated. If the gap is still wrong, remove the actuator and turn the screw a bit more. (Turning the screw clockwise increases the pump gap, and turning it counterclockwise decreases the pump gap.)

To keep the engine hot, you can remove the actuator with the engine running by pinching the pump bellcrank against its stop while the screw adjustment is made. This is a very important adjustment that gives good starting and running and is well worth many consecutive adjustments to obtain the correct 0.019 in. gap.

NOTE —
This hot-adjustment procedure applies to 1969 models so far as the 0.019 in. gap specification. On later models, a dummy tool is used instead of the engine's own actuator to obtain a pump gap of 0.019 in. This tool varies in specification depending on model year. (See table below.)

Model Year/Application	Tool Length (Dimension A)
1969–1974	27.0 mm
1975–1976 1977 49-state 1980–1981 49-state	27.8 mm
1977 California 1978–1981 all others	29.0 mm

Later thermostatic actuators were capable of an additional 4 mm extension when hot. In those instances where new thermostatic actuators have been used to replace old failed actuators, install an 0.080 in. shim between the actuator and the pump to permit a 0.019 in. pump gap to be obtained.

Long Link. With the engine still hot, refit the long link and remeasure the pump gap. It should now measure 0.035 to 0.051 in. Adjust the long link as necessary. A gap of 0.045 in. seems to give good results.

Throttle Plates. Attach the manometer and synchronize the throttles by obtaining equal vacuum levels in the front and rear pairs of inlets. A spring-loaded upside-down screw is located on the throttle shafts between cylinders two and three. Screw it in or out to change the balance between the front and rear cylinder pairs.

Idle. To adjust the idle, first warm the engine to operating temperature. Remove the air cleaner-to-equalizer hose (for the 1969 model, four separate idle circuits are adjusted; for all other models, a single mixture screw is adjusted). Connect an accurate electronic tachometer. Loosen the adjuster lock screw and turn the idle-air adjuster until the highest possible engine speed is reached. You may be able to turn the idle-air adjuster by hand. If not, a pair of pliers can be used providing the tube is not marred or deformed. When the idle is properly adjusted, retighten the lock screw (not too tight) and reconnect the hose.

If a CO meter is available, adjust the idle to give 1% CO at not lower than 600 rpm.

Temperature Compensator. Set the seasonal temperature compensator to the temperature of the work place (typically, always N for "normal;" C is "cold" and F is "freezing"). With everything but the air box in place and on a warm engine, blip the throttle, observing how smoothly the engine comes off idle.

NOTE —
If your car ran satisfactorily before and you found all the gap and link settings to be reasonably correct, grit your teeth and resist the temptation to mess up a good thing. However, if everything else was wrong, it's likely that the mixture was adjusted to try to compensate and it will have to be reset if the engine is to run correctly.

Mixture Adjustment. There are two methods, depending on year. On 1969 models this is a simple 10 mm locknut and a set screw. On later models, mixture adjustment is provided by screwing the fuel cutoff solenoid in or out, after you loosen its locknut. Wes Ingram has observed that the fuel cut-off solenoid is usually screwed in between 9 to 10-1/2 turns from the starting threads. He suggests that the solenoid should be screwed in 10 turns from full-out as an initial setting, if that is in doubt.

While blipping the throttle, begin to lean out the mixture: on '69 models, turn the screw counterclockwise; on later models, turn the solenoid clockwise. Continue leaning the mixture until the engine exhibits a coughing, fluffy response while revving up from idle and can barely be made to run above idle. This is the lean reference point. From here, very little enriching is needed, usually only half a turn, to obtain a proper mixture. The idea is to set the engine as lean as possible and still permit it to rev up smoothly from idle. When you're satisfied with the adjustment, tighten down the lock nut, refit the air box, and begin the extended road test, which provides final mixture adjustment.

An alternate procedure to the blip-the-throttle technique has been developed since Joe's articles. Set the engine to a fast (2500 rpm) idle using the manual throttle or by blocking the accelerator linkage with a wedge to obtain about 2500 rpm. Then, turn the solenoid to lean the mixture until the revs begin to fall. From this position, turn the solenoid towards rich until the revs reach their highest point, then turn the solenoid towards lean 1/16 turn and lock it in place. Return the engine to normal idle speed.

Mixture Check. After warming up the engine thoroughly, hold cruising speed for several miles and then "cut clean": simultaneously switch off the ignition (don't lock the steering), disengage the clutch and coast to the side of the road. Unscrew one or two spark plugs and look at their ceramic insulators. You're after a light gray-brown color. If the insulator is white, the mixture is too lean; if it's black, the mixture is too rich. Adjust the mixture richness as described above by no more than 1/8 to 1/4 of a turn. Repeat the test drive and recheck. Continue repeating and adjusting in very small increments until the desired color is obtained on the spark plug electrode.

Alternately, have a shop set the CO level to 1% at idle.

This completes the adjustment procedure.

Fig. 6-22. *Mixture adjustment is provided by position of fuel cut-off solenoid. Loosen castellated locknut and screw sole-* *noid in or out. There are small marks on top of solenoid to keep track of adjustment, or you can use position of spade lug.*

Fig. 6-23. *The author's mixture adjusting tool, fabricated from a muffler clamp. Tips fits into slots in castellated nut. A heavy washer, cut in half and welded near the tip, provides extra torsional strength.*

2.3 SPICA Troubleshooting

If you're satisfied that the above adjustments are correct and the car still will not run properly, further troubleshooting is necessary. The procedure is divided into two parts depending on whether or not the car will start.

One other point first. Though fuel injection is supposed to eliminate heat-related fuel delivery problems, the SPICA unit on some Spiders has an annoying failure mode in ambient temperatures over 90°. The pump itself heats the fuel when it compresses it, of course. The system dumps excess fuel back to the gas tank. If the ambient temperature is high enough, the tank can't act as an adequate heat sink to cool the fuel so the fuel eventually boils when its compressed by the pump, leaning out the mixture and causing the car to stall. A half-hour of sitting lets everything cool down to the point that the car will run again.

The "home-brew" cure for this kind of failure is to spray the bottom of the tank with aluminum paint to reflect road heat away from the fuel, assure that all the fuel lines have the proper (original) diameter, and that the filters are not clogged. For some reason, this is a Spider-only problem. Coupes and sedans aren't affected. Alfa's fix for the problem was to install a submersible fuel pump inside the tank that fed the main pump at 3 psi.

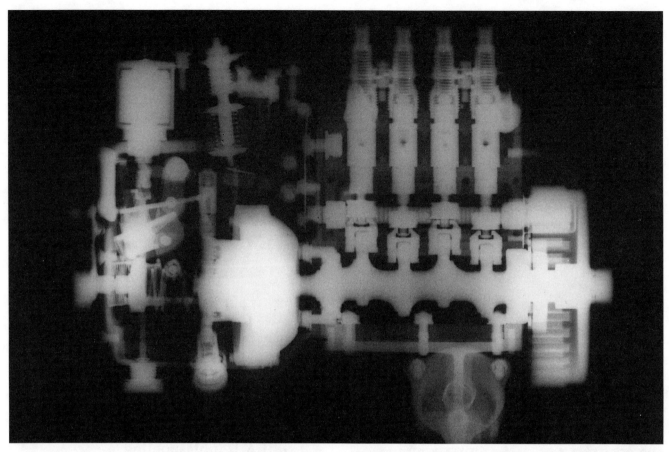

Fig. 6-24. *This X-ray of a 1750 SPICA pump was made by Joe Benson, certainly one of the most knowledgeable Alfa enthusiasts around. (Original print courtesy Ken Askew)*

Missing Plastic Cap Over the Pump Gap Screw

In the chapter on buying a used Alfa I cautioned to deduct about $1000 from the value of an Alfa if the little plastic cap over the pump gap screw has been removed. Everyone—well, almost everyone—who knows anything about Alfas knows that cap is inviolate, so you can safely leap to the conclusion that the person who removed it had no business even opening the hood—and lord knows what else he did, etc.

Several years ago one of the chapter newsletters of the Alfa Club published a long article detailing the proper procedure for setting the pump gap. It required removal and partial disassembly of the pump, and the use of precision measuring instruments. My belief is that if you go to the trouble of removing the pump, you probably should have it worked on by someone who really knows what it's about.

There is a work-around for a misadjusted pump gap screw, part of which comes from Wes Ingram and part from me. Wes has observed that an undisturbed pump-gap screw extends equally on both sides of its threaded boss. So, the first step is to remove its locknut completely and, counting the threads on either side of the boss, set the screw so it extends equally on both sides of the

boss. Then, lock it in place with its locknut and set the pump gap as instructed, following Joe's recommendations under **Hot Adjustments**. The car should start and run with this adjustment. Now, in the following steps, we'll fine-tune the adjustment and finally reset the pump gap screw.

The pump gap establishes the position of the 3-dimensional cam at closed throttle. That is, the actual gap distance is most critical when the engine is idling. The gap itself is adjusted by the screw under the thermostatic actuator, so if you can get that screw adjusted properly, you won't have to worry about the pump gap. This presumes that the engine is in tune, there are no air leaks, and the timing has been verified as being absolutely correct. The reason for double-checking the timing is that the engine idle speed changes with timing, and we are going to adjust the thermostatic actuator screw to obtain a proper idle speed. If the ignition timing is off by any amount, the exercise is futile.

If you're using a dummy actuator (see earlier Fig. 6-21) verify that the actuator has the proper length (dimension A—see table on page 132). If you're using the thermostatic actuator with a Bourdon tube, the engine must be at operating temperature. Turn the screw the actuator bears upon (just below the line labeled B in earlier Fig. 6-20) until you get the proper idle speed. If you're using a real actuator, this means installing and reinstalling it several times, probably. The advantage of the dummy actuator is that you can adjust the screw with the dummy in place. When things seem OK, install the real actuator (the one with the Bourdon tube), shut things down, and leave the car overnight.

In the morning, start it up. You should have an immediate fast idle, near 2000 rpm for a moment and then a slowly falling idle speed as the engine warms up. If there is not a fast idle, or if the idle is below about 1500 rpm at start-up, shut off the engine immediately to keep the coolant from heating up (and thereby extending the actuator tip). The odds are that the screw under the actuator to be turned counterclockwise a bit. Remove the actuator and give it about 1/2 turn.

Reinstall the actuator (both screws, please) and restart the engine. If the engine is still dead cold, you should notice an increase in idle speed over the initial start. If you wish, you can shut the engine off and readjust the screw some more, until you're getting a good fast idle near 1500 rpm (since the engine is warmer, the desired fast-idle speed is lower).

Let the engine warm up fully. It should settle out at an idle speed of about 850–900 rpm. If it doesn't, readjust the actuator screw to get that speed, reinstall the actuator, shut it down, and wait until the next morning to verify fast-idle speed when the engine is again cold.

After several mornings, you should be able to get a good fast idle when cold and a proper hot idle speed around 900 rpm. At that point, you can reposition the pump gap screw so it is 0.019 in. away from the throttle bellcrank with the long link disconnected.

Experience has taught that the SPICA unit is very forgiving, otherwise this roundabout approach to pump gap setting would never work.

Car Will Not Start

If you're working on a car that wouldn't start even before you started working on the injection, your problem is a bit more difficult and may well not involve the injection at all. Here are a few things to check:

Getting a good strong spark?

Distributor timed right? Sure it's not 180° out?

Fuel pump operating correctly so fuel warning light goes out?

Are the cams timed correctly? There are eight different ways to put cams on their timing marks. Several ways will bend valves, some are harmless but will cause backfiring and prevent starting, and only one is correct: with piston one at TDC and with the distributor points open, the forwardmost lobes of both cams must point outward, away from each other, and their timing marks must line up with the line scribed in the front camshaft bearing top half.

Does the engine have reasonable compression in all cylinders?

Have you replaced the front and rear fuel filters recently?

Only if all these items check out correctly should suspicion turn to the injection pump. Here are the major items to investigate:

1. Is the cold-start solenoid functioning? Check it for continuity and shorting with an ohmmeter. If the solenoid is faulty, the engine may still be started by flooring the pedal to flood the engine while cranking, then releasing the pedal. As the engine clears from the flooding (continue cranking) it will begin coughing and can be restarted by delicate use of the throttle.

2. If the engine appears to be flooding, remove the cold-start solenoid wire, being careful not to let the loose wire terminal touch anything grounded. Try cranking again, with the throttle closed.

3. Are there any gasoline leaks around the metal injection tubes or tube fittings? The pump delivers only a small amount of fuel on each cycle and the tiniest of leaks from a cracked metal tube or loose fitting will starve the engine.

4. The retaining spring holding the pump's compensator link up against the altitude compensation bellows may have broken, causing an extremely lean condition. The pump thinks it's at the top of Pikes Peak while actually receiving much denser air. To check, remove the inspection plate from the rearmost end of the pump and see if the compensator link spring clip is in the correct notch of the notched lever for the barometric pressure in your area. See Fig. 6-25.

Rev the warm engine and release the throttle fully to check which notch is selected. If the spring wire is not

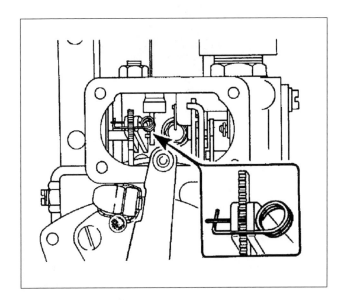

Fig. 6-25. Compensator link spring clip discussed in step 4. Clip is part of linkage used to adjust for barometric pressure.

engaging the proper notch, the barometric sensor should be screwed in to engage higher notches, or screwed out the engage lower notches.

If the compensator link has dropped to the very bottom of the notched lever, the spring has broken and the pump must be removed and dismantled for repair.

Compensator Link Spring Clip Position

Typical Barometric Range	Notch
29.9–30.7	7
29.1–29.9	8
28.3–29.1	9
27.6–28.3	10

Fig. 6-26. *Barometric pressure sensor is carried by triangular plate visible on top of SPICA pump. Cold-weather adjustment (the N, C, and F of pre-'74 pumps) is actually a barometric compensation.*

Car Starts but Runs Badly

If you can get the engine to fire at all, pull one spark plug and inspect its tip to determine if an over-rich or over-lean condition exists. The information will help you limit your diagnostic efforts. Be warned that an excessively lean mixture will cause the plugs to appear wet with fuel, just in the same manner as an excessively rich mixture will.

1. If the car starts and runs well cold but gets more and more sluggish as it warms up, develops a sooty, smoky exhaust, and eventually stops running, you have a classic case of a stuck cold-start solenoid plunger. If you're out in the wilderness, there are several quick temporary cures that can get you home.

The first is the tried and true "give it a sharp rap" technique. The plunger that sticks is towards the bottom of the pump directly below the cold-start solenoid. See earlier Fig. 6-20. Rap the pump in this area with a small plastic mallet or equivalent with the engine running. If you manage to free the plunger the change in engine condition will be sudden and dramatic.

If this fails, you can remove the large brass plug (hex head) on the bottom of the pump directly below the cold-start solenoid. This gives direct access to the offending plunger—put a short screwdriver up into the hole and push upwards to free the plunger. Be prepared for a gush of oil out the hole and try to lose as little as possible, replacing it with an equal amount of clean engine oil. Or, you can remove the side inspection plate and push the rod up with a screwdriver. See Fig. 6-27. Remove the cold start solenoid wire before attempting to restart it or it will simply stick again. The permanent fix requires removing the pump, taking out the plastic plunger, and removing 0.002 to 0.003 in. from its outside diameter. (See page 140.)

2. If the engine starts well and runs very strong above 3500 rpm but is weak and sputtery and spits back through the intakes a lot at lower revs, the engine intake cam may be too far advanced. This problem is unique to injected cars and only one vernier hole too much advance can produce this condition.

3. Be sure the injection pump is timed properly. With the sheet-metal timing cover removed, rotate the engine by hand until the timing pointer is opposite the I

Fig. 6-27. *Removal of rearmost side inspection plate reveals cold-start solenoid plunger.*

Fig. 6-28. Timing mark on the fuel injection pump pulley has been highlighted with a drop of white paint. Mark on the pump housing is barely visible, but it is exactly lined up with pulley mark. Later pulleys have a forward lip that makes belt removal a major effort. Some enthusiasts machine off lip to facilitate maintenance.

mark on the fan belt pulley. Remove spark plug #1 and verify that the intake valve is open. If it is not, rotate the engine 360°, verify that the valve is open, then replace the plug. The mark on the front of the fuel injection body should line up with a (barely perceptible) corresponding mark on the pump's drive pulley. A tolerance of 0.2 in. is allowed in either direction. If the pump is out of time, remove the notched belt from the pump pulley, put it in time, and replace the belt.

4. Check the altitude compensator setting through the inspection port.

5. If you have replaced the distributor rotor you may have used a rotor with a built-in suppression resistor to replace one without suppression resistance. If your engine already had resistance ignition wires or add-on suppressors, adding the suppression rotor will give a very inadequate spark. Also, be aware that suppressors break down and can seriously degrade engine performance. The best combination is solid-core ignition cables with quality terminals and a suppression rotor.

2.4 SPICA Pump Repair

There are some pump faults you can fix yourself, and some that are beyond even the most experienced owner or mechanic. The most common internal failures that can be reasonably repaired at home are: cold-start solenoid adjustment or sticking plunger, a broken altitude compensator spring, a sticking pump control lever, a broken compensator link spring, a broken cam follower, or a sticking gear rack.

Failures of the pump section are very rare. Fortunately, though the repair of any part of the pump is not

beyond the capabilities of a good machine shop, such repairs are uneconomical.

Take a deep breath and buy a rebuilt pump. New ones are no longer available.

All the repairs discussed below involve pump removal and disassembly, but neither procedure is particularly complicated.

Pump Removal

Before the pump can be dismantled, it must be removed from the car. This is probably the most difficult and irritating part of any pump repair. The following procedure is suggested:

1. Remove the front drive belt shield (three 10 mm hard-to-reach nuts).

2. Rotate the engine to line up the injection timing marks. There is a mark on the crankshaft fan belt pulley and another mark on the pump pulley. Both must be set to their corresponding pointers. See earlier Fig. 6-28.

3. Slip the cog drive belt forward off the pump pulley. Alfettas require removal of the pump pulley itself in order to remove the belt. Remove the air pump belt and radiator plumbing.

4. Loosen the injection tube nuts on the top of the pump using lots of penetrant. Be very careful not to twist the tubing. If the nuts are really tight, use a drift to shock the nuts on opposite sides of the tubes: place the drift on the top of the nut and rap it a good one with a hammer. Loosen the tubes from the pump. Factory wrench A.5.0164 is recommended.

5. Remove the two 10 mm bolts that fasten through a stamped bracket to the rear face of the pump.

Fig. 6-29. This wrench has been bent in order to facilitate removal of pump mounting nuts. Nut shown is hardest one to get at. Many short wrench swings are needed for removal.

6. Remove the four nuts holding the pump mounting flange to the engine front cover. Access is from under the car. The right wrench here will work miracles.

Fig. 6-30. Same angle as previous photo, to show mounting studs with fuel injection pump removed. Notice also clean circle where O-ring seals pressurized oil passage into pump. Drain hole is just barely visible at top of mounting flange.

Pump Disassembly

Before proceeding further, plug all openings in the pump and very thoroughly clean the exterior with solvent. It should be spotless to prevent any contamination or grit from getting into the delicate works.

Disassembly of the pump is astonishingly easy and requires only the removal of six screws that hold the castings together and the removal of the spring clip that rides on the notches of the notched lever reached through the rear access panel.

I strongly suggest the following precautions:

Provide a spotlessly clean work area for your now spotlessly clean pump. Newspaper is suitable.

Remove the inspection ports first to help build up your courage

Make a sketch showing the hole on the compensating link through which the notched lever spring clip installs before you remove the clip. See Fig. 6-32. One hole off in either direction results in a hopelessly rich or lean condition. This is especially critical on 1969 pumps. There are only three positions on later units.

Tap the castings gently to convince them to separate, and then do not be surprised when the six rpm-sensing balls come rolling out.

Fig. 6-31. Pump has been cleaned and inspection ports removed. Fuel cutoff solenoid and mounting also removed.

Fig. 6-32. *Notched lever spring clip. Clip holes are arranged front-to-back in this angle of view but are not visible. In this instance, clip is in forward-most hole.*

Fig. 6-34. *Front of the logic section showing race for rpm-sensing balls. Above race is long arm that carries spring used to connect with barometric sensor.*

Fig. 6-33. *The two castings just beginning to separate.*

At this point, you may want to consider lapping the check valves that are captured inside the fittings on the top of the fuel injection pump. See Fig. 6-36. These check valves keep fuel from draining out of the coiled lines between the pump and the injectors. If the fuel does drain, some time is required to pump new fuel to the lines and the car starts slowly. A puller must be fabricated to remove the check-valve assembly. The valves are lapped to their seats using a very fine compound.

CAUTION —
Because the procedure introduces an abrasive material into the fuel system at a most sensitive point, scrupulous cleanliness is essential.

Fig. 6-35. *A closer view of front of logic section. Notice mousetrap springs used to locate levers in logic section. Long snake is altitude compensator retaining spring.*

Fig. 6-36. Check valve at top of fuel injection pump. Removal requires a special tool. Most owners should leave them in place.

Fault Correction

Cold-Start Solenoid Adjustment. The symptoms resulting from a misadjusted cold-start solenoid will be somewhat vague and difficult to diagnose, but the engine will never seem to have good clean starts although all other systems are working normally and all other adjustments are correct.

A special wrench will be needed to get at the solenoid lock nut. With the engine fully warmed up to operating temperature, remove the wire to the solenoid. Hook up a tachometer capable of reading a 100 rpm change in engine speed. Attach one end of a jumper wire to the battery and momentarily touch its other end to the solenoid terminal. Engine speed should drop between 75 and 100 rpm. If the engine speed drops more than 100 rpm or stalls, the setting is too rich. If there is no change, the setting is too lean. Once you've obtained the proper mixture strength, lock the solenoid in place and then retest. Sometimes, tightening the 24-mm locking nut changes the adjustment.

Sticking Cold-Start Solenoid Plunger. This plunger, which is made of a nylon-like material, has a one-way flap valve and moves in an oil-filled cylinder. If there is any water in the fuel injection pump it will freeze in cold weather and disable the plunger. Also, water will cause the bore to corrode, so it's important to keep the pump oil filter changed and also, from time to time, to drain oil from the unit through the brass plug at the bottom of the casting. Remove the plunger with its rod attached through the bottom access in the casting. Chuck the assembly in an electric drill and, with a file or other suitable tool, turn down the outside diameter of the plunger about 0.003 in.

Altitude Compensator Retaining Spring Broken or Detached. This is the most common internal failure

of the pump and is usually relatively easy to fix. The typical cause is water in the fuel injection pump that causes the spring to corrode and break. Symptoms appear immediately rather than by gradual deterioration. The engine will seem to have at least two fouled plugs but will still give strong smooth performance at higher rpm and full throttle. You can verify the failure by removing the triangular plate on the very top of the pump, which on pre-1974 cars housed the seasonal adjustment device. Directly forward of the opening in the top of the pump, positioned vertically is, or should be, the retaining spring. See earlier Fig. 6-35.

If the spring has broken at its upper attachment where it threads into the machine screw, you can effect repair without removing the pump from the car. Back out the attaching screw on the top of the pump and be sure to catch the small spring fragment still on the screw so it doesn't drop into the pump internals.

With the screw removed, the spring can be grabbed through the triangular plate opening, pulled up, and

Fig. 6-37. Three-dimensional cam has been removed from rear casting of logic section and its follower hangs free. Below it and slightly to left is control lever input shaft. Just to left of lever is cold-start solenoid rod, which leads into an oil dashpot at very bottom of pump.

centered under the attaching screw tapped hole and the screw then rethreaded back into the pump and finally into the spring coils. Don't feel bad if it takes a few tries.

If the spring has broken further down or has come adrift from its lower attachment on the compensator link, the pump must be disassembled and the spring replaced. New springs are not available from Alfa, so a substitute spring must be obtained of approximately equal value. A gunsmith knows how to wind new springs.

Control Lever Sticking. The control lever shaft is supported in a long bushing in the pump casting. Although there are seals on the shaft to prevent contaminants from entering the pump, there is a length of exposed shaft and bushing outside the seal where contaminants, especially salt spray, can build up and bind the shaft.

The control lever arm will be quite difficult and very gritty to rotate. A little cleaning and lubrication are all that's required. A bit of penetrant around the shaft may be sufficient, but stubborn cases require the pump be disassembled so the shaft can be removed from the casting to be cleaned and lubricated properly.

Altitude Compensator Link Spring Clip Broken. If the spring clip breaks or falls out of the linkage due to mis-installation, the main rack controlling fuel delivery would snap to the full-lean position, probably cutting off fuel delivery completely.

If the spring is broken or has fallen out, it probably will rest harmlessly at the bottom of the logic unit and a replacement spring can be safely fitted with the pump still in the car. Removal of the triangular plate and barometric capsule from the top of the pump eases access. A bright light, inspection mirror, and several small needle-nose pliers or hemostats will do the job.

Cam Follower Failure. Any outright breakages of the cam follower or its linkages will have to be repaired by welding or fabrication of new parts.

If the ball bearing at the tip of the cam follower is lost, there will be sudden and extreme lean operation under all conditions of temperature, throttle, and rpm. The reason is that the cam follower drops by the height of the exposed portion of the lost ball bearing.

Rotating the pump control lever will produce a scratchier internal sound than normal. The condition must be repaired before the cam follower scratches its way through the 3-dimensional cam. Since new replacement cams are not available, repair can be attempted using a fine-grain metal-type epoxy to fill the damage, followed by careful sanding and polishing to restore the original surface contour.

If the original ball is lost, a replacement substitute can be obtained from a large bearing-supply shop.

Sticking Rack. This failure mode was not included in Joe's original articles, but appears, after a number of

Fig. 6-38. Cam follower resting on the 3-dimensional fuel cam.

Fig. 6-39. While diagnosis and repair focuses on pump logic section, the end of the line for all that linkage is a rack, shown here. Rack motion rotates four sleeves that control how much fuel is trapped in cylinder to be injected into engine. Rack has been known to stick if pump oil is not renewed and filter is not changed regularly.

additional years' experience with SPICA units, to be a somewhat common failure.

Over extended use, especially when the oil filter in the injection pump is not replaced regularly, rust will form on the rack and keep it from returning to idle. Since the throttle plates are closed at idle and air enters the engine through the idle circuit, the engine may idle properly but will immediately go rich as soon as the throttle plates are opened. The richness is less problematic at wide-open throttle simply because the engine is more able to run excessively rich at high revs than at part-throttle speeds.

The sticking rack can be freed by removing the inspection plates and flushing the interior of the pump with a thin lubricant such as WD-40 while the rack is moved through its travel.

Reassembly

Be very careful of the cam follower as the castings are pushed back together. With the pump upright, the tip of the follower will be below the edge of the 3-dimensional cam. Forcing the castings together will bend the fragile follower. Either fabricate a removable wire clip to elevate the follower or simply reassemble the pump upside down.

The balls in the speed sensor also will tend to push the sensor and fuel cam assembly rearward, thereby letting the balls out of their individual tracks if the pump is assembled horizontally. Since keeping the drive sprocket end lower prevents this, the best reassembly configuration is with the pump upside down and the drive sprocket end pointing downwards at about 45°.

Through the inspection ports, ensure that the forked altitude compensator link straddles the vertical notched lever and that the two concentric coil springs that preload the speed sensor have correctly seated. Finally, before the castings go completely back together, the internal rod from the control lever must be guided between the two projections on the 3-dimensional cam, thereby linking throttle input to cam rotation.

Since the 3-dimensional cam freely rotates until engaged by the control lever rod, this can become an annoying thread-the-needle operation. An assistant, a flashlight, and a length of stiff wire to position the elusive cam are all quite helpful. Be very careful, so the pump doesn't eventually become rotated from its upside down position endangering the cam follower again.

Once the control lever rod is positioned, the casting halves can be pushed completely together, letting the pressed-in dowels take care of alignment.

2.5 Rear Fuel Pump

The fuel pump located at the rear of the car near the fuel tank has several configurations depending on the year of manufacture. This is an expensive unit, and usually fails by leaking. The Bosch units are sealed and not repairable. These units have a stamped metal cover swaged in place. Marelli pumps can be disassembled and repaired with little difficulty since their design is very straightforward.

Note that a fuel-injection fuel pump cannot be replaced by a pump that is designed to feed a carburetor. A carburetor expects pressure ranging from 2.5 to 3 psi; a fuel injection system, on the other hand, needs pressures greater than 25 psi. Pumps for SPICA fuel injection (and Bosch) typically cost around $150.

SPICA Fuel Pump Part Numbers

Part	Part Number
Brush	105.04.021.00/05
Front O-ring	105.04.021.00/04
Large O-ring	105.04.021.00/03
Thrust washer	105.04.021.00/01

The high-pressure fuel pump runs constantly as long as the ignition is switched on. These pumps occasionally leak. As noted above, there are repair parts for some of the pumps, but the sealed Bosch units have to be replaced rather than repaired. Several models of pumps have been fitted over the life of the SPICA system. Some have two ports and some (later models) have three. All are roughly interchangeable, but you may find that after you have substituted another pump, the red low-fuel-pressure warning light won't go out.

System fuel pressure is regulated by an orifice placed in the outlet fuel nipple of the fuel injection pump. See Fig. 6-40. By reducing the size of this orifice, system pressure can be increased. If your warning light won't go out, simply insert a 1 mm orifice in the return fuel line (the rear fuel line) coming from the fuel injection pump. You can use a Solex carburetor 100 main jet. The outside diameter of the jet will make a leak-proof seal with the inside of the rubber fuel hose. Just be sure the jet is pushed far enough into the hose so you can slip it completely onto the fuel injection pump nipple.

Fig. 6-40. Fuel system pressure is maintained by a restriction in fuel return fitting on fuel injection pump. Small-diameter hole (arrow) is just visible here.

3. Bosch Electronic Fuel Injection

Unlike the mechanical SPICA system, the Bosch electronic fuel injection system is very well known in the industry. Its principles are applied under license to a large number of European, Japanese, and domestic cars. Though one is available, a specialized computer is not necessary to diagnose the Bosch electronic system properly. The entire system can be checked using only a fuel pressure gauge and a good-quality VOM (volt-ohm meter). There are some tests below under **3.2 Bosch System Checks**.

There are some important warnings when testing electronic components. While an integrated circuit (IC) chip is very durable under normal operation, it cannot stand any surprises. As a result, if it receives a signal (voltage) to which it's not accustomed, it will usually die faster than you can move your hand. There are two things to guard against: spikes, or voltage greater than normal, and voltage applied where it isn't expected.

Along the same lines, if you try to jump-start a Bosch-injected car and get the polarity wrong, you can fry not only the alternator diodes, but the Bosch brain as well. Be very, very certain of the battery polarity before hooking up jumper cables.

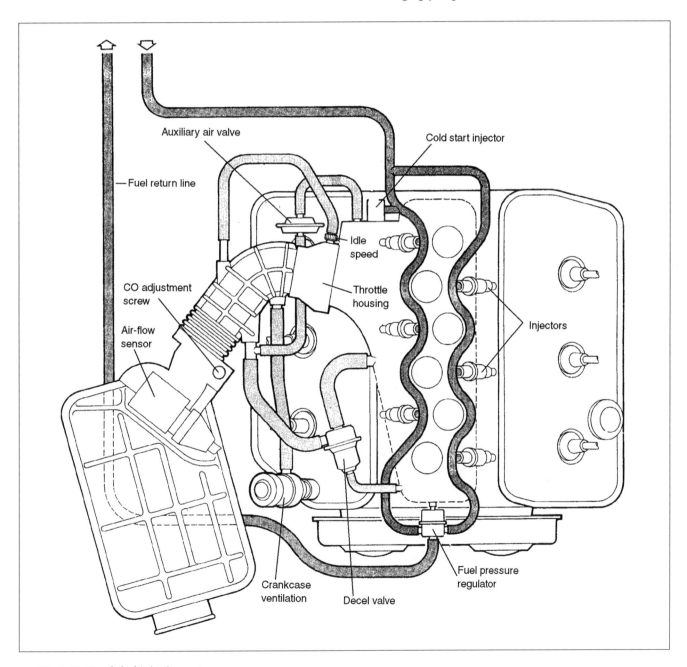

Fig. 6-41. Bosch fuel injection system.

CAUTION —
Never connect or disconnect a component when it is operating. Turn the ignition off before plugging or unplugging solenoids and sensors. Disconnect the battery leads before unplugging the main ECU harness.

Putting voltage where it isn't wanted is a bit more subtle. You need to remember that the normally harmless ohmmeter sends out 1.5 volts to check for resistance. As a result, if you're trying to test for continuity and happen to connect the ohmmeter to wires that go into the Electronic Control Unit (ECU), you can fry the chip that wasn't expecting the 1.5-volt surprise. Unless specifically directed to do so, never try to test the resistance between two wires that lead directly into the ECU. For the same reason, never test an oxygen (Lambda) sensor for continuity.

CAUTION —
Never use an inexpensive volt-ohmmeter to test electronic components. Use only a voltmeter, ohmmeter or multimeter with high input impedance (10 megohm or greater). The electrical characteristics of other types of test equipment may damage electronic components.

If you think a procedure says to test resistance between two wires at a connector leading into the ECU, re-read the procedure carefully to be certain it is not the connector on the component that is to be tested instead of the connector at the end of the wire from the ECU.

3.1 Bosch Failure Modes

There seem to be no typical failure modes on Bosch systems. The oxygen sensor is its most fragile component and should be replaced every 50,000 miles, but its failure is hardly noticeable until you try to pass an emissions test.

If pressed for the most likely failure, I would say it was poor electrical connections due to corrosion or dirt in an electrical connector. Many electronic fuel injection problems can be cured simply by unplugging connectors and cleaning them, and making sure that all wiring is dry and tight. If you encounter trouble, suspect a faulty ground connection first.

The runner-up failure mode would be a sticking air-flow sensor door. It's a simple matter to check the flapper door by carefully moving it with your hand to check for sticking. If you do need to check the door, turn the ignition switch on and then listen for the electric fuel pump to start as you open the door. If you verify that the fuel pump isn't running with the flapper door open, the pump or its wiring is bad.

For troubleshooting purposes, it's necessary to connect a fuel pressure gauge in the fuel line to the cold-start injector. A service bulletin from Alfa in 1982 covers the replacement of the fuel line to the cold start injector. The original fuel line comes out the back of the fuel rail and makes a tight loop up to the cold start injector. This line is too short to be used with a fuel pressure gauge and should be replaced with a longer line.

Fortunately, a Bosch system hardly ever goes bad and you should conclude there is a problem with it only after every other cause has been thoroughly eliminated.

3.2 Bosch System Checks

It's important before beginning to test the EFI system to verify that the engine is in otherwise proper tune. It must have good compression, be properly timed, and not have any major mechanical problems (such as bad rings or valve guides) that would cloud (pun intended) the issue.

The first check should be the intake ducting from the air-flow sensor to the throttle housing. Poorly installed clamps, loose hoses, or cracks in the ducting can let additional air into the intake. This air is unmeasured by the air-flow sensor, causing a lean mixture and all sorts of problems, from hard starting to a rough or high idle.

The next check is to verify that the fuel pump is delivering fuel at 2.3 to 2.7 bar (34–40 psi). Attach a T-fitting at the point the fuel supply line connects to the fuel rail. Measure system pressure with the engine running. If the pressure is too low, pinch the fuel return line so it is momentarily blocked and note the fuel pressure reading. If the pressure rises above 40 psi, the fuel pressure regulator is bad; if it does not rise, the fuel pump must be replaced.

The testable electrical components of the EFI system are either sensors (senders) or solenoids. Sensors can be

Fig. 6-42. *Showing the installation of a fuel pressure gauge.*

tested for continuity or resistance using an accurate, high-impedance VOM. The plugs (the connector on the wire coming from the ECU) for some sensors may be checked for voltage. There are seven solenoids that may be tested: six injectors and the start valve. Solenoids are tested by unbolting them and watching for fuel spray (remember that the start valve only works when the engine is cold). Watching for the spray is a dangerous procedure because you're working around atomized fuel, which will explode easily. See Fig. 6-43 for a typical Bosch wiring diagram and ECU connections.

Checks With a VOM

You should obtain the Bosch EFI maintenance manual for your specific model if you plan to do any testing. The terminals identified below may change on later models (specifications subject to change, etc.). For tests below made at the ECU connector, make sure the key is off before you disconnect anything. Note that by making tests at the ECU connector, you are testing both the component and the wiring, so either could be bad. There is no way to directly test the ECU. If everything else checks out but there is still a running problem, then the ECU may be bad. The following are typical tests for a 1982 GTV6.

Air Flow Meter. The resistance between terminals 36 and 39 of the air-flow meter connector should show infinite resistance (open circuit) with the flap closed and no resistance with the flap fully open. With the ignition switch on, depressing the flap slightly should start the fuel pump.

ECU Power. The main connector to the ECU is a 35-pin connector with pins 1–18 down one side and pins 19–35 down the opposite side. Pin 17 is system ground. With the ignition switch off, there should be battery voltage at pin 10.

Fuel Pump Relay. The resistance between ECU terminals 17 and 20 should measure 52–78 ohms.

Start Signal. With the starter cranking, there should be system voltage between ECU terminal 4 and system ground (pin 17).

Coolant Temperature Sensors. Resistance taken between terminals 13 and 17 should be 7–12 K-ohms at 14°, 2–3 K-ohms at 68° and 250–400 ohms at 176°.

Thermo Time Switch. For these tests, a cold engine has a coolant temperature below 86° and a hot engine has a coolant temperature above 104°. On the thermo time switch, terminal G should have a resistance to ground of 25–40 ohms cold and 50–80 ohms hot; terminal W, no resistance cold and 100–160 ohms hot, and resistance between terminals G and W should be 25–40 ohms cold and 50–80 ohms hot.

Cold Start Injector. When cranking a cold engine (and the thermo time switch is OK), there should be bat-tery voltage across the injector harness terminals. The windings of the injector should be 4 ohms.

Full-Throttle Switch. ECU terminals 3 and 18 should cycle between a closed and open circuit as the switch is activated. If the two wires at the switch are connected together on a running engine, rpm should increase slightly.

Injectors. The windings of each injector solenoid should show a resistance of 2–3 ohms.

3.3 System Modifications

The mixture on emissions-controlled engines is set for best emissions, not power. Modifications to the fuel system are usually directed at enriching the mixture.

It's necessary first to disconnect the oxygen sensor. Otherwise it will force the ECU to provide the proper air-fuel ratio regardless of most modifications. Indeed, the oxygen sensor has enough corrective powers that the stock system can be used on engines that have been turbocharged. Disconnecting the sensor means, of course, that the engine is no longer emissions-legal.

Since the amount of fuel delivered in the Bosch system depends in part on its delivery pressure, increasing system fuel pressure will enrich the air-fuel ratio. The system has a pressure regulator that is controlled in part by manifold vacuum (pressure rises with rpm). It's possible to fiddle with the regulator, or you can increase pressure simply by narrowing the restriction in the fuel return line downstream of the fuel rail.

The system can also be fooled into richness by telling it the engine is colder than it actually is. This requires placing a potentiometer in series with the brown wire that goes from the ECU to the cCoolant Temperature Sensor II. The advantage of this modification is that, if the potentiometer is located on the dash, you can dial in just the mixture richness you want.If handled stupidly, however, you can foul plugs, destroy mileage, pollute the earth, and—possibly—melt your engine.

The most sanitary approach is to fit the modified ECU that is available from Alfa Heaven.

One common modification to the Bosch L-Jetronic system is directed to improving its responsiveness. The air-flow sensor is capped with a square piece of plastic that is glued in place. Under this cap is located the spring that tensions the air-flow door. If you weaken this spring, the door will open more quickly, giving an extra measure of fuel as you accelerate. The tensioning spring can be adjusted. It's held at the desired tension by a cog wheel that itself is held in place by springs. You can turn this cog wheel about 5–7 notches to release tension on the door. Be very careful. This is an easy to ruin an expensive component.

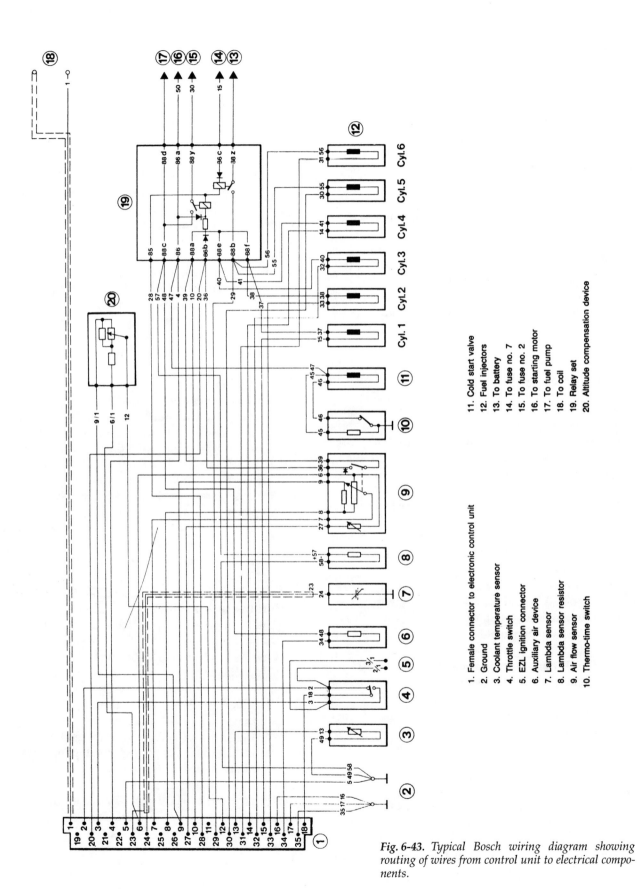

Fig. 6-43. *Typical Bosch wiring diagram showing routing of wires from control unit to electrical components.*

1. Female connector to electronic control unit
2. Ground
3. Coolant temperature sensor
4. Throttle switch
5. EZL ignition connector
6. Auxiliary air device
7. Lambda sensor
8. Lambda sensor resistor
9. Air flow sensor
10. Thermo-time switch
11. Cold start valve
12. Fuel injectors
13. To battery
14. To fuse no. 7
15. To fuse no. 2
16. To starting motor
17. To fuel pump
18. To coil
19. Relay set
20. Altitude compensation device

4. Ignition System

Under normal operating conditions the stock Alfa ignition system is adequate. Not great, but adequate.

I'm aware that a lot of Alfa enthusiasts will argue that the stock unit is perfectly superb under all conditions. Joe Benson could always get enough spark from the stock ignition to start his Veloce in sub-zero cold when I never could. I will admit to running my ignition systems under less-than-optimum conditions. Old points, dirty distributor cap, loose connections, worn rotor and leaking spark plug wires are all faults the meticulous owner should avoid.

I've always felt that insisting on Champion (N5 or N6Y) spark plugs was as much as I needed to do for my ignition system. The V-6 owner's manual lists Silver Lodge surface-ignition plugs, but Champion plugs are my personal preference. I've also suspected that a Capacitive Discharge (CD) system would more than compensate for the habitual state of deterioration that characterizes my Alfas' ignition systems. A case of using technology to overcome sloth.

Electronic ignition systems are expensive and they certainly aren't necessary if you're used to maintaining your ignition system in as-new condition. However, they do give an extra margin of security and for that reason alone I recommend them, especially if you live where the temperature dives below zero. Unless your ignition system is really messed up, an electronic system will not give a significant increase in power nor mileage. The Marelli Plex ignition system is a fine alternative, and a stock feature of more modern Alfas.

4.1 Ignition Timing

Apart from routinely replacing ignition parts, the only other ignition fooling around you're likely to do is to want to add a little more advance to your timing.

The purpose of proper ignition timing is to light off the air-fuel mixture in the combustion chamber at the proper moment so the combustible mixture develops the most power (that is, it burns the most) as the piston travels down the cylinder during the power stroke. This means creating a spark a few degrees before the piston reaches top dead center (TDC).

Unless you're using a dynamometer, the shape of the best ignition curve is beyond discovery by experimentation. The practical solution is to set the maximum

Fig. 6-44. This looks like a stock Giulia engine until you start counting wires from distributor cap. GTA used two spark plugs per cylinder to assure complete combustion.

ignition advance for maximum engine speed so the engine nearly pings. That is, advance the ignition until the engine pings at full throttle and then retard the ignition slightly. This will not optimize mid-throttle ignition, but as I've already suggested, that setting is beyond finding out unless you're heavily instrumented.

Alfa provides ignition timing marks on the flywheel of Giulietta engines and on the front crankshaft pulley on later engines. The static timing mark (F for "fissa" or fixed) is set with the engine off; the maximum (M, for "massimo") mark is used to set timing at 5000 rpm. The P (for "punto") mark, if you see it, is for TDC; if your car has SPICA fuel injection, there will also be an I ("iniezione") mark for setting the injection pump timing.

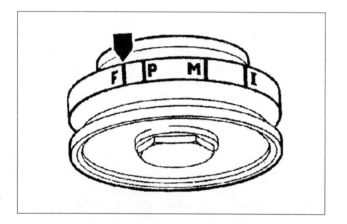

Fig. 6-45. Ignition and injection timing marks for 49-state 4-cylinder Alfas. Some California cars have F and P combined at a single mark.

In practical terms, you may set timing either statically (using the F mark) or dynamically (using the M mark). If you have a strobe that uses battery power (not the neon type that only connects in serial to the spark plug wire) then you should set the timing dynamically for maximum advance. In theory (that is, if the springs and weights in the distributor are stock), if you set static timing, then dynamic (maximum) timing advance will be correct. The most important value, however, is maximum timing advance, so if you can, set the timing using a strobe light at 5000 rpm.

Whether you use static or dynamic timing, you must first adjust the points so they are open about 0.018 in. when the rubbing block of the point set is exactly at the top of the lobe of the distributor shaft. The proper value of the opening will be given in your shop or owner's manual. The actual dwell angle is 60° ± 3°. If you

have a dual-point distributor, the top set of points is gapped to 0.017 to 0.019 in. and the secondary, or bottom set is gapped anywhere from 0.014 to 0.021 in.

To set maximum timing, first mark the M with a piece of chalk or with Liquid Paper so it's easy to see. Hook up the strobe (two wires to the battery, and a clamp over the #1 spark plug lead), warm up the engine and then have a helper hold it at 5000 rpm while you point the strobe at the timing marks. Be careful. The strobe makes all the moving parts appear to stand still, including the fan blade. If the timing mark is not correct, stop the engine and loosen the distributor clamping bolt enough so you can just rotate it slightly in one direction. You're after a one-or two-degree rotation, not halfway around the block. Start the engine, and recheck timing with the strobe. If the engine won't start, you turned the distributor too far. Turn it back a bit. Repeat the entire process until the M timing mark is opposite the timing pointer at 5000 rpm. Only very experienced mechanics dare rotate the distributor with a running engine. If you're not (seriously), don't.

Tighten up the clamp on the distributor so you can't rotate the distributor by hand. It's not necessary to tighten the clamp further than that.

If you don't have a timing light (and it's the most accurate method) you can get by with static timing the engine. Static timing (the F mark) is set with the engine off but the ignition on. With the ignition off, put the car in gear and rock it forward until the F mark appears opposite the pointer. Take the car out of gear without moving it further, set the parking brake, and turn on the ignition (be careful not to engage the starter).

Attach one lead of a 12-volt light to the electrical terminal on the side of the distributor. You can verify you have the right terminal because the wire already connected to it goes to the coil. Attach the other lead of your light to a good ground. What you'll do is to rotate the distributor until the light just comes on (that is, the points just close). Loosen the distributor holding clamp and rotate it in one direction until the light just goes out (the points are open). Then, carefully rotate the distributor in the opposite direction until the light just comes on. Make a mental note of the position of the distributor and repeat the process. The light should just come on at the same position both times. Repeat the process as many times as necessary to satisfy yourself that the distributor is positioned so the points are just opening. Tighten the holding clamp so you can't rotate the distributor.

Chapter 7

Transmission and Drivetrain

THE DRIVELINE IS, properly, everything that makes a car go, including the engine. Since the engine has been discussed at length in the previous chapters, we can begin studying the Alfa driveline just aft of the engine. This may be an esoteric distinction to the novice. "Just aft" can easily appear to be part of the engine, for some of the parts we're going to talk about now are so firmly bolted to the engine that they may seem a part of it.

Fig. 7-1. *A rogue's gallery of transmissions. From front: 750-series 4-speed, early 101 4-speed column shift, 101 5-speed, 105 5-speed, 2-liter 5-speed, and 2600 5-speed.*

1. Flywheel

The purpose of the flywheel is to store energy. Since a 4-cylinder engine manages a power stroke only every 180°, the weight of the flywheel is used to keep the engine rotating steadily between power strokes. Clearly, the heavier the flywheel, the more energy that can be stored—and the smoother the engine.

The difficulty comes from the fact that the heavy flywheel has more inertia. It is harder to accelerate and decelerate a heavy flywheel than a light one. Thus, an engine with a heavy flywheel will be less responsive to changes in throttle angle. For this reason, many modified engines run with a lightened flywheel to improve their responsiveness. Note that a light flywheel does not create more horsepower, it just feels that way. There is an argument that a lighter flywheel will improve a car's cornering ability by allowing a faster engine response to oversteer, but I think this is a debatable point considering the modest power available to steer an Alfa with its throttle.

I do want to emphasize that an engine with a light flywheel is a whale of a lot of fun. It may not be worth quite the cost of fabricating an aluminum unit, but that is perhaps a very subjective judgment.

You can lighten a stock Alfa flywheel by taking it to a machine shop to have it cut on a mill. You want to cut on diagonals across the flywheel so that you're left with six pedestals standing where the clutch pressure plate bolts up. The pedestals should be radiused to relieve stresses where they join what remains of the flywheel. Be sure to rebalance the flywheel after it's been cut. Using this approach, about six pounds can be taken off the flywheel.

There is one drawback to this modification: the heavy lip of the stock flywheel helps shield the clutch from any oil leaking through the rear main bearing seal. With most of the protective circumference of the flywheel cut away, the clutch will be quickly soaked if the rear seal leaks. Also, and this goes almost without saying, the craftsmanship of this modification is critical. A sloppy job is a certain hand grenade. An Alfa engine running with a lightweight flywheel, even an aluminum aftermarket flywheel, should also run with a scatter shield or bellhousing scatter blanket.

All Alfa flywheels look similar, but there are differences in size that render many noninterchangeable, especially in maintaining the mesh between the starter and ring gear. Giulietta and Giulia flywheel/clutch assemblies are smaller than the 1750/2000 units. Moreover, the bolt mounting pattern changed with the Alfetta series so that earlier flywheels cannot be mated to the Alfetta crank. In passing, I should also note that Alfetta cranks don't have a pilot hole for the transmission input shaft, so you can't use an Alfetta crankshaft on earlier 2-liter engines. The surest way to verify flywheel interchangeability is to count the number of teeth on the ring gear.

The Alfetta ring gear plate at the rear of the engine serves as a light flywheel, but its main function is as a mounting boss for the front donut. Since this plate is quite heavy, drilling it for lightness would save a pound or so of rotating mass. I've never heard of anyone modifying it or the Alfetta flywheel, which is located in the transaxle unit. Remember that the rotating mass of the entire driveshaft also serves as a flywheel of sorts.

2. Clutch

The clutch is a connection between the engine and the rest of the driveline. When it is disengaged, the engine is free to run without moving the car. When the clutch is engaged, the engine can turn the drive wheels (through the transmission and differential).

A clutch is composed of two main parts, a driven plate (the clutch disc) and a driving plate (the pressure plate). The pressure plate spins with the engine since it's bolted to the flywheel. The clutch disc, which is captured between the pressure plate and the flywheel, is attached to the input shaft of the transmission. See Fig. 7-2.

The pressure plate acts very much like a vise. Instead of a vise's screw mechanism, it uses heavy spring pressure to clamp the clutch disc firmly to the engine flywheel. The clutch pedal is simply a linkage with enough mechanical advantage to overcome the clamping action of the pressure plate. For a split second before engagement, the clutch disc slips, making the transition between disengagement and engagement somewhat smooth. Additional smoothness is gained by placing heavy springs between the clutch disc and its hub.

On most Alfas built after 1969, the clutch is hydraulically operated. Repair of the master and slave cylinders is straightforward. Many shops refuse to rebuild either brake or clutch cylinders (that is, hone the cylinder and replace the rubber seals) for reasons of economics (it's cheaper to buy a new/rebuilt one) or safety (if it fails, don't sue me). If you're somewhat experienced mechanically you should have no qualms about rebuilding a hydraulic cylinder as long as you are clean and careful with the rebuild. You'll save about 75% do-

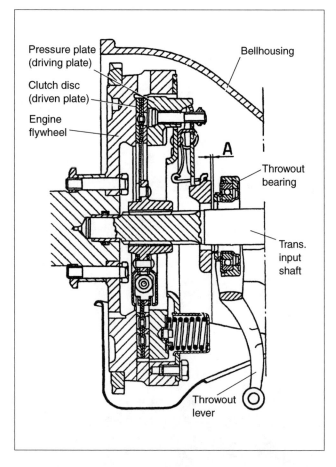

Fig. 7-2. *This 750-series Giulietta clutch dates from 1954, yet it is representative of clutches fitted to 1993 spiders. Clearly, little has changed in clutch technology. For that matter, clutch diagram applies equally well to prewar 6C2500 cars.*

ing a rebuild. Clearly, if you have any doubts about your ability, don't do it for safety's sake.

Alfa clutches are trouble-free if they are treated properly. American clutches have a built-in "idiot factor" to withstand habitual hill-holding, or slipping the clutch just to keep the car from rolling backwards. Do this in an Alfa for more than a few seconds and you'll eventually wind up replacing both the pressure plate and clutch disc and perhaps refacing the flywheel. The proper way to treat an Alfa clutch is to make the transition between in and out as decisively as possible.

Clutch Troubleshooting

That said, your clutch will eventually fail from normal wear. The surest signs that the clutch may need replacement are "chatter" when you release the pedal (also caused by bad motor mounts), and clutch slippage: the engine turns faster with the clutch engaged, but the car doesn't seem to accelerate at a similar rate.

Hard shifting is a possible sign that the clutch disc is binding on a rusty transmission input shaft, though this symptom may also be caused by failing transmission synchronizers or a binding shift linkage.

2.1 750- to 105-Series Clutches

Clutches through the 105 Series were operated mechanically. The 1750 introduced hydraulic clutch operation, making transmission swaps between pre- and post-1969 cars a considerable challenge unless you retained the original bellhousing. Also, the Giulietta/Giulia used coil springs in the pressure plate. Later units used a diaphragm spring plate to clamp the clutch disc.

Clutch Linkage Adjustment

Mechanically operated clutches should be periodically adjusted to maintain some free play at the pedal before the throwout bearing engages the pressure plate springs. The actual amount of free play is not nearly so significant as the fact that there is some. Adjusting free play may also cure slippage or hard-shifting problems.

If you have trouble in determining clutch free play, just crawl under the car and grab the clutch cable. You should be able to pull it back towards the rear of the car a bit by hand. If you can't feel any slack in the system, you should introduce some by readjusting the linkage.

In very salty areas, the clutch operating mechanism may bind because the bushings seize up in the cast aluminum housing bolted to the underside of the body. This problem can be alleviated by installing a grease fitting that will keep the surface of the bush lubricated.

Clutch Replacement

The easiest procedure for replacing the clutch is to remove the engine and gearbox as a unit from the car. This may seem unreasonably major, but, having personally tried removing the engine and transmission individually, I can assure you that the removal of both items as a unit will prove to be the fastest way to get the job done.

It is extremely difficult to remove an engine alone because there is not quite enough room in the engine bay for the pilot shaft of the transmission to clear the clutch throwout bearing. Removing a transmission by itself is not particularly difficult. What is tough is trying to get it back into the car with the engine in place. The home mechanic is most likely stuck trying to force the transmission in place while it's resting on his chest, and before his arm muscles begin to protest. Clearing the exhaust system and getting the shift lever into the transmission tunnel hole, lining up the splines between the clutch disc and the input shaft, and then finally rotating the transmission so it engages the mounting studs on the engine is likely to prove a nearly impossible task for all but a world-class weightlifter (a transmission jack only isolates you from feeling exactly where the transmission is).

Having said all that, I shall now tell you how to replace the clutch on a 750/101-series car by simply removing the transmission. The secret, which I owe to Pat Garrett, is in eliminating the difficult and tedious alignment of the splines as you slip the transmission shaft into the clutch. With this one task eliminated, it's a lot easier to get the transmission into place and lined up with the mounting studs on the engine. This technique won't work on transmissions with solid bellhousings. You need the large inspection plate area to be open. Also, this is not a technique that I recommend for the neophyte.

Getting the transmission out is—in theory—simply a matter of unbolting it from the engine at one end and from the driveline at the other end. I have found that several models of Alfa, most notably the Giulia Super, defy you to lower the transmission from the car without also disconnecting much of the clutch and brake assembly. And, I can assure you, all transmissions make it a test of manual dexterity to disconnect the speedometer drive.

You should support the rear of the engine so it doesn't put too much torque on the motor mounts. An unsupported engine will come to rest against the firewall, which is leaning too far back for my comfort. Put a short 2x4 between the head and the firewall to protect the throttle linkage bellcrank and to keep the engine somewhat level.

If you remove the driveline at the rubber donut, assemble two or three large radiator clamps end to end and tighten them around the donut to retain the proper bolt spacing pattern for easy reassembly. See the section on donut replacement later in this chapter.

Before trying to replace the transmission, mark the clutch pressure plate so it can be reassembled to the fly-

Fig. 7-3. *Two types of clutch pressure plates used on Alfas. On the right is the old-style unit used on Giulias and Giuliettas. Newer type, introduced on 1750, is on left.*

wheel exactly as it came off (maintaining its balance). Remove the pressure plate by taking out the six attaching bolts. You can leave the clutch disc inside the flywheel.

Look at the three large levers that bear against the clutch spring assembly. Their fulcrum is very near the outer edge of the assembly, but there is a short extension on each lever outboard of its fulcrum. You'll want to compress the clutch plate and then block it in its compressed position by putting wedges under the outboard ends of the levers.

To do this, fabricate three large U-shaped wedges that are about 3/8-in. thick. The large staples used to install heavy Romex wire in a building can be used for the wedges. Put the pressure plate on the floor so the throwout bearing faces up and then stand on the throwout bearing to compress the spring assembly. Have a helper place the wedges, then step off the bearing. The levers should be wedged so the spring assembly is still compressed. See Fig. 7-4. What we have here is a virtual set mousetrap. Handle it with respect.

Fig. 7-4. *One of Pat Garrett's wedges in place for installing the clutch.*

Using eye protection, very carefully reassemble the pressure plate to the flywheel with the six attaching bolts screwed in one or two threads only. You'll note that the clutch disc will now move freely inside the pressure plate because the springs are wedged in a disengaged position.

Refit the transmission to the engine. The transmission input shaft should be very lightly lubricated and should easily locate and mate with the clutch disc. You can help the union by putting the transmission in gear and turning its output shaft slightly as you slide the transmission up against the engine. Bolt the transmission to the engine securely.

Now, reach through the clutch inspection opening and begin tightening the six pressure plate attaching bolts. Turn the engine over by hand so you can tighten the bolts evenly. As you tighten the bolts, the pressure plate will begin to make contact with the clutch disc. As it is bolted home, its three levers will be forced further open and the wedges will fall free. Just be sure all three wedges have been removed before you button everything up.

If you replace the clutch, also replace the throwout bearing. The driveline may not need a new throwout bearing right now, but if you reassemble the old one, it will be sure to fail next week.

2.2 Alfetta Transaxle Clutch

The Alfetta carries a massive clutch assembly at the rear of the car, just in front of the transmission. See Fig. 7-5. This location allows replacement of the entire assembly without removing either the engine or transaxle from the car. In fact, the clutch and transmission gears can all be removed without removing the main casting from the car.

I've always been a bit nervous of messing with the clutch, and especially the transmission gears while the main casting is underneath the car. Unless the underside of your Alfa is steam cleaned before you start, there's always danger that some glob of road dirt is going to infect your pristine, rebuilt transmission gearset just as you ease it back into place underneath the car.

While it is attractive to be able to renew the clutch and transmission on an Alfetta without removing the main casting, I do not personally recommend it. Removing the transaxle requires the extra effort of dismounting the two driveshafts and disconnecting both clutch and brake hydraulic lines, but it has the distinct advantage of cleanliness.

A generic description of clutch removal is part of the transaxle disassembly procedure given below.

2.3 V-6 Clutch

All GTV6s used a dual-disc clutch which provided superior clamping pressure. The Milanos reverted to a single-disc design to make clutch operation smoother. While some enthusiasts have concluded that there is no virtue in fitting an early dual-disc clutch to later cars, fitting a Milano clutch to a GTV6 saves about half the parts cost.

Alfetta racers (with the 4-cylinder engine) may want to use the heavier-duty V-6 clutch assembly. I've not heard of this swap personally, but it seems like a good one if your engine output stretches the limits of a clutch's capacity.

Fig. 7-5. *Alfetta transaxle removed from car showing location of clutch at front of assembly.*

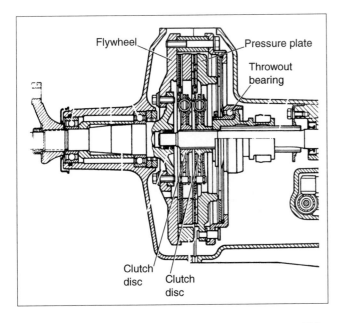

Fig. 7-6. *Dual-disc clutch used on early V-6 coupes. This assembly sits on nose of transaxle, which is to right in drawing.*

3. Transmission

It helps, before starting out, to understand the operation of an Alfa transmission. There are three main sections to it: an input shaft, a mainshaft, and a layshaft.

The input and mainshafts look like they're one piece when you first open up the transmission while the layshaft "lays" beneath the two. The input shaft has a single gear on it that is used to drive the layshaft. The gears on the layshaft, in turn, drive the gears on the mainshaft. See Fig. 7-7. The mainshaft gears are free to rotate on their shaft, which is what connects to the driveshaft (or, on transaxle cars, the final drive).

Shifting gears means nothing more than locking the selected gear to the mainshaft. The locking process is accomplished using a sliding ring. In order for the sliding ring to engage the gear, the speed of the ring must match the speed of the gear. A small expanding clutch (the synchronizer) is attached to the gear to bring it up to the speed of the synchronizer ring. It is the failure of the synchronizer ring to bring the gear up to proper speed that causes a hard or grinding shift.

In the Alfa transmission, there are three of these locking ring assemblies and they are all identical. By moving any one either forward or backward you shift the 1-2 pair, the 3-4 pair, and 5-reverse. A lockout system assures that the transmission engages only one gear at a time.

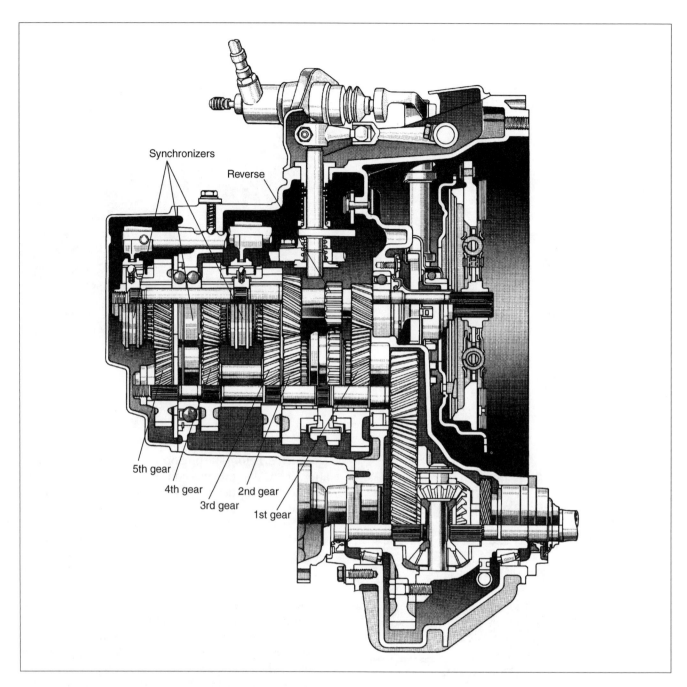

Synchronizers

Reverse

5th gear

4th gear

3rd gear

2nd gear

1st gear

Fig. 7-7. *164 transaxle combines transmission and final drive in one case, but operation of transmission and synchronizers is the same as all other Alfa transmissions.*

Second-Gear Synchronizer Failure

This is the most common problem you'll find in an Alfa. Offhand, I can't remember driving a used Alfa in which I couldn't beat the second-gear synchros. Now, there is a proper way to shift an Alfa transmission: always pause for a heartbeat in neutral and double-clutch coming down through the gears. It's not only kind to the synchros, but it'll impress the passengers with your mastery of the art of driving. But it's not uncommon to find an Alfa that will not take second gear without a crunch, no matter how skilled you are.

Alfa transmissions are very sturdily built. In fact, the gears in the transmission are so sturdy that their inertia almost overcomes the synchronizers. For some reason, this is especially true of the second-gear synchronizer set. After a few thousand miles, you'll begin to realize that second gear engages harder than the rest, and occasionally emits an embarrassing grind even when you're careful not to speed-shift.

There are two fixes for this problem. For short-term satisfaction replace the synchros; for a permanent fix, the individual gears must be drilled to reduce their mass. This solution is so expensive as to be prohibitive, but you at least ought to know that there is an "ultimate" fix. The Alfa club magazine many years ago reported on this solution and showed a sequence of photos of the modification, which worked like a charm. Of course, each gear should be balanced after it is drilled. As I say, expensive.

Synchronizer replacement is not a casual repair. If you've never had a transmission apart, be prepared for what appears to be at least several miles of gears. An engine has a lot of parts, but most of them are unique in their appearance. As a result, most anyone can put an engine together after all its parts have been thrown in a cardboard box. The same can't be said for a transmission. So many transmission parts look the same that a good deal of organization is required to keep them in their proper place.

Normal transmission oil has additives to improve its ability to lubricate under the extreme pressures where two gears mesh. These Extreme Pressure (EP) additives tended to glaze the surfaces of the new (Porsche-type) synchronizers introduced on the 101-Series transmissions, so non-EP lubricants (Shell Dentax 90) were mandatory. The problem was that non-EP lubricants were (1) oddball to almost every transmission application and (2) available only in 55-gallon drums. Thus, when you finally did find the proper lubricant, you bought a lifetime's supply or made a group purchase with a bunch of other Alfa owners.

Some owners substitute heavy (40- or 50-weight) engine oil and a can of STP for non-EP lubricant. In 1968, as I recall, Alfa made an unannounced change in the moly surface of the synchronizers so regular manual transmission fluid would work just fine. The old-type synchronizer rings have radial grooves while the newer type have a rough, textured surface. You should fit the newer-type synchros when rebuilding one of the older boxes. There was a change in the size of some parts of the synchro mechanism, so make sure the synchro assembly works freely on the bench before installing it into the transmission.

3.1 Front-Mounted Transmissions

The 750 Giulietta used a unique gearbox that was essentially a tube into which the transmission gears were pulled. This box requires a special tool to extract or install the gear sets and its care should be entrusted to a skilled Alfa mechanic who has the tool. Your friendly AAMCO man won't be of much help. If you plan to pull one of these transmissions apart, be prepared to become an expert on the matter.

A 750-series transmission can be replaced with a later (101-series) unit, but the swap requires fitting a 101 flywheel, lowering the transmission mount, and cutting a new hole in the floor for the shift lever.

In 1959, unannounced, Alfa introduced a transmission that we initially referred to as the ZF-style. It is now referred to as the 101, in keeping with the first three digits of the part numbers for the box. This transmission is split longitudinally so that it may be opened and easily worked on using common tools. The same basic transmission is also used in the 2-liter and 2600 cars.

The first of the 101 Giulietta transmissions were 4-speed units. For a while, it was popular to raid the 2-liter parts bin for the fifth gear set so you could drive

Fig. 7-8.   A sobering thought: all the power of a V-6 engine flows through this slender (21mm—less than an inch) splined shaft, located dead-center in rear clutch housing. Shaft to right, almost as robust, serves the shift linkage.

Fig. 7-9.   A prototype 750-series gearbox owned by Tom Zat. This is a column-shift unit.

CHAPTER 7

around in a 5-speed Giulietta. The Giulia ended this little game, for it was equipped with a proper 5-speed box. The ratios were different from those used in the 2-liter, however, so it's not proper to say that the Giulia and 2-liter boxes are identical, though they are interchangeable.

Front-Mounted Transmission
Removal and Disassembly

While the transmission can be removed from the car by withdrawing it from the engine, it is in my experience much better to remove the engine and transmission as an assembly, as noted in the clutch discussion above. While it may seem wasteful to disconnect all the plumbing from the engine only to reconnect it again, the hours spent with aching muscles under a car trying to coax a transmission to line up its input shaft and mounting studs is a one-time lesson that is memorably convincing (an exception is noted above).

Typically, in removing the engine/transmission, one item is left attached with sometimes puzzling and occasionally disastrous results. This is again a tip from experience: double-check that you've disconnected the tachometer drive, the ground strap near the starter, a choke cable if fitted, the fuel lines, the throttle linkage, and the shift lever.

On the transmission, the backup light switch wires need unplugging. The speedometer drive is a very awkward item to remove and the transmission cannot be lowered appreciably to help in unscrewing its attaching ferrule. Water-pump pliers are the best way to turn the ferrule free. After one or two short turns to loosen it, try to unscrew the ferrule using your fingers. The following procedure is significantly abbreviated from what is available in shop manuals. If you need step-by-step information, a repair manual is essential. My feeling is that, if you need advice on how to remove the transmission, its overhaul is probably beyond your experience level.

Once the transmission is resting comfortably on a dolly, wash it with a degreaser so its exterior is as clean as you can possibly get it. Otherwise, much of the grit on the outside of the transmission will end up inside. Collect some cardboard boxes to hold assemblies such as the synchronizers, shafts, and bearings. A few sandwich bags will keep very small items from hiding under the bottom flaps of the boxes. Use a large Magic-Marker to label the boxes and bags.

The front bellhousing is removed first and then the rear output flange.

Slip the transmission into third, remove the shift assembly, unbolt and remove the rear tail section and then the 5-reverse shift linkage, including its rail.

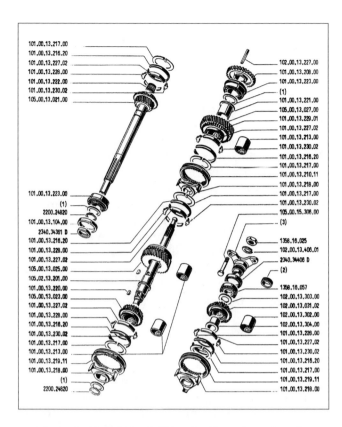

Fig. 7-10. *Mainshaft of 101-series 5-speed gearbox. Input shaft is shown at top. Mainshaft itself is in center cluster of gears. Bottom row also slips onto mainshaft.*

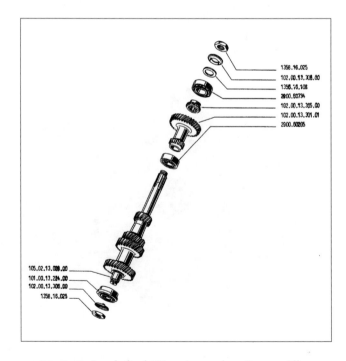

Fig. 7-11. *Layshaft of 101-series gearbox. Reverse idler gear assembly is not shown.*

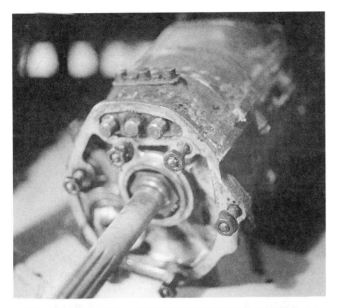

Fig. 7-12. There are two types of 101-series gearboxes. Top one has exposed shift selectors: three rods extending horizontally from the end of the case. On bottom 101-series gearbox the shift selectors are not visible in their bores.

Remove the bolts holding the two sides of the transmission together. The oil filler plug should be on the top when you pull the halves apart. Don't pry on the mating surfaces. Tap the castings with a hammer to separate them.

Remove the interlock mechanism that keeps you from selecting more than one gear at a time. Examine it to see how it works (clever!). Hint: make up a dummy wooden dowel to ease dismantling.

Take a deep breath, grab the ends of the input and mainshafts, and lift the assembly free of the layshaft and housing.

Put the assembly on a bench and pull slightly, separating the input from mainshaft. This will dismantle fourth (on the input shaft) from third gear (on the mainshaft) and the synchro mechanism for the two gears will be exposed.

Take as long as you need to study the synchronizer so you understand its operation completely. The main parts of the synchronizer are the outer sleeve that engages the shift fork, the synchronizer assembly itself (an expanding clutch made up of two sets of concentric segments that engage two floating stops), and the gear, which rotates freely on the shaft. See Fig. 7-13. Do not proceed until you can assemble the synchronizer assembly confidently. You should be able to pass the next-door-neighbor test: he brings you a shoebox with all the parts of three synchronizers thrown in together, and then trips coming into your garage, spewing the box's contents on the floor. Quickly hand him back the assembled units.

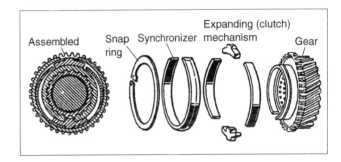

Fig. 7-13. Detail of a typical synchronizer set. Outermost sliding collar is not shown.

Before you begin to dismantle the gears, you should check for both axial and radial play. A gear should rotate smoothly on its shaft without wobble. Hold each gear with your fingers and try to move it back and forth or rock it on its bearing. If there is enough motion for you to see it, the gear bearing is worn and should be replaced. You can compensate for excessive axial clearances by selecting thicker or thinner shims. Now is the time to note whether different shims will be needed, but be careful not to plan so you set the box up too tight. Getting a box too tight is a common mistake. The transmission needs some amount of clearance to work in its heavy oil.

Also, if the gearsets you're looking at are blue from wear, pitted, or broken and the bearings are obviously bad, then you may want to save the effort and expense of repair and simply get a better, used gearbox (this won't make good reading in 25 years, I know). I do have to emphasize that rebuilding a gearbox is labor-intensive. I was once talking to an independent Alfa shop owner and I asked him if he ever had any trouble setting up the pinion on an Alfetta transaxle. After a slight pause of embarrassment, he admitted that he had never had an Alfetta gearbox apart even though he was factory trained. He knew it was always more economical for his customers if he got a used one from a wrecked car.

The bearings are pressed into the inner diameter of the gear and come oversize so they have to be bored to fit after they've been pressed into the gear. If you have a lathe, you can fit the bearings yourself, otherwise take the gears and the shaft to a machine shop for the work.

Service manuals detail the procedure for stripping the mainshaft with a press, which most enthusiasts don't own. Instead of a press, you can rent a three-arm puller that will reach from the output end of the mainshaft to the far edge of second gear, plus about an inch. Use the puller, bearing against the sides of the gears, to slide the roller bearings and synchronizer hubs from the mainshaft. Third gear comes off to the front while the remaining gears remove to the rear. Don't try to remove a gear over a key left in place in its keyway. The key will score the bearing surface. The keys, bearings, and spacers that are captured by the gears along the shaft must be marked so you can return them in proper order.

When you have the mainshaft completely stripped, examine its bearing surfaces for scoring. You can polish the surface with #600 sandpaper to remove slight burrs.

As you reassemble the parts be sure to lubricate all the bearing surfaces well and be surgically clean.

Check a ball bearing by cleaning it in solvent to remove all lubricant and dirt trapped in the races. Hold the inner race with your fingers and spin the outer race. Not too fast, you may damage the bearing. Listen for clicks. If you hear any, replace the bearing.

WARNING —
Do not use compressed air to spin the bearing. The bearing may disintegrate while spinning.

One of the challenges of rebuilding an engine is that you really can't know how it will run until it's all back together. A transmission, on the other hand, can be checked for operation while it's going back together. As you reassemble it, check the transmission for smooth synchronizer operation. With the shift forks attached to the shift rails, verify that the forks center the synchronizer operating sleeves between the faces of the gears.

3.2 Alfetta Transaxle

The clutch and gears on an Alfetta (de Dion) transaxle are serviceable without removing the unit from the car. In order to withdraw the clutch and gears, the driveshaft has to be removed and the transmission case wedged downward enough for the gears to clear the body as they're being pulled from the case.

As noted above, I don't recommend working on the clutch or transmission from beneath the car for reasons of cleanliness. Removing the entire de Dion assembly is not an especially difficult operation though it does require opening up the hydraulic lines for the brakes and clutch. I'll begin by giving the under-car procedure, anyway, and then conclude by showing the extra steps required to remove the transaxle from the car.

Clutch/Transmission Removal

This procedure requires the car to be high enough off the ground that you can work beneath it in comfort. Always use jackstands. You must first remove the

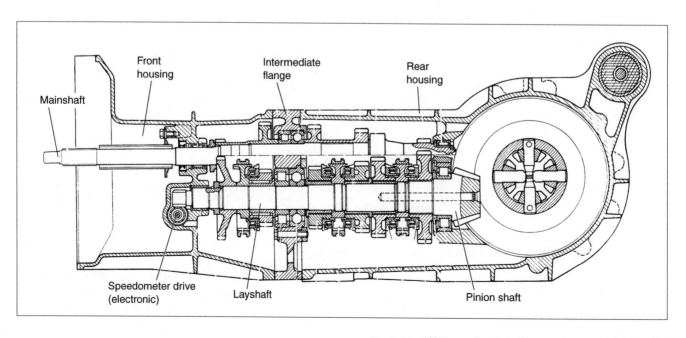

Fig. 7-14. *Alfetta gearbox is built around a narrow intermediate flange that carries main and pinion shaft.*

driveshaft from the vehicle. See the section on donut replacement below.

Remove the six bolts that attach the de Dion tube front support to the body. The support is a square-section fabrication which runs transversely under the car at a point just below the front edge of the rear seat. See Fig. 7-15. In the middle of this support is a single pivot bolt for the de Dion unit. The bolt is clearly identified by a lock tab welded to the support. Leave the single bolt alone. Undoing it won't get you anywhere.

Put a floor jack under the center of the de Dion tube at the rear of the car and very slowly raise it. This will lower the nose of the transaxle. See Fig. 7-16. Fit a spacer between the de Dion front tube and the body to hold the transaxle in a downward position. There's a special Alfa tool for holding the case in its proper downward-pointing position, but a length of 2x4 wood will work as well. See Fig. 7-17. There's no magic to the length of the 2x4 wedge. With the driveshaft removed, simply jack up the de Dion tube until the transmission has tilted down as far as it will go (further jacking causes the car body to begin to lift). Measure the distance between one of the side members of the de Dion transaxle and the underside of the car, then cut a section of 2x4 to fit. My personal preference is to leave the jack in place on the de Dion tube as a safety measure. Release just enough lift on the jack so that the 2x4 is held in place firmly.

Now, clean the underside of the body pan and the transaxle case as thoroughly as you can to minimize dirt falling into the transmission gears.

Disconnect the four front transaxle mounting bolts (two on each side) from their rubber mounts and, using another jack, raise the transaxle off the de Dion support.

Carefully remove the shift selector arm retaining nut from the arm. Don't apply any torque to the selector arm itself or you'll misalign the shift gate. See Fig. 7-18.

Fig. 7-16. We see companion flange for rear of driveshaft. Bellhousing behind it contains clutch. Clutch slave cylinder is visible to left. Transaxle's nose has been tilted down.

Fig. 7-17. De Dion tube and its forward links make a big mousetrap which needs respect. A length of 2x4 and an old cylinder liner can be used to wedge it in its down position. Alfa makes a real tool for the job, but this will do. Overall dimension is not critical.

Remove the four bolts that attach the clutch unit to the transaxle, then pull the clutch unit forward and off the transaxle. See Fig. 7-19. The clutch release lever may stay clipped to its ball stud, but be prepared for it to fall free.

There are three castings that make up the transaxle unit behind the clutch: the front housing, the intermediate flange, and the rear housing. The intermediate flange is a narrow casting sandwiched between the front and rear housings. It carries the main transmission bearings, and both the main and layshafts come with it when the flange is withdrawn.

During transmission disassembly, you want the intermediate flange to stay attached to the rear housing when you pull the front housing away. This is tricky because the front housing and intermediate flange are held

Fig. 7-15. Where are we? Underneath the car, looking back. Rear of de Dion tube has been jacked up to bring transaxle nose down. Its front transverse mounting member is nearest us.

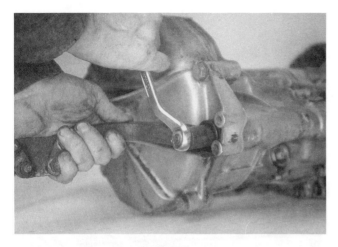

Fig. 7-18. Removing shift selector arm, transaxle removed for clarity. Brace arm with your hand as shown so no torque is communicated to selector shaft, then slip rubber protective boot off selector shaft.

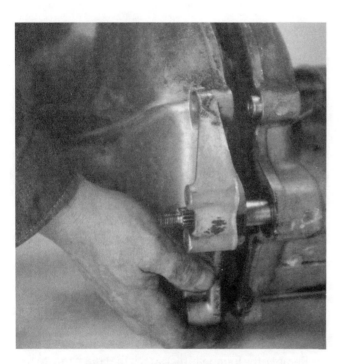

Fig. 7-19. Bolts which attach clutch bellhousing are next removed. Bellhousing should slip off easily. Tap it with a soft hammer to get it started if necessary.

Fig. 7-20. With bellhousing off, throwout bearing can be slipped off transmission input shaft.

Fig. 7-21. A makeshift solution. Transmission intermediate flange needs to be disassembled in proper order. To keep it attached to rear housing, you can make up a clamp using a long bolt and enough washers. Large square washer catches a boss on rear housing, capturing intermediate flange. (There are two bosses on opposite sides of the housing.)

to the rear housing by the same nuts. The secret is to make up a clamp using a bolt and large washer to hold the intermediate flange to the rear housing. See Fig. 7-21.

The speedometer cable is held in place by a lockbolt. Remove the bolt and pull the cable and speedometer drive housing free, then replace the bolt for safekeeping. On GTV6 and later cars, the speedometer is electric and there will be two wires instead of a cable. Pull the rubber boot back from over the sending unit

and pull the wires free. Similarly, remove the two wires from the backup light switch and remove the switch. See Fig. 7-22 and Fig. 7-23.

Remove the nuts holding the front housing and intermediate flange to the differential housing. See Fig. 7-24. Pull the front housing forward, and free of the transmission. Be careful not to drop the reverse idler gear.

Fig. 7-22. Remove speedometer drive housing with pliers.

Fig. 7-23. Backup light switch unscrews.

Fig. 7-24. With intermediate flange clamped to rear housing you can remove attaching nuts for front housing.

Fig. 7-25. Walter Mitty is about to become a brain surgeon. There's no turning back now. One clamp holding intermediate flange is shown clearly at top.

Fig. 7-26. With front housing removed, you can see large straight-cut reverse gear. Input shaft is to left and gear selector shaft to right.

Fig. 7-27. Remove two clamps from intermediate flange and pull carefully. It's better to pull apart transmission with a friend.

Hold the intermediate flange securely, then remove the flange retainer you made up. As you pull the flange forward, the transmission geartrain, complete with the pinion gear, will come free.

When reassembling the flange to the front and rear housings, all mating surfaces should be cleaned thoroughly using a scraper and solvent, then sealed with a silicone gasket material.

Transmission Disassembly

The good news is that most of the things you'll want to do to an Alfetta transmission can be done without special tools or pullers. The bad news is that the transmission gear stack-up controls the depth of the pinion gear, and any operation which would change that setting shouldn't be attempted by normal folk like you and me.

Fig. 7-28. Great design: Alfa's transmission comes out in a single piece, including three shift selector rods, layshaft (top gearset), and mainshaft (bottom gearset with final drive pinion at far right). On mainshaft, first gear is rightmost, largest gear
and fourth gear is tucked next to intermediate flange, just visible behind 3-4 shift fork. Fifth and reverse gears are on other side of flange. Gear synchronization is taken care of on mainshaft.

Before disassembling the transmission, clean the mainshaft thoroughly and make a list of the parts that need to be replaced. By driving the car, you should know before the transmission comes apart whether or not the synchronizers should be replaced (when in doubt, fit new ones). Verify the condition of the bearings by spinning them carefully and check the gears for any wobble which indicates bushing wear.

The pinion is set against the ring gear by the stack-up (the sum of all the thicknesses) of the components along the mainshaft between the mainshaft intermediate bearing and the rear roller bearing. Changes to the fifth-gear assembly will not change stack-up, nor will changing just the synchronizers. If any of the critical parts which control the stack-up dimension (bushings, splined synchronizer hubs, and roller bearings) have to be replaced, then the rear bearing spacer has to be exchanged to regain the original stack-up dimension.

That dimension is measured using a pillar gauge and dial micrometer, two items most owners will not have in their tool kits. Thus, if your examination of the mainshaft reveals a need to exchange a part, you'll want to take the shaft to a shop that's equipped to make the measurement. Just taking a shaft apart and putting it back together is all right, but if you replace a part which affects the stack-up, then leave the details to someone with the right equipment. For pity's sake, don't take a disassembled mainshaft to the shop: they'll have no accurate way of knowing what the stack-up was.

Much of the nitty-gritty detail of transmission disassembly given above for the 101-type gearbox applies to the Alfetta transmission: the synchronizers and gears selectors work just the same. There are nuts on the ends of both the main and layshafts, and removing these nuts allows you to strip the mainshaft of its gears and synchronizers, and to remove 5th-reverse gear on the layshaft. A press or puller is required only if you need to remove the ball bearing adjacent to the pinion.

I really don't need to give the disassembly procedure step-by-step, in part because the construction of the transmission enforces an invariably sequential process and because if you need that much help you shouldn't be doing this in the first place.

Complete Transaxle Removal

Begin by following all the steps for clutch removal, above. When you reach the point at which you can tilt the nose of the transmission down, begin following the procedure given here.

As a checklist, you will already have removed the six bolts which hold the de Dion front mount to the body, tilted the nose of the transaxle down, and then removed the four bolts that hold the front of the transaxle to the front mount. In point of fact, only one bolt now remains to hold the transaxle in place, not counting the

hydraulic lines, wires and axle shafts. At this stage, the transaxle is fairly secure, but over the next several steps it becomes increasingly able to fall out of the car. Work with care and don't position any part of yourself under the transaxle.

Remove both axle shafts where they attach to the inboard disc brake hubs. The shafts are attached using 7-mm Allen-head screws which may be very difficult to break free. See Fig. 7-29. It's important to use a high-quality Allen wrench here, even if you have to go out and buy it: if you round out the hex, you will have to grind the screw head off in order to remove the axle shaft. If your prized collection of Allen wrenches came off a surplus table you'll probably bend the wrench into a pretzel before getting even one of the screws free. A top quality Allen wrench on the end of a 3/8-in. drive will let you put real muscle behind it.

Then again, the Allen screws may be finger-tight on your car. But the ones I've done have all been knuckle-busters which break free with the sharp report of a .22 rifle. Once free, the axle shafts can hang of their own weight. They will need to be held free of the transaxle, however, during its actual removal. If you want to wire them up, have them point toward the front of the car.

Fig. 7-29. *Allen-head bolts that hold axle shaft to transaxle are on tight. Work carefully and without Vise-grips.*

Remove the hydraulic lines to both the clutch and the brakes. See Fig. 7-30. In both instances, you'll have to disconnect the lines where the rubber hose attaches to the body. This operation is usually possible using brake fluid as a penetrating oil and a regular open-end wrench: if you have any trouble at all, get a proper hydraulic-line wrench to avoid rounding off the corners of the flats.

It's important not to twist in half a metal hydraulic line using brute force. It's common for hydraulic lines to seize to their companion nuts so that, when you try to undo a nut, the line twists (and usually ruptures). Watch the line carefully to assure that the nut rotates freely. The junction we're talking about is supported by a metal tab welded to the body. See Fig. 7-31. Use a second wrench on the locknut nearest the tab to counter the torque you apply when removing the fitting for the rubber hydraulic line.

The hand brake cable loops over the top of the transaxle. Remove the adjusting nuts at the end of the cable and then pull it completely free of its linkages and the body. See Fig. 7-32.

As you're removing it, notice that the hand brake cable runs through a sheetmetal boss in the body over the top of the transaxle. Slightly to the rear of that boss and dead-center near the top of the transaxle housing is the single 17-mm bolt which attaches the transaxle to the body. If you were to pull that bolt free right now, the transmission would simply fall to the floor. Fair warning: the transaxle is a formidably heavy unit for one

Fig. 7-30. *Clutch slave cylinder is held with two mounting bolts, one of which has been removed in this photo. Third bolt attaches to rubber mount.*

Fig. 7-32. *Hand brake adjuster on an Alfetta transaxle is tucked up where no one in his right mind would ever find it. That's the price of inboard disc brakes. When removing transaxle, adjusting nut has to be removed and cable withdrawn.*

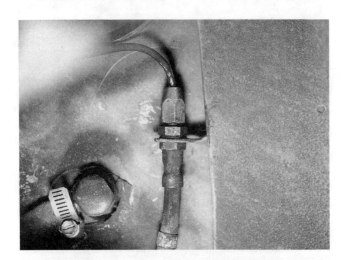

Fig. 7-31. *A typical hydraulic line connection between a metal tube and a rubber one. Be careful not to twist a rubber line. To disconnect, unscrew large bell-shaped fitting (top) while bracing bottom rubber fitting with a wrench. Then, remove locknut (center).*

Fig. 7-33. *The last act before transaxle becomes capable of crushing you. Work carefully here because there is real danger. Transaxle must be safely supported before this mounting bolt is withdrawn.*

Fig. 7-34. *With any luck, you'll have raised car high enough to get transaxle under the de Dion tube. Note how transaxle is resting on one trolley jack while another is lifting de Dion tube.*

person to manage, even with two trolley jacks and some scrap 2x4 wood wedges to help accommodate its weight. Thus, you're working with a heavy unit in a limited space and from a very awkward position. If the transaxle should slip and fall, it certainly will crush whichever part of your body is underneath, so work very slowly and very carefully.

Put a trolley jack or transmission jack under the transaxle to slightly support its weight. It's important to have jacks on casters because you'll need to move the transaxle around while it's supported. You must not try to lower a transaxle if you have only a screw jack or a bottle jack. Ideally, you'll have a transmission jack which allows you to secure the transaxle to the jack and keep it from slipping and falling.

Remove the 17-mm nut from the bolt that holds the transaxle in place. Press the bolt out using a screwdriver (don't try to pound it out: you'll only mess up the threads, and don't try to wiggle a pinky through the hole for fear of not getting it back).

With the bolt removed, carefully lower the transaxle from the car. The length of the transaxle will just barely clear the inside of the de Dion triangle, providing the nose of the transaxle case is elevated and the case is rotated slightly off its front-back axis. Replace all rubber mounts on the transaxle whenever the unit is removed. Don't forget to bleed the brakes.

4. Donut Replacement

On later Alfas, the donuts have given a surprising amount of difficulty. The rubber donuts tend to tear apart, throwing the driveline out of balance and causing vibration which only hastens the catastrophic destruction of the donut.

It appears that on some early Alfettas, the center bearing mounting bracket spacers were installed incorrectly. On these cars, there should be 7 mm between the rear bellhousing mount and the driveshaft (the allowable range is 6–8 mm). If there is a different measurement, the shaft is probably out of alignment and the

Fig. 7-35. *Alfetta donut-itis. This is what to look for if your Alfa vibrates. Driveshaft has been removed here to show tear clearly.*

rubber joint is stressed enough that it will break. Fix that problem by reshimming the center bearing. Alfa had a recall on the Alfetta driveline, but the problem persists. An original alignment problem may have started the legend, but many fixes and even more torn donuts later, the legend lives on.

The donut itself is not overwhelmingly expensive at $80 to $110 per, but if it lets go at high engine speeds, the bellhousing, which is expensive, can be ruined. I don't mean to imply that an Alfetta is likely to throw a donut from under the car after a ride around the block. They are relatively durable, and some Alfettas have traveled for 50,000 miles or more with few difficulties. In the hierarchy of potential problems on the Alfetta driveline, however, the failed donut ranks number one.

I've never heard a satisfactory explanation of why the Alfetta is so hard on donuts. My personal feeling is that wear is exacerbated because the donuts rotate at higher speeds than on conventional driveshaft-behind-the-tranny designs, though Don Black is quick to point out that the GTA driveshaft speed (the engine did 9000 rpm) exceeded that of the Alfetta with no problems. In the Giulietta/Giulia design, the driveshaft speed was reduced from engine speeds by whatever ratio was se-

lected in the transmission. In the Alfetta, the donuts rotate at engine speed, and fast starts with the car cause both heavy loads and high rotational speeds. Using conventional universal joints seems like an obvious fix, but John Shankle has tried such a conversion without success. The solid joints work well enough, but there is an unacceptable level of vibration in the car. Clearly, the rubber donuts do absorb a great deal of shaking. Alfa is not unique in using them; Ferrari and Lotus use them with no difficulty.

There are eight distinct styles of Alfetta donuts: some have round mounting holes and some are oval. See Fig. 7-36. The center bearing support is also different for different years. Make sure you get the exact ones for your car.

(While the following description and procedure applies specifically to an Alfetta sedan, it also applies generally to the GTV6 and Milano.)

Put the car up on four jackstands at least 18 to 20 inches off the ground. Verify that it is solidly supported. Always use jackstands. There will be some shaking and pulling during this procedure and you need to verify that you won't be able to pull the car on top of yourself.

Fig. 7-36. *Rear driveshaft donut looks like this. Donut's metal inserts are folded metal: other donuts have solid inserts. There are eight different styles.*

Lower the exhaust system by removing the suspending straps. It isn't necessary to unbolt the head pipe from the exhaust header on most models. Then, remove the heat shield that obstructs the center portion of the driveshaft.

Mark the front and rear mounting yokes and driveshaft sections with spots of paint so the front and rear sections of the shaft can be reassembled exactly as they came apart. If either piece is rotated in relation to the other, the driveshaft will be out of balance. It's also desirable to mark one flange on the driveshaft and the engine and transmission companion flange so that the driveshaft will be reassembled exactly as it came off.

Examine the frontmost donut carefully. There are six attaching bolts. You want to remove only the three nuts that hold the donut to the flywheel. See Fig. 7-37. The weight of the nuts, bolts, and washers affects balance, so keep track of their positions.

Fig. 7-37. *You will have to block driveshaft from turning when loosening three 17 mm donut attaching nuts. Here, a screwdriver is wedged into rear donut to keep it from turning. Front donut nuts are a bit hard to get to, even though there's a big cutout in the motor mount.*

Rotate the driveshaft so that one attaching nut is centered in the access hole. Put the car in gear and apply the parking brake to keep the driveshaft from turning, then remove the nut and washer. Rotate the driveshaft and repeat the procedure for the other two nuts. There is very little maneuvering room along the driveshaft tunnel and it's necessary to remove the entire shift linkage to get the driveshaft out. Begin by unbolting the two nuts securing the center driveshaft support and remove the support so the driveshaft hangs free in the middle.

Now get inside the car and remove the shift lever knob by pulling it off. Crawl back underneath and remove the two bolts on each end of the shift linkage's front and rear links and then remove the shift linkage shaft. Three bolts and a sheetmetal screw hold the shift lever housing in place. Carefully mark the position of the shift lever housing by scribing a line around it, then remove the attaching bolts/screw and lower the housing out of the way, being careful not to rip the shift boot.

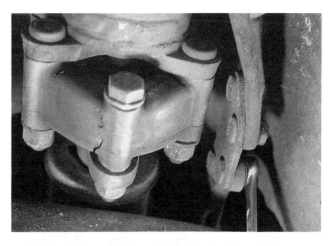

Fig. 7-38. *Rear shift link is removed by taking off two forward nuts. With front and rear fasteners removed, shift link is removed from car. Note crack in donut: though still serviceable, it should be replaced.*

Fig. 7-39. *Removing front shift link. Protective rubber boot has simply disappeared from this car.*

Fig. 7-40. *Shift lever housing leaves a clean tracing of its position. Try to return it to its original position when you reinstall it. Housing must be removed in order to take out driveshaft.*

Remove the two rear engine support attaching nuts so the engine can be moved slightly to help withdraw the driveshaft. The donuts are held in slight compression by their attaching bolts. When the donuts are slipped off the bolts they will expand slightly. To remove the donuts from their attaching bolts, it's important to ease them off in equal increments along the length of the bolt so all three come off at about the same time. If you can reuse one of the donuts, fit one or two hose clamps together and tighten the clamp around the good donut to compress it so it will be easy to slip back on its studs.

Working at the back of the driveshaft, remove the nuts which attach the donut to the companion flange, push the attaching bolts back, and pry the donut free of the flange.

On the V-6 driveshaft there is a pinch-bolt connector which allows shortening the driveshaft just enough to work it free. On earlier Alfettas, the transaxle may have to be tilted down to get enough front-back room for the shaft to be pried free. At the rear of the Alfetta, then, support the transaxle crossmember with a jack and then remove its six attaching bolts. Slowly lower the front of the transaxle with the jack until the driveshaft can be withdrawn from the transaxle yoke. When the driveshaft is free, hold the end of the driveshaft below the transaxle crossmember, return the crossmember to its original position, and reattach it with the six bolts, finger-tight.

At this point, the driveshaft should be free of the car front and back and held in only by the fact that it is resting on the center sheetmetal cross-brace (near the cast-aluminum support that carries the rear ends of the torsion bars).

The driveshaft is removed toward the rear. The biggest (literally) hurdle is to work the donut past the rear

engine mount and the top of the transmission tunnel. To make the opening as large as possible, loop a piece of rope around the rear mount and fit a long pry bar (such as a shovel handle or a 2x4) through the rope so an assistant can pry down on the mount using the underside of the car body as a fulcrum (probably the stiffening member which runs next to the transmission tunnel). See Fig. 7-41. With your assistant pulling up on the lever, work the driveshaft to the rear until it clears the rear motor mount and then pull if free of the tunnel.

Fig. 7-41. *Looking forward along driveshaft tunnel, biggest hazard is muffler hanger flange on mounting boss of rear motor mount. Driveshaft will just barely clear this, even when it is pulled down as far as possible. Lots of grunge here. End of a shovel handle and a length of rope are used to pry rear motor mount down so driveshaft will clear.*

Remove and replace each of the donuts as necessary. Leave the metal band around the new donut for the time being.

Inspect the rubber portion of the center support bearing for any tears or sagging and rotate the bearing by hand to check for smooth operation. If the rubber is torn or the bearing doesn't rotate smoothly and freely, replace the unit. If the center bearing and donut are OK, leave them alone.

There are spherical bearings at the front, center, and rear of the driveshaft. Inspect them for uneven wear.

During reassembly, the bearing surfaces should be covered with a moly paste.

The next step is to reinstall the driveshaft. Use the piece of rope around the engine rear mount and pry bar to gain room past the rear motor mount boss. Push the driveshaft in so the donut mates with the mounting studs on the crankshaft. Make sure the driveshaft goes back in the same position as it came off: here's where the paint marks pay off.

The front cup-shaped bearing occasionally gives some alignment problems on installation. You should be able to fit it into place using a long screwdriver while an assistant turns the engine by hand (that is, slowly). Push the driveshaft forward just enough to hold the bearing against the drive plate and then use the screwdriver to press on the bearing to center it as it's being rotated.

Fig. 7-43. "Offering up" (as Brits say) new donut to rear companion flange. Metal compression band is left on. Watch it: donut can be installed wrong. Small cleats on three donut mounting bosses fit into recesses on companion flange (arrow).

Fig. 7-42. Front of driveshaft has an alignment bearing.

With the front donut in place, fit the rear donut to the transmission companion flange. Slip the center support bearing onto its mounting studs. Thread the nuts onto the bearing support to draw the driveshaft into a straight line. Make certain that the spacers are in place for the rear motor mount and then thread the motor-mount nuts onto the studs. Check the clearance between the driveshaft and the motor mount. As suggested above, it should be about 7 mm. If it is not, now is the time to add washers to bring the driveshaft into alignment. With the fasteners all finger-tight, start the engine and let it idle for a moment to let the driveshaft align itself. Then stop the engine and tighten the nuts to the bearing support and rear motor mount and then secure the donuts with their self-locking nuts.

After the donuts are bolted up, clip the compression rings and remove them. Then, reinstall the shift housing and linkage.

Fig. 7-44. Tab A into slot B: Another view of pilot bearing on an Alfetta driveshaft, and where it fits into flywheel.

5. Driveline Balance

This procedure refers to all cars that have a rear transaxle and a driveshaft with three rubber donuts; that means GTV6 and Milano, too.

Check for driveline balance by sitting in the car and revving the engine with the transmission in neutral. If the car feels like a massage machine, you know the driveline is out of balance. You can feel for less dramatic vibration by placing your fingers on the driveline tunnel or, if you're finicky/brave, put your finger on a stationary part of the center steady bearing and have an assistant rev the engine a bit.

We come now to a very subjective judgement. I have heard tales of incorrigibly out-of-balance driveshafts, even after they have been returned to a professional balance shop several times. One of my good friends brought over such a car. He had been through rotating the driveshaft and even had the driveshaft rebalanced by another shop after the first failed to meet his expectations. I took his car for a drive to experience the out-of-balance driveshaft. The owner was almost beside himself with frustration over the vibration.

I couldn't feel a thing. Further, I've put an Alfetta driveline together using new donuts but no marks at all to guide me (the driveshaft came out of another car) and the thing seemed to me to be in perfect balance. Now, I am either incredibly insensitive to vibration, or there is hope for Alfetta drivelines. I hold the latter philosophy. If you haven't knocked any balancing tabs off the driveshaft, then I maintain you can get the thing put together in acceptable balance just by using care.

Let me put it another way: the engine is balanced quite independently of the driveshaft, so we know it's going to be OK. The transmission doesn't require balancing because it's not the kind of thing that vibrates. That leaves just the driveshaft as a rotating member, and those welded-on tabs tell you that Alfa has already put it right. They may well have balanced the front and rear sections together, so that's why you should take care to paint matching marks so it goes back into its original position.

I've never experienced any of the balance troubles I've heard so much about. I have a GTV6 with one-third of the rear donut missing, so I know what kind of vibration that gives (not much). Is there just one guy out there who had trouble and complained to everybody?

To conclude, don't be overwhelmed by the challenge of replacing a driveline donut. If things vibrate just too much after you put it back together, I'll tell you now how to fix it.

If the shaft vibrates more than you think it should, you can rebalance it to your own taste using simple worm-screw hose clamps. The screw part of the clamp makes a fine balance weight, so simply unscrew the

Fig. 7-45. A make-do balancing act. Weight of clamp screw is enough to put driveshaft back into balance in many cases. Trial and error is only procedure. Note white reference mark above the clamp.

clamp and fit it over the front section of the driveshaft where it won't foul against anything as it revolves. Tighten it down around the driveshaft and then put a paint mark where the clamp is as a reference point. See Fig. 7-45. Start up the engine and note the vibration. Needless to say, you shut the engine off before trying to adjust the hose clamps.

With the engine turned off, then, loosen the clamp and rotate it 180° in relation to the paint mark. The vibration will be either more or less. If it is more, rotate the clamp back toward the paint mark by 45° and try again. Keep trying until the lowest possible vibration point is reached and you're through. If the vibration is less in the 180° position, then add another clamp so its weight is adjacent to the screw section of the first clamp. Move the second clamp around trying to achieve the least vibration. Add/move clamps as necessary. You get the idea. You can keep balancing for as long and as precisely as you like.

You may also be able to eliminate driveline vibration by loosening all the fasteners associated with the driveline, then idling the engine for a while before retightening everything.

Since we don't have Weber carburetors to rebalance any more, we now have Alfetta driveshafts.

6. Differentials

The differential assembly is probably the most trouble-free item on an Alfa. Unless it loses enough oil to damage the bearings or gears, virtually nothing ever goes bad with it. That is just as well. You'll notice above, in talking about taking down an Alfetta transmission,

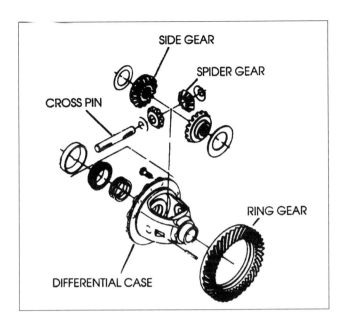

Fig. 7-46. Exploded view of a typical differential assembly. Alfetta differential has stub axle shafts connecting to the inboard disc brake rotors.

that I don't recommend upsetting the ring/pinion setting or trying to put it right. I strongly advise against owners trying to repair or replace a differential or a ring and pinion set. It is not because there are too many small parts or that the theory of operation is obtuse. The skills (and tools) required to set up a ring and pinion are simply beyond hobbying.

Just so we're clear: the ring gear meshes with the pinion gear. Attached to the ring gear is a cage which carries the differential, itself made up of two spider gears which mesh with two side gears. The spider gears let the side gears rotate in relation to each other, but the whole assembly acts as a solid element when both side gears are turning at the same speed in the same direction (that is, going straight down the road). The side gears are attached to the axle shafts.

About the most you can do with a differential is to figure out what's going wrong with it so you can save the real mechanic some diagnosis time. A ring and pinion gearset announces that it is going bad with a whine that rises and falls with the speed of the car. The character of the whine will change depending on whether or not you are accelerating or decelerating. Bearings in the rear axle can also whine, but, when they go bad, they typically clunk and leak oil.

The differential unit itself will announce its failure by protesting in turns. When driving straight ahead, the unit is under virtually no load and even bad differentials will run silently. If the rear end grinds or clunks when turning a corner one way but not the other, the probable cause is a bad wheel bearing and not the differential.

750- to 105-Series Differentials

In practical terms, if something in your rear axle assembly is going bad, you should price an entire assembly from an Alfa dismantler against trying to have yours repaired. The low incidence of failure makes used rear ends quite attractive.

Replacing the entire assembly is quite straightforward and self-evident. The body needs to be jacked up and supported with jackstands so the rear wheels are free of the ground by about six inches. Then, support the axle at the differential with a jack. Remove the limit straps, shock absorbers (on 750 and 101 chassis; on the 105 chassis the shocks can stay in place), companion flange bolts to the driveshaft, trailing links at the axle end, the locating joint at the center of the axle, and the hand brake cable where it operates the bellcrank on the axle housing. You may remove disc brake calipers to avoid opening the hydraulic system. Drum-brake rear ends are most easily removed if you disconnect the flexible brake line that runs between the body and the rear axle.

Reattach the road wheels and lower the jack slowly until the springs have lost their compression. Roll the axle straight back from under the car. If the differential assembly happens to rotate to an upside-down configuration, you'll leave quite a trail as very sticky oil leaks out the little J-shaped breather on the top of the axle housing.

Final Drive Ratios

Alfa has had several final drive ratios over the years and, if you're changing anything in the rear end, you might want to consider an alternate ratio. Sometimes, the advantage of improved mileage or performance makes up for the inaccuracy of the speedometer such a swap creates. The most popular swap was between the 4.10:1 ratio of the Veloce and the 4.56:1 of the standard. The standard got more acceleration and the Veloce got a higher top end. All ring gears for the solid-axle cars had 41 teeth; pinion gears are available with 8 to 11 teeth.

Limited-Slip Differentials

Some earlier Alfas, like the GTA, and all Spiders after 1981 came equipped with limited-slip differentials; on 2-liter and later cars, the limited slip is standard. Alfas between 1972 and 1979 had a 4.55:1 ratio and 1980 on had a 4.10:1. There are climates where a limited-slip differential is a definite advantage and, of course if you're racing, a limited-slip will give better traction.

CHAPTER 7

The standard differential allows one wheel to rotate faster than the other while turning corners. As long as both tires develop traction with the road, power from the engine is distributed fairly evenly between the two driving wheels. But as soon as traction is lost, thanks to the differential the car launches into one-wheel-drive. In snow, you throw lovely plumes of powder while digging yourself in until spring.

A limited-slip is set up to create what amounts to continuous traction within the differential. Small clutches attached to the differential cage apply the same kind of force as wheel traction to the differential gears, and they are never completely free to rotate in relation to each other. As a result, both rear wheels always receive power. The greater the clutch area in the limited-slip differential, the more equal the power distribution to both wheels. The standard set-up is to allow a 25% difference in rotation, though swapping plates can achieve a 47% "lock-out" ratio.

There are instances where a limited-slip differential causes a car to be less stable. Lost traction on ice is harder to regain simply because the differential doesn't "know" it's lost it. In most applications, however, a limited-slip unit is a distinct advantage. Virtually any Giulietta-based Alfa can be fitted with one salvaged from a later model. As with ring and pinion replacement, the switch should be handled by an experienced technician.

It is advisable, since there are clutches in the limited-slip units, to change the differential lubricant occasionally to flush out any debris. Experience has shown that the limited-slip units can start to fail at about 50,000 miles if the lubricant is never changed.

Chapter 8

Chassis, Tires, and Brakes

ALL ALFAS MASS-PRODUCED since 1951 have been constructed with unit bodies. That is, there is no separate frame on which the body sits. What I first want to discuss here for a moment is what rust means to a unit body.

Structural Rust

This is vital information if you're driving a Spider in the "rust belt." I once owned a Giulia Veloce Spider in Michigan that had a near-concours body. Yet its underpan was so badly rusted I finally had to stop driving the car. There is a hierarchy of rust damage to the unit-body Spiders that can be used as a guide when restoring or purchasing one. Rusted external rocker panels are almost entirely cosmetic. Typically, they will rust first, followed by a hole in the floor pan where the driver's right heel fits. Neither problem is serious and both can be cured by patching with new sheet metal.

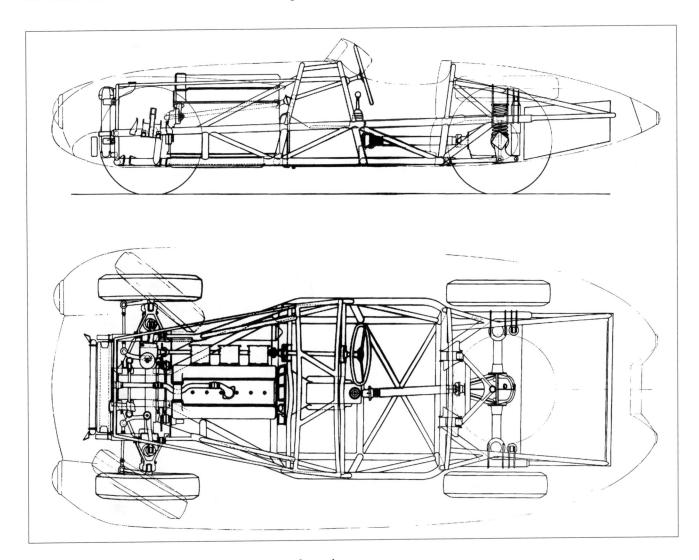

Fig. 8-1. *Disco Volante tubular chassis was most advanced chassis structure of its time.*

The next level of rust is when the outboard welded-on stiffening members on 750 and 101 Spiders are attacked. These members, which run parallel to and just behind the rocker panel, have large circles punched in them for lightness. They literally tie the front and rear suspensions together. If they rust away, they must be rebuilt to ensure a strong body. The terminal stage of rust is when the front attachment points for the rear suspension trailing arms pull free from the body. When this happens, the body is so completely rusted it probably cannot be rebuilt economically.

It is very rare for the lower front suspension attaching points to rust on a unit body. One reason for this is that the front of the car, especially the engine compartment, is usually bathed in a continuous mist of oil from leaking engine gaskets. In spite of this, Alfettas have a tendency for the front-suspension stiffening member to rust. This is the sheetmetal brace that runs up the inside of the fender well in the engine compartment.

In patching a sheet metal unit body, I recommend using a Metal Inert-Gas (MIG) welder to minimize overall heat warpage. If a gas torch is used to run continuous seams of weld, enough local heat will almost certainly be generated to warp the body so the doors no longer close properly. If you're concerned about spot welds holding, you can tack the new metal using a gas torch and then run a series of short beads, alternating

Fig. 8-2. *A close-up of the inner structure of an Alfetta sedan bodywork. Unit body structure gives strength with lightness.*

Fig. 8-3. *This Bertone-designed coupe has been powered by engines ranging from 1300 cc Giulia to 2-liters. Its strong unit body has been virtually identical regardless.*

Fig. 8-4. This Giulia Spider has a unit body: that is, no frame. There are stiffening members underneath, but they are subframes and only assist in rigidity of main structure.

from place to place to distribute the heat as evenly as possible over a large area.

If you've done major repair to a unit body, always have the wheel alignment checked.

Undercoating

It's important to distinguish between rustproofing and undercoating even though there's a tendency to lump the two goals together. Rustproofing includes galvaneeling, electro-deposition of rust inhibitors, and spraying chip-resistant paint on the lower body panels to slow down corrosion. All these steps are taken before anyone buys the car. Owners can add rustproofing only by spraying or painting additional protectants over the sheet metal and behind removable decorative trim. One goal of rustproofing is to protect areas between panels where water can be trapped and cause rust.

Undercoating is a heavy black petrochemical spray that helps protect metal from rust because it offers thick, resilient protection from chipping or cracking. It also offers some additional sound deadening. If water seeps behind undercoating however, rust will propagate without the owner even knowing it.

A car that is only undercoated will rust rather quickly. The older Alfas needed owner-applied rustproofing around the doors, trunk, and wheel wells. Modern Alfas are much better rustproofed, but still benefit from regular antirust care.

I'll finish this brief introduction to unit bodies with an Alfa anecdote. Several years ago when I lived on a farm I invited the Detroit chapter of the Alfa club out for a picnic. Joe Benson and I talked about some kind of event to enliven the party and we hit on trashing an old Giulia Super body I had previously stripped...using a nine-pound maul. The maul is similar to a sledge hammer in that it has an almost three-foot handle but one side of the head is a blade for splitting wood. Now, one would think that such an instrument would make quick work of sheet metal. The first few blows on the Alfa proved otherwise. I thought I would cut through one of the windshield posts for starters. That post remained attached in spite of a protracted day's efforts by partying Alfista. Before long, it became clear that there was very little we could do with the maul that would damage the Alfa any more than a few dents and perforations. Though most of the partygoers tried to inflict some kind of memorable damage to the body, it resisted quite successfully. Alfa makes great sheet metal.

CHAPTER 8

1. The 750/101 Chassis

For its first mass-produced car, the 1900, Alfa chose the absolutely conventional suspension system of the day: twin A-arm front suspension and a solid rear axle with coil springs all round. This was the same basic design that was also used on the 2-liter and the Giulietta (well, almost: the earlier cars used kingpins on the front A-arms while the Giulietta used ball joints). Front A-arm suspension creates a parallelogram suspension system (as seen from the front) that helps keep most of the tire in contact with the road over the entire up-and-down travel arc of the wheel. The A-frame construction gives excellent resistance to front-back bending forces.

There is a significance here in Alfa's unwillingness to introduce independent rear suspension (IRS) on the Giulietta. Porsche, as well as Renault and even Corvair had made much of the virtues of IRS. Alfa itself had used IRS with great success on the 6C2500 cars. But, it chose to retain a solid rear axle, not only for the Giulietta, but for its heirs as well (until the 1975 Alfetta introduced a de Dion rear suspension).

Alfa's conservative suspension has taken a lot of heat from the motoring press over the years. Most reviews, however, end in the acknowledgment that, while Alfa's design may not be state of the art, it is probably better behaved than any IRS setup, as the GTA demonstrated.

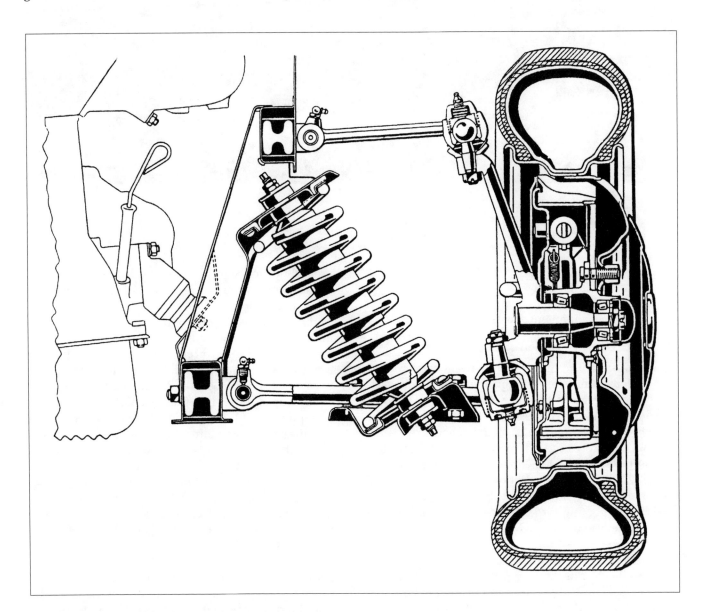

Fig. 8-5. *Conventional, but very refined, Giulietta front suspension featured upper and lower A-arms with ball joints on steering axis.*

What made these early Alfas so good and your typical American sedan (which used essentially the same layout) so bad? The secret of Alfa's success was attention to small details. In the first place, unsprung weight was kept low by the careful design of components to obtain strength and lightness at the same time, including the use of aluminum brake drums. On both the 750 and 101 series cars, limit straps and an antiroll bar on the front suspension kept body roll and weight transfer within the limits of the fairly soft suspension. A fabric limit strap on the rear was also a standard fitting. Boulevard ride was excellent and, though body roll was pronounced, the cornering capability of the cars was superb.

It's important to have good limit straps on the 750/101 cars. Many Giuliettas run around with broken limit straps. As a result, they heel over too far in a turn and transfer more weight to the outside wheel than it can handle. The result is poor traction in a turn.

If you hear any clunks from the front just as you apply the brakes, that's a sign that the bushings in the A-arms probably need rebuilding. These bushes wear slowly enough that you grow used to the deteriorating suspension and hardly notice it. As a result, you'll find that new A-arm bushings will make a (suddenly) dramatic improvement in handling. The only special tool required to re-bush a Giulietta/Giulia front end is an adjustable reamer for sizing the installed bushes. Heim "LS" series uniballs are both cheaper and permanent, but more noisy in operation.

A coil spring rear suspension requires linkages to locate it in both lateral and longitudinal directions. On the Alfa, front-back location is handled by two traction rods that are attached to the underbody at a point just behind the front seat backs and to the rear axle at about the point where the coil spring sits. Side-to-side motion is taken care of by a large triangular member that attaches to the body at two points and to the rear axle on top of the differential housing. The two traction rods and triangular member do a fine job of keeping the rear axle in place for all but the most demanding race applications.

If you're racing a 750 or 101 car, you should strengthen the traction-rod attaching brackets on the rear axle by welding rectangular plates onto them, thereby boxing the bracket.

GTA racing Alfas lowered the rear axle's roll center by adding a sliding member between the differential housing and the body of the car. The sliding block on the differential was captured in a vertical channel bolted to the body to allow up-and-down motion but absolutely no side-to-side movement. This modification would be of interest only if you're a serious vintage racer, but is relevant to all Alfas with solid rear axles.

2. 105 Chassis

With the Giulia lineup, Alfa changed its front suspension to a modified A-arm layout that was both lighter and cheaper to produce. The lower member of the front suspension remained a true A-arm but the upper member was formed by two rods, one almost transverse and the other a trailing arm that connected to the body at a ball joint located at the front of the wheel well. The advantage of this layout was that it widened the base of the upper-arm geometry and increased the rigidity of the front suspension for handling fore/aft loads.

Owners of the new suspension soon found that rough roads would eventually fail the upper joint nearest the headlights. Replacing this joint is an easy task and doesn't require realignment if you measure everything carefully. Just screw in the new ball joint the same number of turns as the old one.

During 1971, a great deal of discussion occurred between owners about the desirability of adding lube fit-

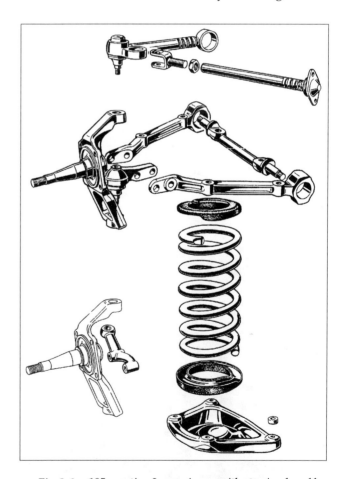

Fig. 8-6.   105 practice. Lower A-arm with steering knuckle is actually out of place in this drawing; it should be bottommost item. Spring fits inside A-arm, and its support bolts to arm. Top suspension elements form a very widely splayed A-arm. This is right-side assembly, seen from front of car.

tings to the lower control arm bushing on Alfa front suspensions. If these bushings seized from rust, the suspension could actually be torn from the chassis, so the discussion was not just academic. You may find that your Alfa has had grease fittings added. If it has, remember to apply chassis lube from time to time.

Generally, wheel alignment is not a task for the owner, though it can be done if you have a flat cement floor and a lot of patience. It's a lot easier just to have your car aligned by a professional shop that knows what it's doing.

Many owners modified the upper transverse link by cutting it in half and threading the tubes so the overall length could be easily modified to adjust camber. This is a desirable modification, especially if you're racing

The rear suspension geometry of the 105 chassis is essentially the same as the earlier chassis, but executed in stamped metal rather than being fabricated from tubes. The springs bear on the trailing link just in front of the rear axle, however, and not directly on the axle as in the 101 chassis. As a result, rear axle assemblies are not easily interchangeable between 101 and 105 series.

Fig. 8-7. *Don't see many of these any more: a grease fitting. Nipple pointing to right keeps splines lubricated on a Giulia Super driveshaft.*

Fig. 8-8. *A photograph of a Sportiva prototype. Phil Hill gives Fangio a ride.*

3. Derivative Chassis

The GTA, Montreal, and Junior Zagato chassis are all 105-series chassis, as are the GT and GTA Junior, Giulia TI, and TI Super. Generally, anything noted above also applies equally to those cars.

The Tubolare Zagato, while it uses many components from the 105 cars, is constructed of a space frame and also has independent rear suspension. This aluminum-bodied car is strictly competition-bred, though it does make a very exciting road car. Prices of TZ Alfas are so high that it is unlikely a novice will be involved in working on them. Typical of most race cars, there are few specifications. The cars are intended to be set up according to the experience of their owners. By all reports, the rear suspension of the TZ was not outstandingly successful, and needs constant attention to be race-worthy. One of my friends poured several knowledgeable years' effort into getting his TZ's rear suspension just right. The next owner, also knowledgeable, confidently reported to me his concern over how poorly the rear end had been set up. There is a fact of life: no two farmers plow a field the same way.

4. 116 Chassis

The 1975 Alfetta introduced an entirely new driveline that answered long-standing criticisms of Alfa's "antique" suspension practices. If anything, the Alfetta's driveline is technological overkill. To improve weight distribution, the transmission is moved to the rear in unit with the differential. A de Dion tube is used with inboard brakes. As a result, the rear suspension has very low unsprung weight and virtually perfect geometry. See Chapter 2 for more information.

5. Suspension

5.1 Stabilizer Bars

Stabilizer, antisway, or antiroll bars are simple devices based on a surprisingly involved interaction. They are nothing more than a stiff connector that assures that the motion of one wheel affects the motion of the other.

All Alfas have front stabilizer bars. Fitting a heavier one reduces the tendency of the body to lean in a turn.

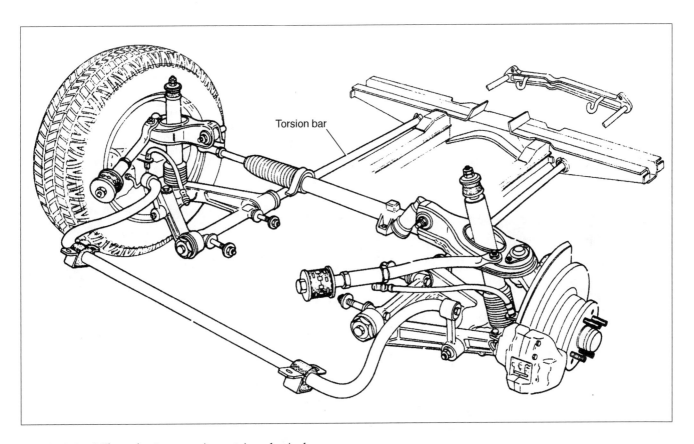

Fig. 8-9. Milano front suspension retains classic lower A-arm and uses a modified version of the Giulia (105) upper layout. A significant difference from Giulia is torsion-bar springing.

Fig. 8-10. *A close-up of splined torsion bar connection to lower A-arm. To adjust ride height, torsion bar is pulled free of the locating socket and rotated (not quite as easy as it sounds).*

This is desirable at high speeds because, by lowering body roll, you also reduce weight transfer as the body leans over on the outside wheel. Because the outside wheel carries a smaller load it can maintain traction in a harder turn than would otherwise be possible.

The purpose of an antisway bar, then, is to partially cancel the effect of independent suspension. As a result, when you fit a heavier antisway bar you give up some of the advantages of independence because the wheels are less compliant with the road. In effect, you reengineer the riding characteristics of the car. Now, since Alfa spent quite some time engineering a good mix of roadability and ride, changing the suspension is likely to make the car less desirable in some way. A heavier antisway bar causes a notably harder ride because a bump taken by one wheel is communicated more to the other wheel.

5.2 Springs

Clearly, the alternate method of reducing body roll is to fit stronger shocks and springs. Quite a selection of springs have been fitted to Alfas over the years, and stiffer aftermarket springs are available from specialty suppliers. There is no secret in judging the relative stiffness of springs. The diameter of the wire used and the number of coils are the two variables you can use for comparison. Using stiffer springs to decrease body roll helps retain the advantages of independent suspension

Fig. 8-11. *This is ABS system pump and pressure reservoir, located on driver's side inner fender.*

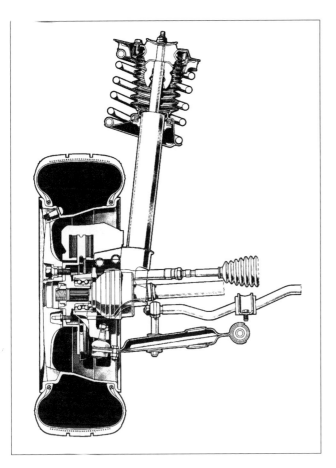

Fig. 8-12. *Front suspension of 164 utilizes a Macpherson strut, which is a combination spring/shock absorber that also acts as a suspension member.*

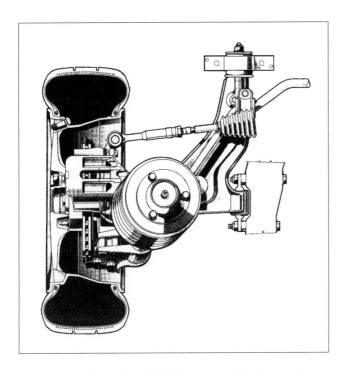

Fig. 8-13. *A top view of 164 front suspension shows extreme offset of lower A-arm.*

Fig. 8-14. *Author's cache of springs for Giulias and Giuliettas. Notice how different they are, considering that basic suspension geometry has hardly changed over the years.*

and allows you to use stock shock absorbers. This is the approach Toyota Racing Development uses for its competition conversions.

You can fit shorter coil springs to get a lower ride height. This modification improves the car's appearance, reduces its center of gravity and promotes greater stability. It also causes the aluminum fins of the oil pan to come perilously close to the pavement so that even small bumps are potential sump-scrapers. You can shorten stock springs by cutting off a coil. Usually, only one coil should be removed. By removing a coil you'll also stiffen the suspension, something you may not wish to do. Of course, the suspension must be aligned if the coil is modified, and on 105-series cars an adjustable upper link will be helpful to add camber to the slumping suspension.

5.3 Shock Absorbers

The shock absorbers fitted to Alfa are really very good. If you're in doubt as to whether or not your shocks are serviceable, lean on one fender and then release your weight quickly. The car should bounce once. If it bobs more than that, your shock absorbers need replacing. It's popular to opt for adjustable shock absorbers such as Koni and then put them on their hardest setting. The result, unless you're racing, is just a harsher ride. Shock absorbers are designed to resist quick motions but permit slower ones, such as those caused by body roll. In this respect they add stiffness to the springs at high frequencies while still permitting body roll at lower frequencies. Gas pressure shocks can offer a compromise, with increased control in all situations, without being too harsh.

Occasionally, the mounting nuts for the shocks can come loose. This is infrequent because the large rubber bushes that mount the shock to the sheet metal helps retain torque. A loose shock mount will clunk just like a loose front suspension and is occasionally mis-diagnosed. The easiest test is to grab the shock and try to move it.

6. Steering

Alfas have very accurate steering. American-car drivers, when they first drive an Alfa, are frequently impressed with the responsiveness of the steering because it is devoid of the kind of play most domestic manufacturers seem to prefer. If there is one-inch play in your Alfa's steering wheel before the road wheels begin to turn, then there is something that needs adjustment or replacement. Tie rod ends are an often overlooked point of wear.

Alfa didn't use rack and pinion steering until the Alfetta chassis. The earlier cars use either a worm and sector or a recirculating ball system, depending on model. Both older systems are very similar and those built prior to 1967 permit easy adjustment of steering play.

To check for wear, jack up the front end of the car so the wheels are off the ground. Turn the steering wheel completely through its arc, checking for any binding or grinding noises that might indicate frozen joints or bad bearings. If there is any binding, you'll have to troubleshoot the entire steering linkage before proceeding. Be-

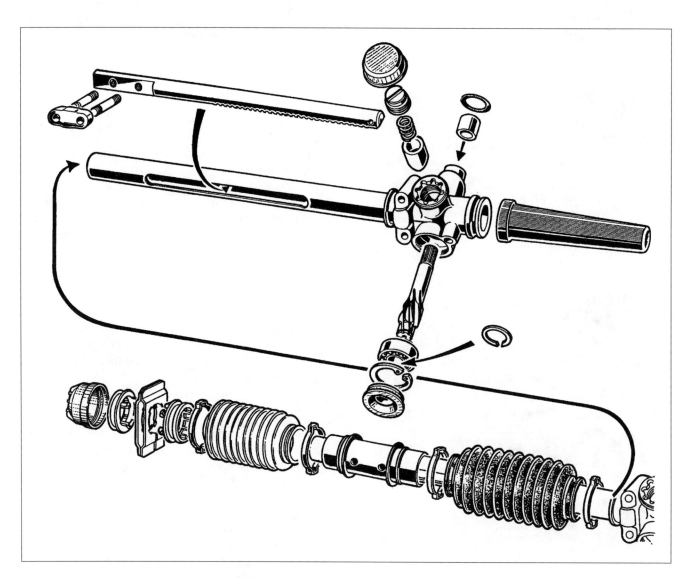

Fig. 8-15. *Alfasud rack and pinion steering is very similar to units fitted to Alfetta, Milano, and 164. 'Sud lacks power assist.*

gin disassembling the steering linkage at the joints closest to the wheels to determine which joint is binding.

If the steering is free, return the wheels to their straight-ahead position. On pre-1967 cars, the top of the steering box in the engine compartment carries a screw with a locknut. Loosen the locknut and tighten the screw until you feel some resistance. Turn the steering wheel through its entire arc again, noting especially any resistance as the wheel comes to its straight-ahead position. Keep tightening the screw on the steering box and testing the steering through its arc until you just begin to sense some binding in the straight-ahead position. When the screw has been turned so that some straight-ahead binding is just perceptible, tighten the locknut. If this adjustment doesn't remove play from the system, the steering box should be rebuilt.

Beginning with the Milano, Alfa offered power rack and pinion steering. The first racks became (in)famous for leaking, but the problem now seems resolved, even though leaking power steering pumps still haunt the model. The Milano power steering system can be retrofitted to the GTV6.

To obtain power steering, the rack housing becomes the cylinder of a hydraulic system, and the rack a double-acting piston. Hydraulic pressure is applied to either side of a seal around the rack to help it move either right or left. The fluid on the non-pressurized side of the rack is exhausted back into the power steering pump reservoir. A valve system connected to the steering meters high-pressure hydraulic fluid to the appropriate side of the rack.

7. Wheels and Tires

Tires are those black rubbery things. Wheels are the metal donut the tires go around. In conversational use, we tend to ignore, or confuse, the terms.

7.1 Wheels

The Giulietta used 15- in. steel wheels, which was the standard sporting wheel size in the late 1970s. The early Giulias with disc brakes carried 15 in. wheels that were 4.5 in. in width. In 1969, wheel width was increased to 5.5 in. with a 14-in. diameter. In 1974, alloy wheels became stock items. Although the Alfetta series cars continued with a four-bolt pattern, the mounting radius was slightly decreased, so Alfetta wheels will not fit earlier cars. Beginning with the 1981 GTV6, bolt pattern changed from four to five.

The early Alfas, that is 750- to 105-series cars, were fitted with tube-type tires. Fitting a tubeless tire to a rim designed for tubes runs the risk of a sudden loss of air during a high-speed turn, so it's a safety measure always to fit tubes to the rims that were designed to take them.

Fig. 8-16. Late practice: Milano Verde carries these alloy wheels for its Michelin MMXV tires. Many cars ape this achievement with decorative plastic wheel covers. This, like most everything else on Alfa, is the real thing.

The TZ Alfas had stock (if anything about a TZ can be called stock) magnesium wheels, distinguished by a flat gray appearance.

You cannot use 14-in. wheels on Giuliettas because the smaller rim will not clear the brake drums. Giuliettas are now beginning to be so valuable that they should only be restored to stock condition anyway. Unless you're racing, stay with 15-in. wheels on Giuliettas.

Many enthusiasts have fitted Minilite alloy wheels to their Alfas, and a variety of other alloy wheels are available on the Alfa aftermarket. These wheels look classy and provide a wider rim, which permits using fatter tires. One easy conversion is to fit the alloy wheels from Ford products that carry 390-mm Michelin TRX radials. The fatter tire puts more rubber on the road, improving traction under dry conditions. The improved cornering provided by these wider tires is significant.

Offset

Offset refers to the relation between how far the end of the axle sticks out to how far the tire sticks out. You can hang the tire outside the car body if you pick a wheel with enough offset.

A large offset results in a wider track and improved stability, but the large offset also increases the load on the wheel bearings and may cause them to fail.

You can increase offset in moderation (about one inch), to improve cornering ability without much risk of damaging anything. The first barrier you'll most likely run into with increased offset is the wheel well itself, especially on the front wheels of the Milano.

Most Alfa bodies allow some additional wheel offset. If you're also fitting wider tires to start with, however, you may actually have to choose wheels with less offset rather than more in order for the tires to clear the wheel wells. You just have to experiment to get the correct mix of tire width, offset, and clearance.

You cannot fit wider tires to Milanos with ABS because the nonstandard tires confuse the ABS sensing system.

Rim Width

As tires have become smaller in diameter, their tread widths have increased, requiring a corresponding increase in rim width to support the sidewall at the proper angle.

If you want the larger patch areas of wider tires, buy wider rims to go with them. But always measure tire-to-body clearance at the wheel wells before making any change. You can change the wheel offset, as discussed above, to obtain the fattest tires and widest rims possible.

Rim Material

First there was wood, then steel, and now alloy wheels. One factor in superior handling is that as little weight as possible should be unsprung. That is, every-thing that bounces up and down with the wheels should be as light as possible. Because they reduce unsprung weight, alloy wheels actually contribute to improved handling. That fact probably has never sold a single alloy wheel. What does sell them is that they look great on a car.

Alloy wheels require special care. Because the alloy is much softer than steel, special tire-changing tools should be used when they're reshod, and you must never run on a flat tire with an alloy wheel or you will grind the rim down to nothing very quickly. From time to time, you also must check for hairline cracks in the rim, especially around the lug nuts.

7.2 Tires

Before you buy a new set of tires, verify positively that they will fit, especially if you're going for a fatter tire and plan to purchase new rims with a greater offset. A cheap way of doing this is to buy a trashed tire of the size you plan to use from a junkyard and then fit it into the wheel well, allowing for any change in offset. Remember that the suspension goes up and down when you're trying to shoehorn a tire into your pride and joy.

The remainder of this chapter is pretty generic. That's because Alfa doesn't make either tires or roads, the two major players in the discussion. Fortunately, information about tires and wheels is essentially public domain. There are really no secrets the customer can be concerned with.

Fig. 8-17. *Though we didn't know it when cars were first released, wire wheels were a legitimate Giulietta factory option.*

It seems that tread design should be clouded in corporate security. In fact, treads are as much aesthetic as they are practical and the performance of a tread design cannot be predicted. Certainly there are secrets in the tire business. The chemistry of tread compounds is especially guarded. But you can learn all you need to know about a tread's compound simply by jamming your thumbnail into it. Some are soft and gummy; they're the ones that stick. But, they also wear out fast. Harder treads, like your year-old eraser, wear like iron and have about as much traction.

More than anything else on a car, tires and wheels are a good-news/bad-news game. Here's the good news: some feature or other has real advantages. Here's the bad news: the same feature or other has real disadvantages. There is literally no such thing as a free ride.

Consider tire size, for example. The advantage of a small tire is that it takes less space from the passenger compartment and so provides greater space utility. The disadvantage is that small tires cannot bear heavy loads.

All the handling of a car is communicated through four small patches formed where the tire touches the road. When you turn, sideways pressure on the car tries to peel the patch off the road. The weight of the car and its suspension geometry work against the sideways forces to keep the patch on the road. That, in a nutshell, is the entire science of suspensions.

Since larger-diameter tires roll on larger patches, in general terms, the larger the diameter of the tire the more performance it can handle. This generalization, in practice, has so many work-arounds that it is hardly defensible, but the suggestion of high performance given by the new 16 in. rims is founded in fact.

Clearly, one work-around, if you don't want to increase the diameter of the tire to achieve a larger patch, is to increase its width. Further, it's more practical to increase patch area using a wider tire than by using a larger-diameter tire. So it's not surprising that the typical 13-in. high-performance tire has a very wide tread. OK: the good news is that you can have small tires, wide treads and great handling.

On dry pavement. Here's the bad news: in wet or snowy weather those fat tires tend to float like a boat, suddenly losing all traction at relatively low speeds. (In bad weather, narrow, tall tires are best.)

You get the idea. Whatever you choose in the way of wheels and tires, you'll bear both advantages and disadvantages. Superior traction? Harder ride. High-performance capability? High-speed price. Aggressive tread? Aggressive noise. Smooth tread? Lower traction. Soft compound? Rapid wear. Harder compound? Lower traction. Lightweight alloy rims? Fragile, even for normal use.

Most people consider wheels and tires in purely cosmetic terms, the details of tread compound, ply construction, and tread pattern being completely ignored. Given the complexity of the subject, it is probably just as well.

Consider the subject from the manufacturer's standpoint: premium characteristics warrant premium prices. Thus, if you simply want the best, you'll simply pay the most. Bargains in the tire and wheel business are about as rare as bargains in the gems and precious-metals markets. Better to enjoy a four-carat diamond than to be concerned with how its facets are cut.

But you wouldn't be a real enthusiast if you didn't want to know more about the subject.

Balancing

When the plies of a tire are laid up and the tread is cast onto it, there may be a little bit more material in one spot than another. The unevenness of material around a tire causes it to be out of balance. A tire's balance is corrected by adding small lead weights to the wheel. Obviously, if the tire is changed, or a weight falls off, then the assembly has to be rebalanced.

Everyone has experienced an out-of-balance wheel at one time or another. The wheel will run smoothly at low speeds but begin to vibrate at higher speeds. If you can stay with it through the bouncy part, you can sometimes drive through the out-of-balance period so that at even higher speeds the wheel will no longer vibrate.

If the vibration simply increases with speed and gets worse no matter how fast you go, then the problem is wheel roundness and not balance. Radial tires go out of round when the tread belts separate from the rest of the casing, so if you have a truly out-of-round radial tire, you must have it replaced immediately.

Alfas with de Dion rear suspension must have the rear wheels balanced as well as the front. IRS or not, it's a fine idea to have all four wheels in balance. The weight of a solid rear axle damps out any but the most extreme vibrations so you never feel them, but the tire is still working at a disadvantage when it's not in balance.

The cheapest way to balance a tire is to take it off the car and set it horizontally on a specialized level. The heaviest part of the tire will tilt to one side. Adding weights to get it to stabilize at a level position will also put it into balance. That is called static or "bubble" balancing.

Dynamic balancing spins the tire, either on the car or off. On-car balancing uses a sensor that picks up the tire's vibrations and a strobe light to show the operator where to add the weights. Off-car dynamic balancing, which is the most accurate, requires mounting the tire to a machine that spins the tire at fairly high speeds.

CHAPTER 8

Most shops will balance a tire in only one plane. That presumes that any unevenness in tire material that causes imbalance is also in only one plane. If you're a purist, you'll want a tire balanced not only on its vertical plane but also on its horizontal plane as well. The best off-car dynamic tire balancing machines will detect out of line conditions in both planes so that weights can be added both to the inside and outside of the rim to help make the tire truly balanced.

If you notice that your tires have been balanced with weights on both the inside and outside of the rim, don't conclude you're received the more exotic balance, however. It's common to add weights to both sides of the rim especially when the amount required exceeds the largest single weight in the parts bin.

Tire Pressures

A tire's sidewall, especially if it is a radial, is very flexible. That ensures that the tire will be able to keep most of its tread on the ground even during hard cornering. The thing that supports the sidewall is air. Thus, inflation pressures have a direct impact on the tire's ability to bear weight. Under hard cornering, an underinflated tire could be pulled off the rim. Similarly, if you know a tire is going to be carrying an especially heavy weight for a short period, either because you've loaded it up for a move or will be racing, higher inflation pressures are appropriate.

In normal use, the range of acceptable inflation pressures is quite broad. Habitual under- and overinflation will affect tire wear. A tire that always runs low will wear its outer edges. Overinflated tires will wear the center of the tread. The maximum safe inflation pressure of a tire is printed on the sidewall. The recommended pressure is given in the car owner's manual. This is an important distinction: the maximum pressure may be 36 psi or more, while the recommended pressure 30 or 32 psi. Unless you have good reason, stay with the manual's recommendation or handling may be adversely affected.

For racing, when every part of the suspension has been tuned to get optimum performance, a difference of one pound inflation pressure will make a measurable difference in lap times. Start with 35 psi and then experiment to find the best match for your suspension.

400 mm Tires

The 1900, 2000, and 2600 Alfas were fitted with a wheel having a rim diameter of 400 mm. This size rim was also stock on contemporary Citroens and Peugeots. Obtaining 400-mm tires may prove a problem, though I'm assured they are still being made. Because of their limited application, it's possible that at some point in the future they will be absolutely unavailable.

An easy work-around to finding 400-mm tires is to weld a new 15-in. rim to the original wheel center. While this approach destroys originality and typically cuts into the vent holes of the center, it is a procedure that lets you select your own offset and rim width. That's altogether good, since one shortcoming of the 400-mm-shod cars was a lack of rubber on the road. On the 2600 with disc brakes, some small part of the front caliper has to be ground away to clear the new rim. You can increase the offset somewhat to reduce the amount of grinding required, but you will end up rubbing the wheel well if the offset is too great. The proper procedure is to measure carefully before you begin welding.

You may choose also to weld on 390-mm rims, which were in vogue a few years ago. The 390-mm rims are an easier modification because they more nearly match the original size.

8. Brakes

Alfa has always had superb brakes to complement its outstanding suspension. People frequently do not know the significance of a really good set of brakes because they have never experienced them. Most passenger cars are engineered to be able to lock up the brakes once. The idea is that if you get into a situation where you have to make a panic stop, you're not likely in the next several minutes to do it again. A cool-down time is inherent in most driving situations. The approach is not very subtle but, considering its popularity, it has proved a very successful one. It's at least cheap, because the brakes don't have to be very large to lock up a wheel, which then loses traction so the amount of work required to keep them locked is significantly reduced.

The biggest challenge in designing brakes for a sedan is to select the proper amount of braking force distributed between the front and back of the car. When you stop, the weight of the car moves toward the front. The rear end rises and the rear wheels bear on the ground with significantly less traction. As a result, all cars are biased so the front brakes have much greater stopping capacity than the rear brakes. On slippery surfaces, this means that the rear end will lock up rather quickly, especially if there are no rear-seat passengers.

It's clear from the discussion so far that the one thing you want to avoid with brakes is locking them up. That is why antilock brakes are so good. When a brake locks the wheel and causes it to skid it is out of control. Two locked brakes on a car put the car completely out of control and usually result in a spin. The moral of all this is to try not to lock up the brakes, ever. Skilled drivers are able to apply the brakes just short of lock-up. The

Fig. 8-18. Late, fully equipped Milanos offered an antilock brake system (ABS). This is a computer-controlled system that moderates brake pressure to keep wheels from locking and skidding. Master cylinder of ABS system is clearly complex, with several electrical connections not seen on standard systems. Large fluid reservoir is notable. System's ECU stores failure codes which can be read using an analyzer.

antiskid system has sensors at each wheel that compare the speeds of all the wheels and moderate the brake system pressure when one wheel suddenly stops turning as fast as the others. Now, this is not a lesson you practice during a panic stop, so a skilled driver must learn how to stop as well as go.

Alfa Romeo ABS

An Antilock Brake System (ABS) was introduced on the Milano Platinum and is offered on the 164. Since ABS (along with EFI and SRS, to throw around some acronyms) is virtually a requirement for the modern hyper-sedan, you can expect its complications to grace any models offered for sale in the U.S. ABS interposes wonderful electronic and hydraulic sophistry between your foot and the brakes. Like SRS airbags, however, ABS could some day mean the difference between life and death.

The Bosch 2B ABS system is installed on the 164. This system uses a hydraulic modulator, controlled by an electronic control unit, to control hydraulic pressure to the brake calipers. The basic input to the control unit are pulses from sensors that monitor the speed of each wheel. If, when you step on the brake pedal, the rate of speed change (not the actual speed) of one wheel differs significantly from the other, then the control unit assumes that wheel is about to lock up. The control unit then activates a solenoid to reduce hydraulic pressure to the offending brake. This allows the wheel to speed up and regain traction.

The solenoid valve simply stops fluid from flowing between the master cylinder and the caliper. Thus, at some point a change in pressure at the master cylinder does not affect pressure at the caliper. I n other words, while pressing harder on the brake pedal will not make the car stop faster, you can slam your foot against the pedal in a panic without worrying about skidding.

Diagnosis of the system requires a special ABS tester. As with electronic fuel injection, if there is an ABS fault caused by some electronic glitch, the cause may be a loose or dirty wiring connection in the system.

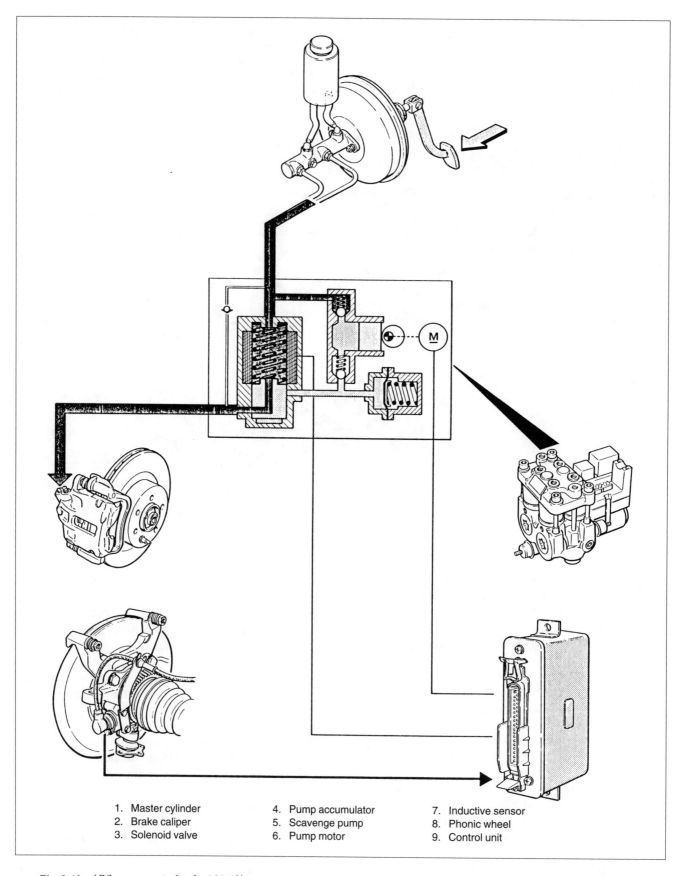

1. Master cylinder	4. Pump accumulator	7. Inductive sensor
2. Brake caliper	5. Scavenge pump	8. Phonic wheel
3. Solenoid valve	6. Pump motor	9. Control unit

Fig. 8-19. *ABS components for the 164 Alfa.*

8.1 Giulietta

The brakes on the Giulietta were quite a conversation piece: large, helically finned aluminum drum brakes up front and radially finned brakes at the rear suggested the car could handle 200+ mph speeds in confidence. When the Giulia first appeared, the front brakes featured three shoes to provide more braking area. These were dropped in favor of four-wheel disc brakes on the 105 chassis (though the 1965 Giulia Spider had front disc brakes and rear drum brakes).

Alfa has used the brake components from a number of companies including ATE (pronounced ah-tay), Dunlop, Lockheed, Bonaldi, and Girling. In my experience, components from any manufacturer are interchangeable with those of other manufacturers only as long as you keep within the system. That is, don't try to use a disc-brake master cylinder on a car with drum brakes. My Giulia Super was originally fitted with ATE disc brakes on all four wheels. When I replaced its rear axle I retained the Dunlop disc brakes fitted to it with no difficulty other than having to shorten the hand brake cables.

Dunlop brakes, especially the rear units, are not liked by Alfa enthusiasts and the tendency has been to refit Dunlop-equipped cars with ATE brakes. Front wheel spindles are needed from an ATE-equipped car along with the rear-axle backing plates (and all the brake mechanisms) for the change.

Drum Brake Adjustment

Alfa front drum brakes use a star wheel adjustment while the rear brakes use a threaded pyramid to expand the rear shoes. If you've never adjusted brakes before, it's worth the effort to pull a drum just to see how things work. Jack the car up and remove the road wheel. Two screws hold the drum to the axle. Use the largest screwdriver possible to unscrew them and be careful not to destroy the slots in the heads. An impact screwdriver, used especially on motorcycles, is the proper tool to use if the screws do not come out easily. The drum will slip free of the brake shoes only as long as it's not cocked to one side.

> **WARNING —**
> *Brake friction materials such as brake linings or pads may contain asbestos fibers. Do not create dust by grinding, sanding, or cleaning brake parts with compressed air. Avoid breathing any asbestos fibers or dust. Breathing asbestos can cause serious diseases such as asbestosis or cancer, and may result in death. Use only approved methods to clean brake components containing asbestos.*

The rear pyramid adjuster has an absurdly small square head for the adjusting wrench. See Fig. 8-21. It is very easy to round off this adjuster so only a vise grip will work thereafter. If the adjuster does not rotate easi-

Fig. 8-20. Adjusting Giulietta-style front drum brakes using adjusting mechanism.

ly, I recommend removing the rear drum and very carefully spraying a solvent such as WD-40 on both sides of the adjuster to free it (be sure to wash the WD-40 off completely with brake cleaner so you still have brakes on reassembly).

Brake shoe replacement requires a lot of physical effort trying to detach and reattach the spring that holds the two shoes against the slave cylinder. Special pliers are available to help ease the effort, but I have never replaced a shoe without working up a sweat.

Adjust all shoes so that they are one adjustment notch away from rubbing against the drum.

After any work that involves removing the drum, I always wipe the shoes and the drum with a brake-cleaner-soaked rag to remove any grease.

Fig. 8-21. Giulietta-style pyramid adjuster for rear drum brakes.

CHAPTER 8

8.2 Giulia

Beginning with the Giulia, Alfa brake systems employed vacuum boosters ("power" brakes) to allow lower pedal pressures. The boosters have proved to be the most failure-prone part of the system. Typically, the large vacuum diaphragm ruptures, requiring increased pedal pressure. In some cases, brake fluid can be sucked back into the intake manifold and cause the car to look as if it's burning oil. Rebuild kits for failed boosters were available for some time but I don't think that is the case any more. That makes serviceable brake boosters for the older cars quite valuable.

On cars with remote booster systems (that is, the booster is not bolted to the master cylinder) it's quite possible to bypass the vacuum booster system providing you're physically able to exert the additional pedal required for safe stops. Simply disconnect the hydraulic lines to the booster and use some new hydraulic line with the appropriate hydraulic fittings to complete the circuit. Always bleed the system after opening it for any reason. I drove a 1750 GTV for many miles with disconnected boosters and found that I quickly became accustomed to the increased pedal pressure. On later cars where the master cylinder is physically bolted to the booster, this modification will not work. On the other hand, I haven't heard of many newer boosters failing.

8.3 Disc Brake Pad Replacement

Giulia, 1750, and 2-liter

Front disc brakes are not adjustable. You can tell the wear on a disc brake by how far down the brake pedal travels. I have found that the discs tend to wear so that no pedal is left on a worn disc. The rear brakes on Alfetta and later cars with the de Dion transaxle are adjustable.

Disc pad renewal is very easy on the Alfa. A pin and stamped steel spring hold the pads in place. Drive the pin out from the outboard side using a sturdy nail with its end ground flat, or a punch. Have new pads ready and pry the old pad back against the piston with a large screwdriver so that the piston is completely compressed. Quickly remove the old pad and slide the new one in place. If you work fast enough, you'll be able to get the new pad in before the piston works its way out far enough to block the pad. If you are too slow, just fit the old pad and force the piston back in again. Be careful not to tear the dust seal around the piston (any more than it is probably already torn).

If you encounter a frozen piston (one that won't move) you may be able to free it by first spraying it carefully with brake fluid (as a penetrating oil—do not use a petroleum-based product), waiting for a few minutes, and then prying on it using the old pad. Force the piston as far into its bore as possible. Then, with the old pad in place, press on the brake pedal to force the piston against the pad and rotor. Repeat the process several times to exercise the piston and lubricate the entire working surface of its cylinder with brake fluid. I've used this process several times to free stuck pistons.

A warning here: a frozen piston usually means there is some rust, either in the caliper bore or on the piston itself. If the piston is rusted, it will likely destroy the seal as you free it, or somewhere down the road with the resultant loss of brakes. In cases of a stuck piston, the safest repair is to replace the caliper.

Never use any abrasive on the piston or its cylinder and clean everything in sight with brake cleaner.

Alfetta Rear Pad Replacement

Replacement of new pads on the inboard brakes of the Alfetta requires adjusting the piston retracting screws on either side of the caliper. The inboard adjuster has a 7-mm bolt head located on the face of the caliper that is perpendicular to the axle shafts. The outboard adjuster consists of a 17-mm locknut and set screw that requires a 5-mm Allen wrench for adjustment. Retract both pistons until the new pads can be slipped in. (The turning direction of the adjusting screws differs, but turning the adjuster in a way opposite to the expected way seems to work most often. All the adjusters work the same way except the inboard passenger-side adjuster, which works opposite.)

Once the pads are in place, readjust the pistons so you can hear the pads just begin to rub against the disc. The factory spec is a clearance of 0.005 in., though 0.004 in. seems to be an optimum clearance.

Fig. 8-22. Three-shoe front brakes fitted to early Giulias.

Fig. 8-23. Inboard disc brakes of the Alfetta transaxle keep unsprung weight low but are difficult to work on. Retaining pins and cross-clip have been removed from caliper in preparation for pulling pads out. Hexagonal inboard piston adjuster is at A, caliper bleeding screw at B.

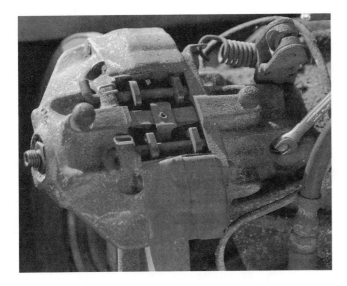

Fig. 8-24. Adjust Alfetta rear brakes using 17mm nut (extreme left, almost hidden inside caliper) and 7mm adjuster (with wrench on it). Spring clips and brake pad retaining pins are clearly visible.

8.4 Brake Fluid

There is a lot of mumbo-jumbo about brake fluids. For ordinary street use, any of the DOT-approved fluids will work fine, though DOT 4 is preferred. I recommend against mixing fluids, and against the use of silicone-based brake fluid unless the car is going to sit for a long time. Silicone fluid does not attract moisture the way mineral-based fluid does, but if you do use silicone fluid, you must clean the system of any mineral based fluid, otherwise the system will rust away.

There is no doubt that some brake fluids boil at higher temperatures than others. If you're racing, then no amount of money is too much to keep your brakes working properly. The fact that a brake fluid is imported does not, of itself, make the fluid any better.

WARNING —
Brake fluid is poisonous. Wear safety glasses when working with brake fluid, and wear rubber gloves to prevent brake fluid from entering the bloodstream through cuts and scratches. Do not siphon brake fluid with your mouth.

CHAPTER 8

8.5 Brake Bleeding

You can tell if the brakes need bleeding because the pedal becomes spongy when you press on it hard. Air in the lines compresses, while fluid doesn't.

Two persons are required to bleed Alfa brakes properly, unless you are equipped with a pressurized drum designed specifically for brake bleeding. A hand vacuum pump can be used at the wheel cylinders to draw fluid through the system, but I prefer to have someone help me by pumping the brake pedal to verify that all the air has been expelled from the lines.

Be prepared to use a lot of brake fluid to do a proper job. You should have a short piece of transparent plastic tubing that you can fit between the bleeder valve and a small reservoir of brake fluid. If you keep the tip of the tube immersed in the brake fluid in the reservoir you assure that no air is aspirated back into the system as the master cylinder is pumped.

It's easy to bleed Alfa brakes improperly. The most common error in bleeding Alfa brakes is simply to move the bubble of air from one line to the other, never really expressing it from the system. To avoid this, begin with the passenger-side rear wheel, then the driver's-side rear wheel, then the passenger's-side front wheel, and finally the driver's side front wheel. The idea is to begin bleeding the slave cylinder furthest away from the master cylinder and to end with the one closest to it.

If you have completely evacuated the master cylinder of fluid, you may have some trouble in getting the system to start pumping. A hand vacuum pump is a satisfactory way to start a system. Attach it to the open bleed nipple and pump it until fluid begins to flow in the line. Hand vacuum pumps are relatively expensive but are becoming essential in diagnosing (non-Alfa) emission-controlled engines.

On Alfas equipped with ABS, simply open both rear bleed nipples and let the system's pump clear the lines of bubbles.

Chapter 9

Bodywork and Interior

FACE IT, YOUR ALFA says something about you. It's that kind of car, just like Porsche, Jaguar and Ferrari (or, for that matter, Plymouth and Chevrolet). Much of what the car says about you has to do with its bodywork. Compulsive types stand out. They're the ones who park their cars on the farthest corner of the lot at an angle across two parking places so no dimwit can impress a spot of paint from his door into the side of the pristine car. They're also the ones who have the location and number of door dings memorized, and each new one collected makes an impression equivalent, to them at least, to a major international crisis.

I have known owners so compulsive that they won't drive their cars when there is any possible threat to the perfection of the body finish. Their cars are truly beautiful to see, but hardly utile. They are objects, but not transportation pieces.

If you're one of those compulsive types, you already know more about body care than I wish to discuss in this chapter. You will avoid car washes with whirring brushes that abrade your car's perfect finish. You will not park in the sun or allow the painted surfaces to experience rapid changes in temperature. You will guard against paint oxidation by regular waxing, keep your plastic and rubber parts flexible with a preservative, and keep your leather seats fed with conditioner. A car cover will reduce dust build-up while your car is safely secured in the garage. Should you drive through salted snow, you will most certainly spend about an hour washing down the underside of your car to guard against the possible onset of rust (being careful, of course, not to force water into the center support bearing on the driveshaft). And any oil spot on any part of your car, including the tops of the transmission and differential, are cause for a general clean-up.

Philosophically, I have no argument with the perfectionist. Alfa is an object, and the sales contract doesn't prohibit you from treating it as a precious gem.

In practical terms, however, I do take issue with those who disable an Alfa as a piece of transportation simply because they are so compulsive. While Alfas (some, at least) are beautiful to look at, their real soul

Fig. 9-1. *One of the more memorable bodies to grace an Alfa is this one by Zagato, on a 1900C chassis. Only a few of these cars were made, and they are now collector items.*

Fig. 9-2. Of all modern production Alfas, this is arguably the most beautiful. John Julien's 1900CSS Touring coupe reflects lines of Giulietta, introduced shortly before this body style appeared.

can be glimpsed most clearly nearest the limits of their physical abilities. To many, testing the limit of an Alfa's ability will seem like abuse. It is not, but driving in a four-wheel drift with the engine touching red-line does put at risk some of the jewel-like surfaces of the car.

1. Exterior

1.1 Cosmetic Rust

I discuss structural rust in Chapter 8, and this chapter deals with cosmetic rust, which can be ignored if you're not particular what your car looks like. Unchecked, it will become structurally significant, of course, but most Alfas can withstand a lot of rust without losing their structural integrity.

Certain climates—generally temperate zones such as the "rust belt" in the U.S.—are more conducive to rusting a car body. For very many years, Alfas were built with no regard for conditions in the rust belt. As a result, many early Giuliettas have simply rusted into oblivion. For years, Alfa has reported improved anti-rust features on its cars; as late as 1979, results of these efforts were hardly perceptible.

Fig. 9-3. This rust, around rear window of an Alfetta sedan, is both common and, for its quantity, terminal. Twenty years from now, when cars have become rare and valuable, this will be regarded as a "slight" amount. But as long as the cars are as plentiful as they are now, this amount of rust costs more to repair than the car's worth.

Don't be misled by the rust-belt terminology, however. Alfas rust in dry climates, too. As I mentioned earlier, many Alfa purchasers unknowingly took delivery of rusty cars, no matter where they lived. These cars sat in the port of entry for months because sales were slow.

While they sat, they were bathed in salt air. As a result, they were well on their way to destruction before they were delivered to unsuspecting owners.

Giuliettas and Giulias rust at their rocker panels and around their rear wheel wells. Alfetta coupes rust around the rear window and at the top of the front shock tower. The 1750 and 2-liter cars rust around the windshield. In all cases, these rust points are caused by water trapped between two metal panels. Short of cutting out the rusted portions and replacing them with new sheet metal, nothing can be done about the problem.

All Alfas can suffer sheetmetal corrosion near the battery. Older cars that carry the battery in the trunk usually have some portion of the trunk floor missing.

The easiest cover-up for nonstructural rust is to spread some fiberglass over it. Unless the rust is first ground away, it will simply keep on working underneath the fiberglass, eventually popping the repair free from its bond. Even a little bit of rust trapped under fiberglass will propagate to eventually destroy the bond. As a result, the essential technique for applying fiberglass is to grind the metal to an absolutely even, shiny surface and apply the fiberglass as quickly as possible to avoid trapping any water. A rust solvent is an effective and modern aid.

Fiberglassing is a messy procedure that requires no special skill until you get to finishing the surface. Just be sure you grind away all the little cavities of rust before you apply the mesh. Use wood clothespins to hold the mesh in place while the resin cures, then grind the pins away as you finish the surface. For a final smoothing, use a fiberglass filler paste.

Purists will eschew fiberglass for lead. Modern fiberglass is a viable body material but lead is clearly both a better and a more expensive solution.

In any case where rust endangers another item, or attacks a structural member, do not use fiberglass. Weld in new sheet metal to maintain proper supporting strength.

Replacement rocker panels are available for early Spiders. No wonder: the originals all rusted out long ago. You can use fiberglass mesh to cover large rocker-panel holes. If the rocker panel is seriously rusted, check the chassis sill stiffening member that runs just behind it. If that member is seriously rusted, you'll

Fig. 9-4. More terminal rot around a rear window. Metal has rusted completely away, and car is good only for parts.

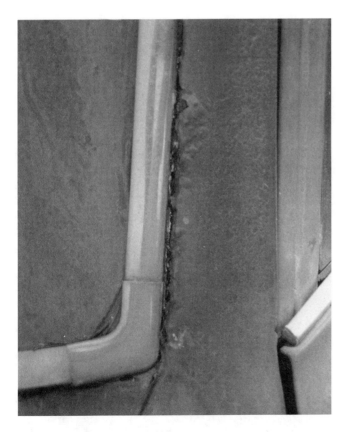

Fig. 9-5. Rust around a front window. This is an affliction on virtually every Alfa sedan from 1750 on, and is very expensive to repair. Alfetta coupes share the same problem.

Fig. 9-6. Rust bubbles around rear fender of a Giulietta coupe. This is very common and easily repaired.

want to rebuild it by welding in new sheet metal. That process is a major structural repair to an Alfa, and goes far beyond the cosmetic discussion of this chapter. Having said that, I need to remind you that, especially when working on a Spider, it's important to have the chassis on a jig so you can check that you're not deforming the basic structure when welding. The amount of heat generated to replace a structural member is enough to warp the chassis. You may end up with doors that won't close or, if closed while the work was done, won't open.

1.2 Painted Surfaces

Paint is a layered medium. Nearest the sheet metal, there is a layer of primer that provides a good bond between the paint and metal. Italian body builders tend to use primer as a body filler material as well as a bonding agent. The primer can be sprayed in a very thick layer and then sanded down to achieve a smoother surface. The problem with this approach is that primer has a slightly different coefficient of expansion from sheet metal; over the years, as the body heats and cools, the primer will lose its bond from the paint and begin to flake. Since paint has much more film strength, the result is a system of cracks that makes the surface of the paint resemble an alligator-skin handbag.

Giulietta coupes suffered from alligatored paint. The only real solution to the problem was to take the surface down to bare metal and refinish the entire car using a thinner layer of primer. Virtually any Giulietta running today will have been taken down to bare metal at some time in its life, so the problem is not as important now as it was in the 1960s.

The color coat of paint is sprayed over the primer. This coat is actually very thin, and measured in thousandths of an inch (mils). A proper lacquer paint job has a 3-mil color layer; less than that produces a paint job that will wear too quickly; more than that runs (pun intended) the risk of chipping and alligatoring.

There are two basic varieties of paint: lacquer and enamel. The new enamels are based on some fairly exotic chemistry and involve urethane or acrylics. Lacquer is a very hard and thin medium capable of superb finishes. It cannot hide small imperfections and requires great preparation (mostly sanding) to achieve a satisfactory surface. It chips easily but can give a mirrorlike surface because it can be sanded. If you're going for concours, you're going for lacquer.

On the other hand, enamel is a thicker and softer medium that requires much less surface preparation and resists chipping but gives a rougher final surface. When it is polished, the surface of the enamel is actually removed, so a good polish job is almost literally a new paint job. A really good enamel job can be as smooth as lacquer, but most will show a telltale orange-peel surface that cannot be sanded. Virtually all new cars have enamel surfaces, and all price-sensitive shops use some form of enamel, never lacquer.

Fig. 9-7. *A PininFarina factory hard top is a rare item that adds to value (and comfort) of your Giulietta or Giulia.*

With both lacquer and enamel surfaces, it's possible to spray a clear top coat to give the impression of a thicker, more lustrous surface. With clear-coat, the color coat is a special paint that gives a matte finish. A metallic paint also helps to give the impression of a "deeper" surface. Small metallic particles (generally, aluminum) are mixed into the paint just before it's sprayed. The particles are suspended in the paint and reflect light to give an impression of depth. Because of their reflective properties, the metallic flakes make the paint appear lighter than normal and must be mixed very uniformly to give a uniformly colored finish. A bad metallic paint job looks splotchy.

Recently, manufacturers have started mixing translucent mica chips into paint to give a pearlescent glow to the surface.

All the attempts at achieving a "deep" finish—clear-coats, metallics, and pearls—are a substitute for the legendary "21 coats of hand-rubbed lacquer" that are supposed to grace the most beautifully painted cars. In point of fact, hand-rubbing lacquer removes it, and the desired final thickness is still close to 3 mils regardless of the number of coats applied.

Acrylic enamel can give the hardness of lacquer and the surface-filling capacity of enamel. Though it is the classic finish, lacquer hardly makes sense any more except for beginners who need a medium that is easily corrected.

Imron is a medium that is very durable. A special chemistry is used with Imron, and the medium cannot be mixed with conventional lacquers or enamels. (Centauri is now replacing Imron as a more modern and less polluting medium.) Enamel, on the other hand, can be sprayed over virtually anything (including rust). Lacquer can be sprayed over other lacquer, but not enamel. If in doubt, you can spray a sealer over the old surface before applying new primer and color coats.

Any enthusiast can paint his own car; the problem is that painting is a skill that can only be learned by practice. The typical outcome is that, by the time you have completed painting your car, you are just skilled enough to begin the job. If you absolutely must do something with your paint job, the most helpful thing you can do is to remove all trim and prepare the surface for paint to the best of your ability. This involves using fiberglass filler and lots of sandpaper. Take your prepared car to a body shop for priming and a final color coat. If you set up the deal before hand, the body shop will give you pointers on preparing the car for painting.

Continued on page 200

Fig. 9-8. *Tom Zat's Zatmobile is a "shaved" Sprint Speciale that goes to soul of this sleek Bertone design by deleting some decorative flourishes on bodywork. It is prettier than original.*

A Tribute to Zagato

The relationship between Alfa Romeo and Zagato goes back to the 1920s. Over the years, Alfa has used many premiere bodybuilders such as Bertone, Pininfarina, Touring, and Castagna, but the most desirable road-going Alfas, historically, have carried Zagato bodies. Zagato's most illustrious Alfas were produced between 1927 and 1934 on Jano's 6- and 8-cylinder chassis.

Just after the war, Zagato kept busy building production bodies for the Fiat Topolino. The 1955–57 Alfa Zagato is based on the cast-iron 1900CSS, Alfa's high-performance version of its first mass-produced car. Not many 1900 Zagatos were made, but they were very popular in Europe for hillclimbs and rallies.

Zagato's most famous series for Alfa began with Giulietta. Early Giulietta Zagatos had rounded tails; later cars had truncated "Kamm" tails. Since Zagato has always been hand-produced, none of the cars are exactly identical.

Unfortunately, Zagato did not create any bodies for the greatest streetable Alfa, the 8C2900B. Its most celebrated series was the Tubolare Zagato (TZ), based on Giulia components but featuring independent rear suspension. These cars were both furiously fast and heart-stoppingly beautiful. Zagato continues its relationship with Alfa; the ES30 "Monster" is a Zagato creation that illustrates that shock, not pleasure, is Zagato's current goal.

One of the earliest collaborations between Zagato and Alfa Romeo is this 6C1500 Spider, seen in factory museum in Arese.

Zagato-bodied 8C cars were virtually identical to earlier 6Cs. This one has its proper fenders removed to lower drag. Slotted "Monza" grille is typical of some 8Cs.

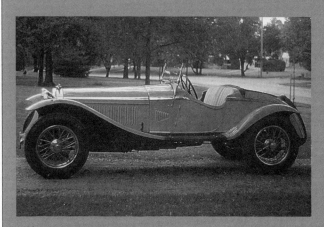

Probably the most famous of all Alfa Zagatos, this 1929 6C1750 Gran Sport combined very best engineering with styling. This was the author's first Alfa Romeo. Photo was taken in July 1961.

The Giulietta SZ was available first with a rounded nose and tail and was called the "football." A Kamm-tailed version resembled the TZ.

Only replicar of its own product by a manufacturer. Giulia-based 1750 4R (Quattroruote) Zagato recalled lines of 6C1750 Zagato. Since modern running gear was used, car ended up much "fatter" than original, but is still a very desirable collectible. Car was a joint effort between Zagato and Italian Quattroruote magazine.

This Tubolare Zagato (TZ) is one of few cars to rival elegant grace of classic 6C1750 Zagatos.

This production Giulia Junior Zagato is unique in having a steel body. Astute observers will note Junior Z was inspiration for Honda CRX.

Zagato's work on rather staid 6-cylinder 2600 Alfa transformed it into a very sporty car, even though weight loss of steel-bodied car was roughly 100 lb. This prototype 2600SZ, at factory museum, has hood scoops, which, fortunately, didn't make it to final production. 2600SZ, like Junior Z, was intended for serial production. Quality control problems cut short 2600SZ production run.

A second series of TZ Alfas was produced in very limited numbers as the TZ2. This plastic-bodied car is lower and faster than TZ, which, to standardize terminology, became unofficially known as TZ1. Carlo Chiti, chief of Autodelta, was responsible for creating TZ2.

"Monster" SZ Zagato.

CHAPTER 9

Sunlight is an enemy of paint. Left in the sun, paint will produce a milky patina of oxided material that cannot be completely removed even when sanded. If your paint surface is weathered and dull, you can use rubbing compound to renew the surface, but be aware that rubbing compound, especially when used with a buffing wheel, will get you to primer or bare metal rather quickly. There are two grades of rubbing compound: red is rather coarse and leaves visible scratches; white is less coarse and intended for finished surfaces.

Wax puts a protective coating on the paint to help preserve the finish. If you want to take the very best care of your Alfa's paint, then wax it three or four times a year after a thorough washing using some form of detergent. The detergent removes the old wax. A hand wash using a soft fabric such as a diaper is the least damaging way to get your car clean. Wet the painted surface first to rinse away most of the grime and then proceed to wash the car, rubbing only lightly. Never rub a painted surface that is dry.

Liquid wax contains a mild abrasive, which helps give a high final shine with minimum buffing but also eats through your paint. A true paste wax is abrasive-free and does not wear the paint. You can smooth small scratches and remove light stains in the paint using the abrasive properties of liquid wax. If that doesn't work, try toothpaste from a tube; it's slightly more abrasive.

Car washes are fine for everyday cars, but they do wear the paint rather quickly. If you've put over $1000 into a paint job, you'll want to care for it without the help of a car wash.

Chrome and stainless steel can be brightened with a light abrasive such as liquid wax. Commercial chrome cleaners are simply slightly more abrasive than liquid waxes. You can polish your way through chrome, but it takes quite a bit of effort.

Cotton-tipped swabs are essential for cleaning and polishing a concours car.

1.3 Dents

There are two kinds of dents in bodywork. The first kind shows a distinct crease where the sheet metal has been re-formed as the result of brushing against a post, bumper, or other unforgiving object. There is very little the average enthusiast can do to minimize creased sheet metal without running the risk of breaking the paint surface. The other kind of dent is a smooth impression that does not include any abrupt change in the metalwork. These dents can usually be eliminated simply by pulling them out with a plumber's helper. Press the rubber part of the helper in the center of the dent and pull out sharply. The metal will usually spring back into place.

You can smooth some dents from behind using a rubber mallet. Press your hand on the painted surface to absorb the blow and strike the panel from behind with the mallet. Don't use a metal hammer because its hard surface will emboss dimples in the painted surface. Generally, when working with sheet metal, the softer and broader the pressure, the better.

Dent pullers are slide hammers with a sheet metal screw captured on the end. You slam the tip of the screw through the sheet metal, then screw it into the metal and use the slide-hammer to "pull" the dent. As you can tell from the description, this does violence to the metal and should only be used when the panel will be refinished with new paint and body filler.

1.4 Touch-up Paints

Right off, you'll never match colors perfectly. Your paint has faded and the stuff from the store hasn't. Also, the overwhelming odds are that, after you've done your best with a touch-up paint, you'll have a very large blob of discolored stuff smeared on your car where once there was a hardly noticeable scratch.

If you want to fix a scratch in the paint that goes to bare metal, a brush-type touch-up paint can do the job. Shake the paint thoroughly to mix the thinner evenly throughout the color, otherwise you'll get an unworkable substance on your brush that will be either too thick to spread or too thin to cover. Dab the paint into place so that it fills the scratch, then let well enough alone until the paint dries. Use white rubbing compound or toothpaste to smooth the final surface.

Spray cans of automotive touch-up paint are appealing but, in my experience, unworkable. They are usually so thin that they run, further spoiling the surface they're supposed to repair. If you must try them, dust the surface slightly with the spray then wait for it to dry before applying the next coat. Keep shaking the can and always spray only when the can is moving. Never hold the can still to build up the paint surface.

1.5 Striping

A tastefully executed stripe highlights some styling line of the car and makes it more attractive. Stupidly applied, stripes detract from the basic lines of the car and shout "amateur." The smoother and more integrated the body, the less opportunity for attractive striping.

No Giulietta should ever be striped. Anywhere.

The 1967–74 coupes benefit from a small stripe that runs around the car at its beltline.

The 1750 and 2-liter sedans can be striped to highlight the slightly concave surface that styles and strengthens their side.

The Duetto and later Spiders should benefit from striping that highlights the "blood trough" that gives the side of the car its character. Similarly, a second-color treatment of the trough seems logical. Neither seems to

Fig. 9-9. *Mag wheels add luster to a beautiful GTV, photographed at one of the Alfa club's national conventions.*

work in practice. That is probably because the scallop down the side is so prominent that it overwhelms anything so subtle as a stripe, and a second color is so dramatic that it makes the scallop gross and unappealing.

Plastic stick-on stripes are hard to apply and painted-on stripes are out of the question for people who don't make their living doing them. If you must stripe your car, use the plastic striping that comes in rolls. I stripe by attaching the beginning end firmly at the front of the car for about an inch and then draw out about a foot (too short to sag) of the material without stretching it. I've found that if I stretch the material in an attempt to keep it running straight, the stuff will curl itself off the paint in a day or two. Hold the plastic stripe just off the surface of the car. Have a friend check to see it's straight and then lower your hand so the stripe settles on the surface of the paint by itself. Use your thumb to stick it against the paint. The real trick is to do the next section just as you did the first without making a visible dip in the stripe. Work slowly and wrap the stripe around the ends of the sheet metal panels for about an inch to keep them from peeling.

If you're working on an horizontal surface or with a thick stripe, you can get away with striping the entire panel in one length instead of working it along in sec-

tions as I've just described. People who have worked with plastic striping kits can do this even with the thinnest of stripes, but I can't.

1.6 Door Protection Strips

In the same family as stripes, but more useful, are the plastic strips that run along the side of your car to protect it from door dings. If you park in a lot, these strips will reduce, but not eliminate, the dings left when thoughtless people open their door into your car. The proper way to apply the strips is to dangle a yardstick next to the car and note where it touches the body when held perpendicular to the ground. It is at this exact point that the protective strip should be applied.

1.7 Trim

Trim is usually installed so that there is no visible clue how it's attached to the body. On Giuliettas and Giulias, chrome trim is attached using studs soldered to the back side of the piece. There are holes in the body to accept the studs and small nuts and washers are used to hold the piece in place. These nuts are usually covered with dirt or underbody sealant. Frequently, when they are accessible, they are sufficiently rusted in place that one or two turns of a wrench breaks the stud off, either from the trim itself, or level with the sheet metal. When you try to resolder a new stud to the trim piece, you usually ruin the chrome surface. The best technique is to weld or braze the stud in place and then have the piece chromed again.

If you're certain to ruin the studs anyway, you can sharpen a small putty knife, slip it behind the chrome to locate the studs, and then use it as a chisel to cut the piece free.

All modern trim is glued in place with adhesive that will almost certainly ruin the paint if it is pulled free. Use a hair dryer to soften the adhesive while you free the trim from the body. Alternately, 3M makes a trim adhesive solvent that is very effective. You can polish the paint once the trim is removed, but be warned that the paint behind the trim piece has not faded like the rest of the car and so will be a darker color. Replace the trim using 0.016 in. neoprene double-side adhesive foam tape from 3M.

1.8 Aerodynamic Hang-Ons

Undirected, air flows under the car and creates lift. The purpose of an air dam is to direct air around the front of the car and force it along its sides. The dam, by keeping air from under the car, improves downward aerodynamic force at speed (on the Montreal, it also overheated the rear axle). "Ground effects" extensions along the side of the car further direct air spilling off the front dam and route it to the rear of the car. Realize that both modifications, these extensions and the air dam, are effective only if they nearly rub against the pavement. Not very practical in everyday driving. A wing at the rear of the car helps create downforce. In the absence of a wing, a slight lip at the rear of the bodywork creates a turbulence that breaks the laminar airflow free from the body, reducing drag.

That's the theory. You can tell from the description that most "aerodynamic aids" fitted to modern cars fall into the same category as tailfins on '49 Cadillacs. The tragedy of the modern aero-stuff is that Alfa has succumbed to it in the interest of increasing sales, most no-

Fig. 9-10. To counteract falling sales, U.S. Alfa execs decided to hang things on an essentially beautiful body. Cars were typically top-of-the-line and desirable for that reason. Their *rarity is another drawing card, but the esthetics of the additions are questionable—here on an Alfetta.*

tably in the series of cars that were the children of Aldo Bozzi (Niki Lauda Spider, Andretti GT, MM GT, and Velocissima GT), and Ernesto Vettore (Quadrifoglio Spider). These aerodynamic fads are unworthy of Alfa.

1.9 Dress-Up Accessories

There is a whole industry of exterior dress-up add-ons to make your car more attractive, comfortable, and convenient. Most traditional American add-ons, such as mud flaps and wire-wheel hubcaps, are not appropriate for an Alfa.

The most common accessory for the exterior is a matched pair of side-view mirrors, which has undeniable utility. For Giuliettas, a front bumper over-rider was commonly fitted to help protect the grille from being violated by someone else's bumper.

1.10 Euro-Spec

American headlight and running light customs are different from European practice. As a result, there is some market for Euro-spec accessories such as Hella headlamps and running lamps and yellow rear tail light lenses. European headlight laws allow brighter illumination than U.S. laws, so you'll find that Euro is also brighter, hence not exactly legal in all states. Similarly, the clear headlamp covers for the Duetto and later Spiders are legal in Europe but not the U.S. These clear plastic covers are attractive and distinctive and are certainly in keeping with Alfa's character.

In a more practical vein, European bumpers on the Alfetta series coupes and Spiders are much lighter. Although they are expensive, the Euro-spec bumpers are recommended if you want to improve your power-to-weight ratio at the expense of crash-worthiness.

2. Interior

No matter how technically sophisticated Alfa is, or how svelte you consider its body, most of your experience in the car happens behind the wheel. That puts the interior in proper perspective.

Alfas have taken a beating for their ergonomics ever since the Giulietta. Reviewers have consistently complained that Alfa's arms-out driving position is awkward and unnatural. The pedals are simply too close to the steering wheel, leaving the driver with the choice either to bend his legs almost double to achieve a comfortable arm position, or (the most common solution) to stretch for the wheel with arms fully extended in order to achieve a comfortable leg position.

You would think, based on the critical reviews, that a large aftermarket industry would have grown up producing either custom pedals or custom steering wheels. Not so. Considering the number of Alfas on the road, either a considerable market is being neglected, or the

problem is not so serious as generally reported. Since Alfa's driving position has hardly changed over the years, it is clear that they subscribe to the second opinion. Satta conducted a study of driving positions and concluded that an elbow angle greater than 120° reduced fatigue.

I concur with Alfa's assessment. I switch cars frequently, and so do not drive an Alfa all the time. When I do return to one of my Alfas for everyday use, I find the position refreshing. That is probably because I'm quite willing to assume the arms-out position Alfa demands.

On classic cars, steering was heavy and large steering wheels were mandatory so the driver could have a workable mechanical advantage. Cut-outs on the doors let your elbows swing into a position that applied most torque to the steering wheel. Modern cars steer lightly and can be quite safely directed without much effort. The arms-out position does help establish a directional line, and I suspect I tend to follow my intended path more sensitively with my arms stretched out than when my elbows are akimbo.

If the Alfa driving position really bothers you, a deeply dished steering wheel is a simple fix. Leave the clutch and brake pedal positions alone. The allowable travel of those pedals has real safety significance. You can change the mechanical advantage of the accelerator pedal linkage to lower the pedal simply by shortening and lengthening the appropriate pivot points along its linkage. Lengthen the bellcrank arm to shorten pedal travel and then set the pedal as close to the floor as you can while still obtaining wide-open throttle.

For Alfetta and newer cars, you can remove the limit stops that are bolted into the adjusting tracks of the front seats to get about two inches more legroom.

For whatever reason, I find that a semireclining position in an Alfa is much more comfortable than a bolt-upright one. Rather than fight the intended position, give it a try. At least you won't confuse the car with an MG or Jag, with the wheel very close to your chest.

2.1 Seats

It's not possible to adjust the seatback rake of the Giuliettas (though the seatbacks can be wedged a bit more upright). In fact, the Giulietta Sprint and Spider seatbacks are quite fragile, and most Sprint seatbacks will have been broken and repaired by welding. The pivot point and brace are placed very close together on the seatback and, especially in the coupe, when you lean over the back, your body puts more pressure on the pivot point than it can stand. As a result, you should always practice getting out of the car and folding the seatback forward in order to get something from the back of a Sprint.

Later Alfa seat designs are much more robust and require no special care. The reclining seatback, which

Fig. 9-11. *Photos such as this should not be undervalued. Twenty-five years from now, some Alfetta Sport Sedan owner may peruse it with a magnifying glass to verify authenticity of a valuable restoration.*

was introduced with the Giulia sedan and coupe, became standard equipment before long and permits some mediation of the requisite arms-out position for those who simply can't stand it.

Sheepskin seat covers—real or imitation—let you save the cost of a reupholstery job for very little effort and no special skill. These covers usually have elastic backs that permit a good fit. So long as the shaped foam material of the seat itself is not damaged, a quality seat cover can be a good alternative to an expensive interior restoration.

You can buy reupholstery kits for Alfas that will give you a serviceable if not original interior. These kits are designed to be installed by the owner and represent an ambitious step up from the sheepskin-cover solution. The biggest challenge in doing your own reupholstery with a kit is getting the new stuff on straight. All the kits come with instructions. Read them first. Don't wait until you've ripped one or two panels and successfully covered the seat with the back material.

If you're restoring a car, however, you'll want to duplicate the original fabric and pattern as closely as pos-

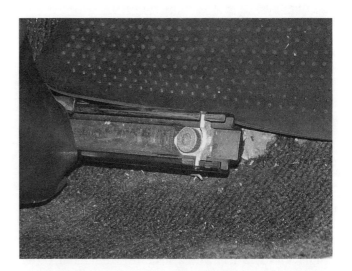

Fig. 9-12. *One type of seat-track stop found on Alfas. The identifying feature is that it's a tab held in place by an in-hex screw. Seat is pushed to limit of its travel to expose the stop. With stop removed, seat can then be moved far enough to expose attaching screw at the other end of the track. Some owners remove stop to gain a little more seat movement.*

sible. There is a very good reason to duplicate the original: aside from being a point-loser in a concours, a nonoriginal interior makes it much more difficult for the next restorer to do a proper job.

The most accurate source of information about interiors is a contemporary factory sales brochure. I need to caution, however, that some brochures show prototype cars, so you can't take them as gospel. If you're restoring an Alfa, try to obtain an original sales brochure for the exact year. Ask other enthusiasts and advertise in magazines such as Hemmings. If you can't find one, get as many contemporary road tests as you can find. The pictures that accompany them are invaluable.

2.2 Seat Belts

There is a limit to how far back you should recline because of seat belt considerations. It's important to sit up enough to avoid "submarining" in a head-on. That is, the force of the impact threads you underneath the lap belt (the shoulder belt will have no effect) and throws your knees into the dash or the steering column.

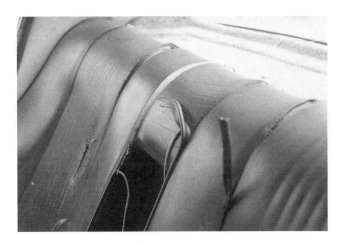

Fig. 9-14. Owners in sunny, hot climates know all about cracked dashes and torn seats from constant heat. Notable in this photo is fact that stitching, and not fabric itself, gave up first.

Fig. 9-13. Milano Quadrifoglio Verde has stock "Recaro-type" seats. Very comfortable and adjustable, but high side bolsters make seats hard to get into.

You want to be erect enough that the force of your body does, in fact, engage the shoulder belt to restrain you from the dash. Therefore, it's important to adjust the shoulder strap so that, in a crash, your body engages the lap belt securely before being restrained by the shoulder belt.

Giuliettas didn't have seat belts. You should fit them for safety even though they are not original. I recommend a 3-in. racing-type belt with a latch that uses a pivoting J-shaped cast-metal hook. The style was contemporary with the car and is certainly rugged enough. Giulia sedans were the first Alfas in the U.S. to have seat belts, and they were a three-point design that included shoulder belts. Some early Giulia belts were recalled because they did not meet proper strength standards. By now, any original Giulia seat belts should probably be replaced simply because the strength of the old fabric is questionable.

The belts fitted to the Giulia Super were similar to those installed in some domestic cars. These belts were clearly intended not to be used, otherwise there is no way to explain whey they were so difficult to adjust, un-

comfortable to wear, and easy to stow. The same motion required to pull the shoulder belt up to slip it on also disconnected it from the lap belt. If you still have this design in your car and the fabric is serviceable, wire the shoulder strap lug to the lap belt end to keep it from being so easily detached. This design, I am convinced, was conceived by domestic manufacturers in the hope that, by making the seat belts unusable, usage would be extremely low and the seat-belt law would eventually go away.

Later Alfa seat belts are a significant improvement and offer ease of use and excellent protection.

2.3 Steering Wheels

Aftermarket steering wheels give your car a distinctive look and, very often, a distinctive feel. It's surprising how the character of a car can change depending on its steering wheel. The absolute classic wheel is a Nardi with three thin spokes and a wooden rim. The grand-prix style Momo wheel, which is small, black and thickly padded, is another popular aftermarket item.

Fig. 9-15. *This interior view of a Giulia TI shows distinctive steering wheel and instrument cluster. Circular tachometer and bar speedometer are a curious combination, but work well in* *this design. TIs are not as common as Supers, and are desirable for their rarity.*

Fig. 9-16. This leather-wrapped steering wheel is standard on Milano Verde. Notice how well its three-spoke design is integrated into dash behind it.

Smaller steering wheels change the mechanical advantage of the system and produce heavier steering effort and quicker steering response. Larger wheels ease steering but seem less precise. You will probably not be able to find aftermarket wheels that permit the horn/light-flash functions of the Giulietta wheel. Beyond that limitation, any wheel with an adapter that fits the Alfa's splines should work OK. Just check that the new wheel doesn't block your view of the instruments.

You may not be able to install the wheel yourself without a steering-wheel puller. These are not expensive, and the installation is not complex. But if you have any questions about your ability to install the wheel, have an experienced mechanic put it on. It's better than having the wheel come off in your hands at an especially awkward moment.

Steering Wheel Lock/Ignition Switch

Beginning with the Giulia Super and GTV, Alfas were equipped with a safety steering wheel lock that included an anti-tamper ignition switch. If you need to remove your ignition switch on virtually any modern Alfa, you're in for a bit of work.

First, remove the two-piece plastic shroud that encloses the steering wheel column. The shroud is held in place by five long screws that are accessible from underneath. The ignition switch is housed in a metal protector and held in place by a headless bolt. On the Giulia, this bolt is fairly identifiable for it sits vertically near the base of the housing. On later cars, the bolt is made more inaccessible and is virtually horizontal, pointing toward the firewall.

In either case, the bolt you need to remove has a domed head with no flats, and is slightly recessed in a protective boss. The easiest way to remove the bolt is to get it loose with a small chisel. Place the chisel so its tip forms a radian on the head of the bolt, and is at an angle of about 45° to the axis of the bolt. Strike the chisel sharply with a hammer so it bites into the top of the bolt and tries to unscrew it at the same time.

If you're careful, you won't have upset the top of the bolt into the threads of the boss and the bolt will unscrew easily with your fingers once started free.

When the bolt has been removed, you can cut a slot in its head with a hacksaw if you wish, thus allowing replacement and subsequent removals with a screwdriver.

Fig. 9-17. This 1984 Spider interior has a slightly different treatment from earlier models, especially row of rocker switches above shift lever and digital clock below. Otherwise, it is the same interior as 1971 model.

2.4 Interior Trim

The goal of the trim designer is to attach panels to sheet metal with as few clues as possible as to how to remove the panels. There are some general techniques used by Alfa. Knowing them will get a panel off about 90% of the time without breaking anything.

The biggest challenge in the interior is removing the dash. After that, door panels offer the next greatest puzzle. Seats are uniformly easy to remove: just unbolt the seat track from the floor. Most carpeting is screwed in place on Alfas. If the carpeting is old and torn anyway, just pull it up trying to keep it in one piece so it can be used as a pattern for the new stuff.

The Giulietta interior is spare compared to the later cars. The dash is a structural member of the chassis and so is not removable. The door panels on the Spider are held in by clips at the bottom and attached by several screws toward the top. The panel is removed by taking off the handles (set screws attach them) and the attaching screws and then lifting the panel straight up to release the clips.

Giulietta trim panels are made of a fiberboard, which is usually warped from being soaked when rain seeps down the window opening. It can be easily replaced at any store that stocks home paneling. In fact, you can do a reupholstery job on Giulietta Spider side panels quite easily. The original Giulietta door pulls are sometimes broken. You may be able to find folding handles in the bathroom section of the store that will make fine substitutes for the handles if originality is not paramount.

Beginning with the Giulia, interiors became more complex and somewhat harder to manipulate.

Giulia sedan and GTV dashes are removable, as are Duetto dashes. The bolts that attach the dash are very hard to get to and sometimes you'll have to remove accessories such as the windshield wiper and heater ducting just to get to the attaching bolts. Typically, there are large bolts at each side of the dash and a series of smaller nuts running just under the windshield that hold the dash in place. The 115 Spiders have four bolts: two on the side and one under each defroster duct. The V-6 uses the same mounting plan, but the Milano requires removal of almost everything on the dash to get it out. In most instances you should try to do whatever work is necessary without removing the dash. Removing instruments or replacing burned-out bulbs is not reason for dash removal.

In hot climates, dash panels shrink and eventually crack in the sun. This process can be slowed by the careful use of window shades or dash or car covers. It can also be mitigated by regular applications of a plastic conditioner, which replace some of the lighter fractions of the hydrocarbons that forms the dash. These components are what fog up your windshield with a hard-to-remove film, and the more conditioner you apply, the more fog you'll get.

Door Trim Removal

Giulia and later door handles are held in with spring clips. You have to pull the U-shaped clip free of its slot in the handle in order to slip the handle free of its splined shaft. Auto shops sell clip removers, but you can make a perfectly serviceable one from a large paper clip. Straighten one end of the clip with your fingers, then use a needle-nose pliers to put a very short hook on the free end. The hook should be about 90°. Anything greater than that makes it hard to engage the clip, and any less than that assures that the clip will slip off before you get it free.

To get to the clip, press firmly against the decorative bezel that fits between the handle and the door. With the bezel pressed as far into the door as possible, rotate the handle until you see the top arc of the spring clip. If you're 180° off, you'll see the two ends of the clip. Hook the clip with your paper-clip tool and pull it completely free of the handle. Then, simply slide the handle off the splined shaft. See Fig. 9-19.

Fig. 9-18. This is a custom interior for a custom Alfa, the Alfasud Sprint 6C. Note that where there would normally be a back seat there is a V-6 engine.

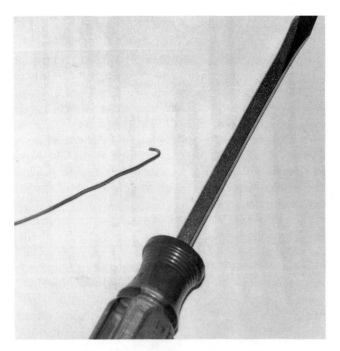

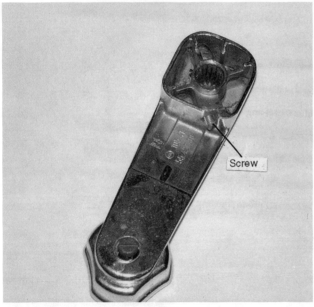

*Fig. 9-19. **Top***: *If you can't figure out how to remove window winder handle on your Giulietta, the bent piece of wire will suggest how. Winder handle is held in place by a spring-clip, which can be hooked and pulled free after a bit of fiddling.* ***Bottom***: *On later models the winder handle uses a set screw as shown on handle for Alfetta sedan.*

The next step in removing door trim is to remove any small Phillips-head screws that attach the panel to the door sheet metal. These screws are frequently rusted in place and may break off when you try to remove them. That's OK, because you'll be able to drill out the broken part after the panel is off. Some Alfas supplement the visible screws with clips that are attached to the underside of the panel and therefore hidden from

view. If you've taken off all the visible screws and still can't work you hand completely around the bottom and sides of the door panel, you've got hidden clips to contend with. Most of these clips use a stepped serration to get a tight fit, and will pop free with just a little careful work prying next to the clip with a screwdriver. If you're working on a door that uses spring clips, you'll probably break the clip with this technique. That's OK too, because it's virtually the only way to discover what kind of gizmo is actually holding the door panel to the sheet metal.

After all the decorative screws and hidden clips are removed, you'll be surprised that the door panel doesn't just lift off. That's because Alfa usually hides two or three screws along the armrest to further support the door trim, especially when it doubles as a handle to close the door.

First, look under the handle to see if you can't locate the attaching screws. For Giulietta sedans and their 1750/2-liter cousins, that's where you'll find them. The GTV6 and Spiders of the same era hide one screw behind the triangular metal insert that highlights the door handle and another behind the silver plastic channel that runs along the length of the handle. On Spiders from the 1980s and some GTVs, one of the attaching screws is hidden behind a trapezoidal metal panel that looks as if it were embossed into the armrest. The part has to be very carefully pried out of its mounting.

After you've located and removed the screws, unscrew the inside door lock handle and lift the panel free.

If you're lucky, you'll find a plastic vapor barrier glued in place behind the trim panel. This is an original-equipment protection against rain damage to the back side of the trim panel and should be saved at all costs. If yours is torn or missing, make a new one out of a 4-mil plastic trash bag and stick it in place with automotive trim cement.

Once the door trim is off, you can have your way with the door lock and window winding mechanisms.

The Giulietta door handle is held on by a large screw and a small stud. The stud at the skinny end of the handle is going to snap before you can remove its attaching nut, so be prepared to drill it out and tap it for a screw.

Those fragile Giulia Sprint (and 2600 coupe) lift-up handles are held in by an attaching bracket inside the door. The upper surface of the handle has a lip that holds it in place on the door. If the handle is broken, it can sometimes be repaired by riveting a reinforcing piece of sheet metal to the underside. Not sightly, but workable.

The later door handles on all Alfas are variations either of the Giulietta or Giulia style and are really quite easily removed. The lock mechanisms that attach to the handles are usually operated by rods. In use, these rods can flex and become inoperative. Once the door is

Fig. 9-20. This door latch is characteristic of the Alfa Spider. It offers some resistance to being pulled away from its mating piece during a severe collision, but is not so robust as "lobster claw" latch used on later sedans and coupes.

apart, study how the lock mechanism works and you'll be able to readjust it to like-new operation.

The door lock itself is usually attached with heavy in-hex or Phillips-head screws. Occasionally, a regular slot-head screw is also used to attach some flange of the lock. It's easily removed. To remove the remaining attaching screws, select an Allen wrench or large cross-head screwdriver, depending on the kind of attaching screw used. In some instances, especially with the Allen wrench, you'll be successful. If the Phillips-head screws don't loosen, you'll need to get buy an impact driver from an auto supply store to remove them. Don't try to force them with Vise-grips on the screwdriver because you'll only ruin the screw head.

Once the lock is freed from its attaching screws, it may be something of a puzzle getting it and its linkages out of the door. Remove as few of the links as necessary, because they all have funny shapes and only go back one way. Roll the window up and down to see if you can get better access to the pieces. And, be prepared to cut your hand or arm on some sheet metal in the process.

With the lock removed, you can actually dismantle the key cylinder, or take the assembly to a locksmith for service. This is the perfect time to have all the exterior locks put to the same key pattern. At this point, also, you may want to renew the fabric track in which the window slides.

2.5 Window Winders

If you've removed the door panel to repair a window winder, you're in for quite a bit of work. The early Alfas used gear-driven winder mechanisms that were very straightforward in operation. As straightforward as they are, the gear-type winders are awkward to remove simply because they have such an odd shape. The other option is a wire-type mechanism that graces the Duetto, subsequent Spiders, and Alfetta sedans.

The wire-type units are indecipherable if they've broken, so study the good winder on the other door to determine how the wire runs. There is a clutch mechanism built into the wire-type winder that is intended to release pressure over a point that would endanger the wire. If the clutch gives up, the entire winder has to be replaced, and this involves unstringing the old wire.

When replacing a wire-type winder, make sure the wire wraps around the winding drum so that it's not tangled between the tracks on the drum.

Electric window mechanisms are much more straightforward, with gear drives. My experience has been that the control switch gives much more trouble than the winder mechanism. See Chapter 10 for troubleshooting details.

2.6 Headliner

Removal and replacement of the headliner on a closed car is beyond the capabilities of the amateur, since it requires removal of the front and rear glass.

2.7 Instruments

Virtually all early Alfas attach their instruments using L- or U-shaped brackets secured by two serrated thumb-nuts. You reach behind the dash, feel your way to the instrument in question, find the two big thumbnuts, and unscrew them. Then, slip off the bracket and then push the instrument forward. Remove it from the front of the instrument panel. If you're removing a tachometer or speedometer, you'll have to remove the drive cable. All the other instruments will have some kind of electrical connection to their sending units.

All, that is, but the oil pressure units on Giuliettas, which are mechanical. The Giulietta oil pressure gauge reads actual oil pressure, communicated to the gauge by copper tubing. The copper tube has to be released with an open-end wrench from behind the gauge before its clamping bracket is removed. Because of the Giulietta's dash design, this is a simple job that only requires sitting upside-down in the driver's seat with your head on the floor (or a similar thoughtful position).

One of the things most repair manuals leave out is the removal of instruments and the dashboard. I don't intend to cover the waterfront here; if you're about to try to take an instrument out, the advice of someone who's done it before on your model will be of great help. And parts books, with exploded diagrams, are invaluable. However, there are some general guidelines.

The first thing to determine is whether the instrument is removed from the back or the front. If there are no obvious fasteners up front (as on the Sport Sedan or GTV6—don't be fooled by the fake screws on the Alfetta sedan), then you have to probe at the rear of the instruments for U-shaped brackets or thumbscrew fasteners. If there are no obvious clues, then press on the instrument cluster carefully, both from the front and rear, and see if you can determine in which direction it moves more freely, or if there are any places where it seems to "hang up." These places are frequently the location of hidden fasteners.

Since the Giulietta, Alfa has slowly improved its wiring design from a bundle of individual, color-coded wires, to an integrated system with multi-pin connectors. The Alfetta coupe and Sport Sedan introduced printed circuit boards on the dash clusters of U.S. models. The instruments and their wiring became an integrated cluster to improve reliability and serviceability.

On Alfettas, removing an instrument cluster is a bit of a puzzle because it's retention is not at all apparent. The secret is two large spring clips, and once you overcome these you can simply lift the panel out. Well, not that simply: over the years, the plastic shroud surrounding the instruments usually slumps and captures them in place. You can make up a sheetmetal tool that holds the shroud out of the way and releases the two clips at the same time, or you can use a bunch of screwdrivers to pry the cluster out.

On GTV6s, remove the instruments from the front after undoing the decorative screws.

The GTV6, Milano, and 164 speedometers are electric, with the sending unit located on the transmission. Both the speedometer and sending unit are prone to sticking with age. You can free the speedometer head by disassembling it and then using a light lubricant.

2.8 Accessory Instrumentation

Alfas have neither ammeters nor—until the 164—voltmeters. I've always thought that was Alfa's basic public relations ploy, for it is certainly an effective way of hiding The Bad News. Any Alfa with a generator needs an ammeter. Hang it under the dash and use at least 14-gauge wire to hook it to the voltage regulator. A voltmeter on alternator-equipped cars will tell you overall system health.

Giuliettas had oil temperature gauges, and I've always lamented that omission on later Alfas. An oil temperature gauge will tell you when its really safe to use

Fig. 9-21. Spring retaining clip location on Alfetta sedan instrument cluster. Another spring is similarly located on speedometer side. You can compress them with pieces of sheet metal, then pull the bottom of the cluster forward to release them. Work carefully: you can break back of speedometer housing if you pull too hard.

full throttle (not until the oil is up to operating temperature), or when you need to pull over and let things cool down.

If you're turbocharging your Alfa or are mildly curious about mixture strength, a pyrometer is a great aid. It's installed on the exhaust manifold and measures exhaust temperature to tell you when things are about to go to meltdown.

I've always been partial to accelerometers. If you use the same freeway entrance every day, the accelerometer is a legal and accurate indication of overall state of tune as you floor it merging into the slow lane. As a one-shot reading, they're entertaining but useless. It's the day-to-day pattern that is really revealing.

All those extra instruments look great even if you never notice what they're trying to tell you.

2.9 Carpeting

As with upholstery, you can buy carpeting kits for all Alfas. Most carpeting is screwed in place. Over the years, the screws rust solidly to the floor pan, so they are unremovable. To further attach carpeting, trim is used to trap the carpet to the contours of the transmission tunnel and sills. This trim is also held in by screws that are likely to rust solidly in place.

For old carpeting, the easiest removal procedure is simply to rip it up in as many large pieces as possible. Beneath, you may find original sound-insulating padding backed by a black asphalt-like material. You may also find a thin foam rubber, which is an easy water-trapping, rust-inducing replacement for the original equipment.

No fabric carpeting is molded. That is, it's basically just flat material cut and trimmed to fit over irregularly shaped surfaces such as the transmission tunnel. If

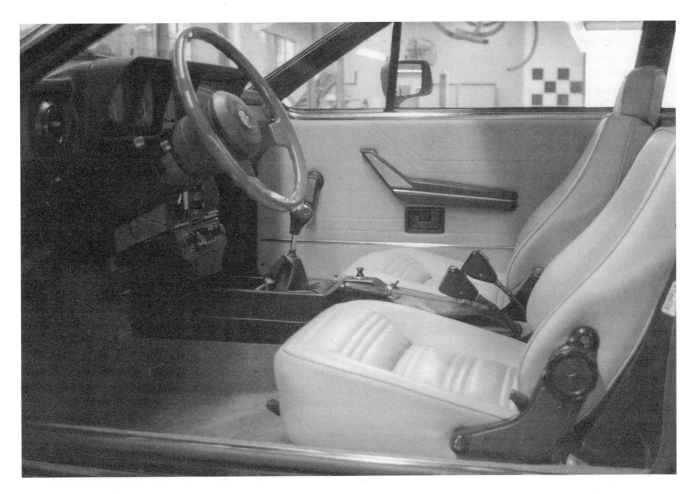

Fig. 9-22. This GTV6 interior was photographed in 1981, the first year of the model. Steering wheel and dash layout distinguish it from 4-cylinder coupe.

you're careful in lifting the old carpet largely intact, you have a ready-made pattern for the replacement that you may wish to sew up yourself. You may ruin a home sewing machine trying to hem the edges of the carpeting, but a rubber-backed (nonfraying) dark carpet can be cut with scissors to make quite an acceptable replacement piece, especially if you're careful to hide as many edges as possible under trim.

A lot of Alfas use molded rubber carpet with unique patterns, such as the Alfa badge, cast in. If the badge is not torn, you can carefully cut it out and stitch it to new carpeting for an interesting effect. Alternately, if the rubber is torn but otherwise sound, use a heavy packing tape on the back of the rubber to hold everything together and carefully fit the carpet in place. A few daubs of trim cement will help keep things together.

If you're restoring a car and the original carpeting materials are unobtainable, use a small roll of commercial ribbed black rubber such as used on stairways to finish off the trunk area, and recover the floors in a single dark color of carpeting that complements the seat color.

3. Windshields

Windshield removal is probably beyond the competence of the enthusiast. Prior to 1972 (on GTV), windshields were held in place by a rubber seal into which an expanding bright trim piece had been pressed. You pry off the trim piece and then pop the windshield free of its rubber seal. Don't try it unless you're absolutely certain your local shop has a replacement windshield they'll be happy to install.

Modern windshields are glued in place. If you can get a piece of piano wire wedged through the glue and between the windshield and the body, you may be able to work the wire around the windshield enough to let it come free. Again, breakage is a rule rather than an exception for the amateur.

Limited-production cars all face the same windshield breakage nightmare. If you have an old Alfa, then presume that your windshield is irreplaceable. You need a willing glass shop as a resource even more than a skilled mechanic.

CHAPTER 9

If your windshield is broken, do not try to remove it. Get the car to the glass shop where the dimensions and curvature of the glass can be measured. Yes, a new one can be made, but at a price guaranteed to keep you from buying popcorn at the movies for a few years. If it's a show car, you can use Lexan to make a replacement. Lexan won't work for a car that is used regularly because it scratches (and isn't legal in most states).

4. Convertible Tops

Giulietta

The top mechanism of a Giulietta Spider is removed with some difficulty. Since the top fabric can be replaced without removing the mechanism, it should be left alone unless some part of it is broken and needs replacing. The problem is that there is a threaded inner pivot rod that is not evident just looking at the mechanism. Initial appearances suggest that about six bolts are all that hold the assembly in place. With the obvious attaching bolts removed, the parts will not come free until this inner rod is shortened by turning with a wrench. Loosen the large locknut on the long bar on which the top pivots, then turn the shaft clockwise using a 15-mm wrench on the squared section of the threaded shaft on which the locknut is threaded (and which is itself also threaded into the long tube that runs clear across the car).

As you turn the shaft, hold the locknut and eventually the rounded end of the shaft will emerge from the body bracket hole about the time the squared end of the shaft disappears into the locknut. However, the rounded end is also a hex, so you can continue to turn the shaft with a 19-mm wrench until the end clears the hole. Remove two short cap screws with a 17-mm socket and remove the assembly from the car.

Duetto and Later

Beginning with the Duetto, convertible tops were attached at the back by a strip held in place on studs. The nuts for these studs are almost certainly rusted on and the studs will snap when you try to remove them unless they have been thoroughly soaked with penetrating oil.

The remainder of the top is held on by glue and a few well-concealed screws. A stiffening wire should run along the flap that seals the top against the side windows. The tension of this wire is adjustable for fit.

The front top bow has a leakproof rubber gasket that is glued in place. The glue used to attach the top is a special nonhardening material that allows you to pull the fabric away carefully without tearing it too badly. If you find that the glue has hardened with age (or the last job was done using the wrong kind of glue) you can try to free the fabric with a hot-air gun or an industrial solvent.

4.1 Top Removal and Replacement

With the above general description, you should be able to remove an Alfa convertible top. It is not an essentially easy task, but with a little care and understanding you can get the job done. Further, removing the top is a lesson on how to put a new one on. Pay attention to details, work slowly, and make notes.

Start by soaking with oil the 23 attaching nuts and their studs at the back of the top. While the oil is doing its work, vacuum the trough along the back to remove dirt and debris. Find the drain hole at either end of the trough near the door jamb and clean it out carefully using stiff wire and a little water.

Carefully remove the small nuts that clamp the top to the back of the car. If you snap a stud, you'll be able to use a sheetmetal screw as a replacement. Remove the chrome trim piece and then slowly pull the fabric free off the studs.

Loosen the set screws on the tension wire that holds the horizontal flap over the window. Loosen the locknuts and remove the adjuster body, the pull the wire out the front at each corner.

Drill out the rivets on the top's vertical post near the tension-wire adjuster.

Begin removing the top from the front bow by unscrewing the handle, latches, and the two chrome trim pieces. Carefully pull the rubber weatherstripping free and set it aside. Note that the edge with the lip faces rearward.

You'll now need to remove a bunch of rivets by drilling them out. Begin with the rivets at the ends of the front bow, then the six that hold the gutter in place.

You can now remove the entire top assembly by unbolting it, but this is not actually necessary.

Find the edge of the top material that will be tucked near the front edge of the top bow and begin to peel it away from the bow.

With both ends of the top free, it will be easier to remove the fabric from the side rails. You already know how to get the fabric free because the technique is the same as for removing the material from the front bow.

Finally, detach the fabric from the top bows. The fabric is glued in place and should be carefully peeled away. The fabric end is probably tucked out of sight between the top and the bow.

Sand all surfaces where metal touches fabric, then paint the surfaces with a rust-resistant black paint. Let the paint dry completely. Check all the pivots of the top mechanism. This is your chance to repair any that have worn, using short, fat bolts.

If you are replacing a factory top, you will encounter aluminum stiffeners and a slightly different flap configuration than for an aftermarket top. The differences are minor so far as procedure is concerned.

Take a moment to read any instructions that came with the new top and take inventory of all the items in the kit. If you need to purchase top cement, go to an auto parts store or an auto paint store and get the right stuff. It's horribly messy, but it will let you do a creditable job that can be undone with little effort if you want to perfect your technique.

Replace any foam rubber that pads the top at the front bow. Then, slip the new top on the front bow. This is the most critical step in the whole procedure. The top must be absolutely square, not only on the front bow, but with the back attaching studs as well. Furthermore, the top must hang evenly over the side windows on both sides. Take several minutes to verify that the top is as evenly fitted as you can manage. Check especially the fit of the material at the ends of the bows and around the curve of the side windows. Rotate the top material around the bow so that the small fabric lip at the front of the top hangs down evenly over the windshield frame. Roll the side windows down slightly to act as an additional guide to help in verifying that the top is squarely positioned.

With chalk, mark the fabric and metal so that you can verify the top's alignment as you proceed. Don't put glue between the top fabric and the foam rubber. Instead, glue the free end of the top material to the front bow, wrapping the loose end out of sight between the top material and the bow. Glue the rubber gasket in place and replace all the hardware. Try the top latches for alignment, then release them and let the front bow hang free.

Now, position the rear of the top along the studs. Carefully verify the top's alignment and attach the chrome clamping strip using sheetmetal screws to replace any broken studs. Begin at the middle of the strip, working to the sides to eliminate any side-to-side wrinkles. At this stage, don't worry about front-to-back wrinkles. The aim is to get the top material taught around the rear circumference. The work gets quite awkward as you progress toward the sides, so take your time.

With the fabric attached at both the front and rear, secure the front bow with its latches and temporarily rotate the intermediate bows in place to verify alignment. Correct any alignment errors now, because this is truly your last chance to do it easily. Don't worry if the top sags slightly between the bows. It'll tighten with time. The important check at this stage is to verify that the top is perfectly square on the car.

Next, with the front bow latched in place, glue the top to the side bows that support the top just behind the side windows.

Release the front latches, string the stiffener wire through the top material, and attach it so it is taught when the top is up. You may have to use a stiff piece of wire with a hook on one end to get the wire threaded through the fabric properly.

The top is now completely installed except for the fabric strips that are glued to the intermediate bows. If the top material is loosely attached to these bows, the top will flutter and roar at speed. Make sure that the top is pulled as snugly as possible against these bows. Slide the bows into place so the two fabric attaching strips hang evenly on either side. Apply glue to the inside of one flap and wrap it around the bow as tightly as you can manage. Pull the second flap firmly around the bow and then glue its end so it tucks out of sight between the bow and the top material.

Expect the new top to be just a bit loose. It will shrink and tighten up in a week or so.

5. Audio and Other Electronica

There is really very little use in fitting a high-quality sound system in a Giulietta convertible, and not much sense in installing one in a Giulietta coupe. Those cars are not meant to be rolling discos. The later Alfas, because they are more quiet, can be fine acoustic containers for exorbitant sound systems.

Early Alfa ignition systems can be effectively quieted using conventional high-resistance secondary ignition wires and a capacitor on the generator. The later systems are quiet enough.

Placing an antenna requires punching a sizable hole in some body panel. Most Alfas will have had a radio at some time or other in their life, so the pain of violating a pristine panel may not present itself. If you must punch a new hole, however, I recommend putting it near the windshield on the left side (looking towards the front). Check first to verify that you can run the cable easily from the hole to the underside of the dash. Watch out for hidden panels that block your access.

Most Alfas have provision for dash-mounting a radio, but the factory holes may not match the shape of the radio you want to install. The older Alfas had a thin rectangular hole to accommodate the popular shape of their day. Modern radios have a much deeper rectangle. Rather than hack away at a Giulietta or Giulia dash, I strongly recommend that you buy a radio that fits the original hole in the first place. You also must check that the depth of the radio will clear the space available behind the dash. Giulia Supers and GTVs are especially cramped behind the dash. In hard cases, or just for an easy installation, consider an under-dash mount.

Monophonic installations are easy because you can usually find some unobtrusive place to hang a speaker. If you want the best kind of mono sound, get one of the small boxed speakers and fit it somewhere up under the dash. You can hang it using mechanic's wire.

CHAPTER 9

Fig. 9-23. Functional simplicity, in best Euro-tradition. The radio, at bottom of "stack," is out of place, ergonomically.

If you must have stereo, tape the speaker wires behind the dash to keep them out of the way. You can install stereo speakers in the door panels. To do this you need to run speaker wires behind the kick panel and through the door jamb. That's a lot of hole-punching and grommet-fitting, and some Alfa doors are hard-pressed to give up space for a decent-sized speaker. You can also surface-mount speakers to the kick panels, being careful not to block the fuse box, hood release, or similar useful items. In closed cars, the rear parcel shelf makes a good surface for stereo speakers. You can run the wires under the carpet and rear seat into the trunk or you can run them underneath the car (following the fuel line, for instance).

5.1 CB Transcievers

I've used CB radios in a number of Alfas with success and pleasure. The antenna should be bumper-mounted even though your transmission pattern will suffer. Except for the antenna, CBs offer no special installation problems.

5.2 Theft Deterrence

In spite of the fact that Alfas seem to survive quite well in Napoli, you may well want to have a theft deterrent system installed in your pride and joy.

I need to say up front that, no matter what the salesman tells you, if an experienced thief wants your car or its contents, nothing's going to stop him short of an alert German Shepherd in your passenger seat. The fact of the matter is that a good thief can look just like an embarrassed owner when the alarm goes off, and, once inside, will disable the system with the same speed as if he had an ignition key. Unless your car is parked in your garage, it's a sitting duck to a real pro.

There are several steering-wheel clamps that are brightly colored and offer a limited degree of maneuverability if the car is stolen. While these devices may be hard steel and practically impossible to cut, don't be lulled into a false of security by them. A thief needs only to cut through the steering wheel itself, bend it back a bit, and slip the device free of the wheel.

One of the big problems of theft deterrent systems is that the industry that sells and installs them is populated by largely unskilled labor. The old saw about the windshield wipers running when you turn on the air conditioning has been recreated frequently by ill-trained alarm system installers. A really sophisticated alarm system is intimately involved with your electrical system. That means that the installers should be similarly intimately involved. At $5 or $10 an hour, they usually aren't. Watch out for hanging wires and plastic splicers that are the telltale signs of shortcutting.

The most common alarm depends on a free-swinging weight that is used to detect slight motion of the vehicle (as when the thief punches out a glass to get inside). This system, which operates exactly like the "tilt" sensor on a pinball machine, is notorious for giving false alarms.

Some systems also use a microphone to pick up the high-pitched impact of breaking glass. Most systems are backed-up by electrical switches on all opening panels.

They may give you peace of mind. That ill-founded illusion is really what you're buying when you install a theft deterrent system.

In point of fact, the most effective theft deterrent is to take the keys out of the ignition. And consider this: locking the doors just means the guy will have to break something getting in.

The upside of Alfa ownership is that they are one of the least-stolen cars available. Too odd and distinctive. My personal solution is to have a very cheap radio installed in the car, nothing of value in the glove compartment or trunk, and all opening panels unlocked.

Nothing here you'd want, buddy.

Chapter 10

Electrical System Basics

ELECTRICAL TROUBLESHOOTING is one of the most frustrating challenges to confront the hobbyist owner. That is probably because wiring has low visual impact. Most disabled electrical systems look no different from functioning systems. The wires don't rupture and a non-working switch can look very shiny, indeed. In contrast, mechanical failures are typically dramatic. There's no mistaking a rod through the block or oil spraying from the head gasket.

Electrical troubleshooting requires a special brand of sleuth. You must understand the system you're troubleshooting and you must use some sort of tester to verify what you can't see with your own eyes. Once you have a sense of electrical troubleshooting, virtually all electrical problems are fixable using only a trouble light or VOM (volt-ohmmeter). I do not say easily fixable, for the two tools you absolutely must bring to electrical troubleshooting are patience and logic. Frequently, a simple short or misconnection will be discovered only after several hour's work. Or, a disconnected wire, lying in plain sight, will go unnoticed until your spouse passes by and asks innocently what it goes to.

1. Wiring Diagrams

With Alfas, an understanding of the system is usually buried in the twisted mass of lines called the wiring diagram, a postage-stamp map at the back of the owner's manual that defies interpretation. If you're in need of the information locked inside a wiring diagram, the first helpful act is to get it enlarged about 200% on one of the new breed of photocopiers. Use the edge of a colored business card to follow a wire through the maze. When the line does a 90° turn, move the corner of the business card to the turn and you'll be able to follow the wiring with little difficulty. The information you're chasing includes the source of the power, the color of the wire used to carry it, and the path of the current back to ground.

Back when there were two-stroke engines in cars (I owned a DKW once) we used to say if it had gas and spark it had to run. Electrical components are something like that. If it has power and a ground it should work. In most cases, after you've verified that a compo-

nent is truly getting power and has a solid ground, if it still doesn't work you replace the component. Very few electrical components are repairable nowadays, thanks to die-molded plastic and sheet metal stamping machines. Nowadays, you don't even repair alternators. The possible exceptions are wiper, fuel pump, and starter motors—though your parts man might look askance at a request for fuel-pump brushes.

Though the electrical systems of all modern cars are so complex as to defy comprehension, they are built on the simple truth already alluded to: a source of power and a path to ground are the only two things any electrical device needs in order to operate.

This basic concept has been used to produce what are called current-flow wiring diagrams. While Alfa doesn't use them, it's important to understand how they work, because you need to draw your own current-flow diagram, at least in your head, if you're going to do any troubleshooting. In a current-flow diagram, the source of power is represented by a line at the top of the chart and the source of ground is a line at the bottom. Everything between is a flow chart of how the components are wired into the car. See Fig. 10-1. What the current-flow layout doesn't tell you is where the components are located nor exactly how the wiring is arranged to get the job done. Only the classic wiring diagram does that.

Reduced to words, an Alfa is wired like this: from the positive terminal of the battery, a large wire goes directly to the starter solenoid. This assures that the starter gets absolutely first dibs on available power. After all, if the car doesn't start, nothing else really matters.

From the starter connection, another heavy wire leads to the fuse box. This is the main lead for electrical power in the car. At the fuse box, electrical power is divided between several categories of circuits, each one protected by a fuse. In early Alfas, the division is fairly straightforward. At one end of the fuse block is a main fuse for the ignition system, which includes the ignition key and ignition coil. At the other end of the block, five fuses handle the running lights, and right- and left-hand high- and low-beam headlight circuits. There is a fuse for the instruments and then one for "accessories," which is a true catch-all that includes the radio you're trying to fix. On later cars, the #5 fuse controls the dash instruments and charging warning lamp.

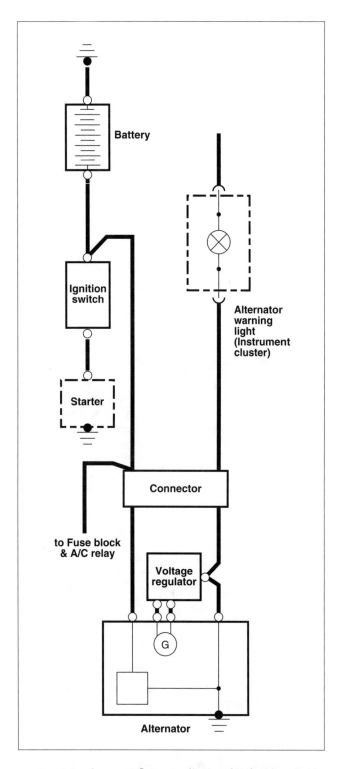

Fig. 10-1. *A current-flow type diagram for charging circuit of 1979 Sprint Veloce. In this drawing, ground is actually depicted at two points, battery at top and alternator at bottom. Battery current flows through alternator warning light to alternator to begin charging process. Generated current flows from alternator to battery.*

1.1 Looking For a Few Good Grounds

The negative side of the battery is usually attached to a short, fat wire that is bolted directly to the body to provide a source of "ground" throughout the entire car. Remember, however, that while most systems have a negative ground, Lucas systems (Giulietta only) use a positive ground.

Plastic parts have created an additional electrical challenge. When bodies were predominantly steel, you automatically provided a ground for the device as you installed it into the sheet metal. Now that both you and the electrical devices are floating in nonconductive plastic, the electrical circuits will require a separate ground wire to work. These ground wires are typically black, and run from item to item to provide a return path for the current. In a modern car, it's as important to verify a solid ground as it is to verify power.

On all modern Alfas, the engine and driveline are isolated from the body with rubber to control noise and vibration. This also means that the engine is not grounded to the car through a solid mechanical connection. To provide a good ground for the engine, a wide braided strap is usually connected between the body and one of the bolts used to attach the starter to the engine. Again, the starter needs all the help it can get (on Giuliettas, it was fashionable to run a heavy wire from the battery in the trunk to a starter mounting bolt).

If this ground strap should break or corrode enough to lose its conductivity, the engine is literally floating without a proper ground. On early Alfas, the ground path is then provided by the throttle linkage. Since this linkage is fairly robust, it is possible to lose chassis ground on a Giulietta and never know it. On Giulias, the throttle linkage is nonconductive plastic, so a loss of engine ground will force all the starting current to pass through the choke cable.

On SPICA fuel-injected cars, the source of ground becomes the thin wire that connects the accelerator linkage to the throttle bellcrank. On these cars, trying to start the engine will cause the cable to heat up and become beautifully incandescent just before torching the surrounding grime. The moral: find the braided ground wire, inspect it, and make sure its connections are bright and shiny on both ends. Use the procedure outlined later in this chapter to verify a good ground.

1.2 Fuses

The charging circuit is typically unfused and gets its current either directly from the battery terminal on the starter or the battery side of the ignition fuse. For a period of time, however, it was fashionable for owners to put a fuse in the main alternator lead, so you should check the fattest wire coming from the alternator to see if a fuse has been wired in and has either blown or is not making proper contact.

A fuse is nothing more than an easily melted wire. If something happens in the circuit so a lot of current flows to ground, the fuse will melt before the wiring, saving a nasty fire.

In any circuit, after the current passes through the fuse, it is then distributed to all the devices protected by the fuse. As a result, in a typical fuse box there will be only a few wires attached at the top of the box but a myriad of wires running out the bottom.

Alfas are notorious for poor fuse connections. The tip of the fuse corrodes against its spring clip and breaks the circuit even though everything looks fine. You can increase the spring pressure of the clips by looping small rubber bands around them. Orthodontists' rubber bands work fine.

2. Troubleshooting

Tom Zat proposes a general approach to troubleshooting that makes sense: take out all the fuses from the box and test across their terminals for battery voltage with a VOM. You will certainly find one fuse with current flowing: it's the one for the clock. If you can disconnect the lead to the clock (and any theft-deterrence system installed), then there probably shouldn't be any voltage across any of the terminals with the ignition off. If there is, you've identified a circuit that needs further investigation to determine how the current is flowing to ground. Generally, Tom's approach works best for fuses on the left hand side of the block. Don't bother testing the fuses that control the parking lights or headlamps, because they don't get current until you turn the proper switch on.

Color coding on the wires will usually indicate where the wire belongs in the wiring diagram. In many cases, especially on old cars that have had their wiring mutilated by several generations of amateurs, it's virtually impossible to tell which circuits are represented by which wires at a fuse block. If you're blowing a fuse and can't locate the source of the problem, disconnect all the wires from the spade connector at the bottom of the offending fuse, replace the fuse with a good one, and then reattach the wires one by one. The fuse will blow just as you attach the offending wire. Secure all the good wires, fit a good fuse, and then work through the electrical system systematically, operating all the devices to discover which ones are inoperative.

If a device isn't working and no fuse has blown, there can be only three possible causes: no power, bad device, or no ground. You can troubleshoot most electrical circuits using a trouble light. Make one using a 12-volt bulb and a socket with two leads. See Fig. 10-2. One is for power and the other for ground. Solder alligator clips to the ends of the leads and you're in business.

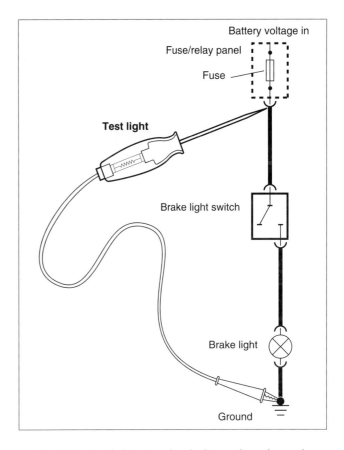

Fig. 10-2. *Test light set-up for checking voltage from a fuse. A test light is the quickest way to check for voltage and ground, but it's also a quick way to kill an ECU. Don't use it on cars with control units on board.*

Verify that there is power to a nonworking device by attaching one lead of your trouble light to the powered side of the device and the other lead of the trouble light to ground. The light will come on if you're getting power. If there is a switch in the circuit, you have to verify that it conducts current when ON, and does not when OFF. See Fig. 10-3.

If you are getting power to the device, next verify ground by connecting one lead of your trouble light to 12 volts and touch the other lead to the ground connection of the device. Make sure that you don't inadvertently connect the test lead directly to ground. For instance, relays are often grounded through one of their mounting screws. The screw itself may be making a good ground connection, but there can also be enough corrosion between the mounting tab and the screw to insulate the relay from ground. Ground connections are almost always lost from a hidden build-up of corrosion. In many cases, the corrosion isolates the part itself from the obvious ground connection. Just because the attaching sheet metal screws goes to ground is no proof that the device itself is grounded. A voltage drop test will help determine ground conditions. See Fig. 10-5.

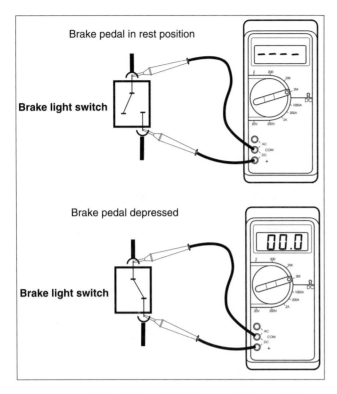

Fig. 10-3. Brake light switch being tested for continuity. With brake pedal in rest position (switch open) there is no continuity. With brake pedal depressed (switch closed) there is continuity. (No voltage is present in this test.)

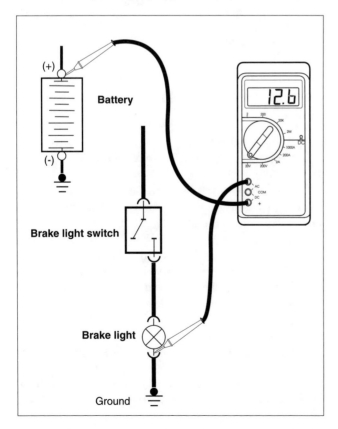

Fig. 10-4. Voltmeter being used to check for ground at brake light.

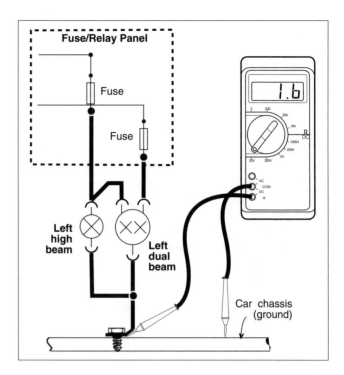

Fig. 10-5. Example of voltage drop test on dim headlights. Voltmeter showed 1.6-volt drop between ground connector and chassis ground. After removing and cleaning headlight ground, voltage drop returned to normal and headlights were bright.

NOTE —

A voltage drop test is generally more accurate than a simple resistance check because the resistances involved are often too small to measure with most ohmmeters. For example, a resistance as small as 0.02 ohms results in a 3-volt drop in a typical 150-amp starter circuit. (150 amps x 0.02 ohms = 3 volts).

NOTE —

Keep in mind that voltage with the key on and voltage with the engine running are not the same. With the ignition on and the engine off, battery voltage should be approximately 12.6 volts. With the engine running above 1500 rpm (charging voltage), voltage should be approximately 14.4 volts. Measure voltage at the battery with the ignition on and then with the engine running to get exact measurements.

Large automotive parts stores that cater to professional mechanics usually carry some kind of short-finder. This is a kit that includes a flasher unit that replaces a blown fuse. The flasher makes and breaks the shorted circuit frequently enough so there is no danger of melting wires. Now, every time the flasher turns ON, and then OFF, a magnetic field is created along all the wires in the circuit up to the point of the short. Finding the

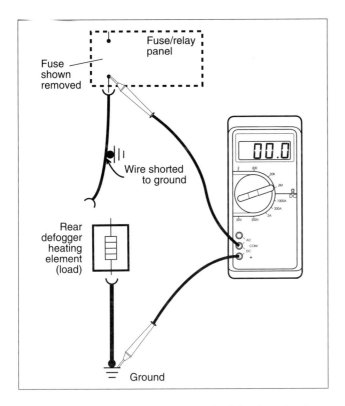

Fig. 10-6. Ohmmeter being used to check for short circuit to ground.

short involves nothing more than moving a meter, supplied with the kit, along the wire. The meter's needle moves each time it senses the wire's make-and-break magnetic field. The short is located at the exact point the needle ceases to move. A short-finder is wonderful when you have to trace wires under carpeting.

Early Giuliettas suffered a lot of starter-switch failures, all of which will have certainly been corrected by this time. If you have a Giulietta with an aftermarket push-button starter switch mounted near the ignition switch, that's why. Later 105 and 115 cars also had starter-switch problems that could be corrected by dismantling and cleaning the switch.

I've run into a curious wiring technique in the GTV6, which will cause no end of difficulty if you're not prepared for it. Up to this point, we've acted as if only one wire to a device carries current and only one goes to ground. But if you test the power window motor circuits, you'll find that both wires carry 12 volts of power. These circuits use relays that switch the ground side in order to reverse the rotation of the motor to raise or lower the window. That is, both wires normally carry power and motor operation involves switching one of the wires from power to ground.

I would be remiss if I didn't say something about electronic fuel injection in a chapter on the electrical system. But, in fact, there's not much to say. You should not use a trouble light or meter to test components un-

less you absolutely know that they can be tested and you have some guide to the results to expect because of the risk of damaging components. The problem is no different in kind than dealing with a radio. It's not appropriate to explain in a book like this how to align the FM section of the radio's tuning circuit, either. With electronic fuel injection, in practical terms you may never be able to identify either the source of power or the ground connection, and current flow may be so brief or weak that it is virtually undetectable using the tools most hobbyists have.

The new Motronic ECUs (Electronic Control Units) will store any detected faults and report them using diagnostic codes that are read by counting the flashes of the "check engine" light. A terminal on the ECU is shorted to ground to start the sequence and the codes are given in the Motronic manual.

There is one truth about electronic fuel injection that applies to the rest of the electrical car, however: the best way to check a suspect component is to substitute a known good one. While this may not be a viable approach with electronic control units, it will work just fine for starters, generators, and most motors.

Generally, if your battery keeps discharging and the headlights don't brighten when you rev the engine, you should simply replace the alternator. If the starter doesn't start, replace it. This is clearly not the cheapest approach but it is certainly the quickest. If you count your time as money it may also prove to be the cheapest in the long run. Thus, if you have verified power and a ground and the thing still doesn't work, your next step is probably to go out and buy a new one.

Now, rebuilding a starter or a generator is fun, providing you can get new brushes and you have the time. You can also replace the diodes in an alternator. Rebuilding windshield wiper motors and fuel pump motors is a pain because they're small and parts are really hard to obtain (you can sand large brushes down to fit most motors if you really need to). Running new wires to replace shorted or frayed ones is diverting, and frequently the only way to fix some problems. Always use stranded wire and tie everything up tidily so the wires don't flop all over.

According to Fred DiMatteo, alternators from 1980-on have attached voltage regulators that can wear out brushes or simply fail. Of course, it's a lot cheaper to replace just the regulator rather than the whole alternator. The older electromechanical regulators can be replaced with aftermarket all-electronic "potted" units.

As noted above, the newer Alfas have much more sanitary and idiot-proof wiring. The most important single advance in Alfa wiring technique was the introduction of the plastic multiprong connector that had been used for a long time in American cars. You can take advantage of this advance while troubleshooting:

if you unplug a connector and the short disappears then you know the problem is downstream of the connector. Finding the connectors may prove a problem because whenever possible they're hidden behind the trim. But the connectors are easy to probe with a VOM and they offer a good reference point for troubleshooting using a wiring diagram. Typically, the wiring harness itself doesn't fail, and as long as you can see the color code on a wire you have a good idea of what it's supposed to do.

One final note: if you're shopping for an Alfa and find that the fuse box is hanging loose, with many disconnected wires and lots of nonfunctioning electrical items, I recommend against buying it at any price. You'll probably never be able to get it fixed short of a complete rewire and you'll learn to hate the car in time. That's no way to feel about an Alfa.

3. Jump-Starting an Alfa

It used to be that you could slap jumper cables on a car to get it started when the battery was low. If you got the cables wrong, the fireworks from crossed polarities alerted you to the fact that the red lead goes to the positive terminal and the black one to the negative, which is ground on all Alfas except the Giulietta. Crossed polarity typically did no mechanical damage.

> **WARNING —**
> • *Battery acid can cause severe burns, and will damage the car and clothing. If electrolyte is spilled, wash the surface with large quantities of water. If it gets into eyes, flush them with water for several minutes and call a doctor.*
>
> • *Batteries produce explosive gasses. Keep sparks and flames away. Do not smoke near batteries.*

As cars have become more electronic, their tolerance for crossed polarities has vanished and it is now true that any inadvertent polarity error will probably do over $1000 worth of damage before you can blink your eye.

That means you can't just attach jumper cables on one of the newer Alfas and hope for the best. Never attach jumper cables to a car without first verifying which terminal on both batteries is negative and which is positive. By verifying, I mean actually seeing a + or – sign or the letters POS or NEG. Don't just go by the wire location, the color of the wire, or what you think it ought to be.

Never jump-start an electronically fuel-injected car from another car with its engine running. The intent here is to minimize the danger of transient voltage or current spikes. For the same reason, never jump-start another car using an Alfa 164. It has a sensitive fuse that blows if you try a jump start.

Before attaching jumper cables, arrange the free ends to make certain that they won't accidentally short out against one another. Typically, this means clamping the positive cable around a well-insulated section of the negative cable. Attach the negative cable first to the good battery and the other end of the negative lead to some part of the engine block on the other car. Then attach the positive cable to the good battery and the other end to the positive post of the run-down battery.

You may conclude from all of this that it is better to tow a fuel-injected car to start it than to try to jumper it. Not so: when you try to tow-start, you flood the exhaust system with raw fuel which, when the engine does start, will probably overheat the catalytic converter. Technology is not always kind.

The easiest and safest way to start a car with a run-down battery is to recharge the battery.

Chapter 11

Performance Modifications

I NEED TO SAY at the beginning of this chapter that Alfa is a car that attracts the kind of owner who is always in search of more performance. You will be able to see in this section that Alfa can be modified to give significantly more power and handling from stock. But I want to caution one thing before getting into the nitty-gritty of mods: Alfa knows what it's doing, and you're up against the real pros if you think you can out-design what's already there.

Race Tuning

Many parts of this book (most notably, the sections on fuel injection modification) are relevant to preparing an Alfa for competition. Whatever you do to improve engine efficiency or cornering ability bears directly on racing.

Everyone has their own idea of the best way of extracting more performance from a car. The fact that there are absolutely conflicting ideas makes race preparation a truly fertile field. It also allows the neophyte to sound as wise as the veteran. One of the jobs of learning about racing is developing a litmus test that distinguishes matters of mistaken opinion from the facts.

I want to emphasize that the subtleties of making a car go fast are still not completely understood, even after one hundred years of effort. Whenever the book is

Fig. 11-1. *Some idea of the sporting dominance of GTA Alfa and its derivatives can be gained from this photo, taken at Brno, Czechoslovakia, in 1970. Tony Ezehans leads in a 1750 GTAm. At least six GTAs are identifiable in following field.*

223

closed on this process of discovery, there will be a single engine design, a single transmission design, and so on. Everything will be, literally, perfect. And, all cars will look exactly alike. It's happier not to know everything.

In the U.S., one guide in the search for ultimate truth is the Competition Advisory Service (CAS) once published by ARI. This information was directed to the racer and gave plenty of authoritative information about futzing with hardware. If you're racing a newer Alfa, the CAS will be out of date. An abridged version of the CAS, originally produced by the Alfa Romeo Owners Club, is available from Alfa Ricambi, and much of its information is included in the author's book on the Giulia.

1. Engine Modifications

The owner who thinks Alfa is stupidly designed to waste power has no business being around the car. And the owner who is certain he can double Alfa's output needs to think carefully about the trade-offs of his heroic ministrations.

Proceed with caution.

Alfa engines develop very nearly one horsepower per cubic-inch displacement, a figure domestic manufacturers occasionally crow about as if it were a nearly miraculous accomplishment. Alfa engines have been developing that kind of power/displacement ratio for a long time (the supercharged 6C1750 Gran Sport developed about 100 hp from about 100 cubic inches in 1929).

The 4-cylinder's race breeding makes it a willing subject for further power extraction, a fact established when Alfa itself hot-rodded the Giulietta engine into a ferocious little fire-breather called the Veloce.

Alfa"s elaborations to produce the original Giulietta Veloce make a literal road map for modifying any Alfa engine. The original Veloce featured twin DCO Weber carbs and individual-runner manifolds for the intake system, true headers for the exhaust system, an electric

Fig. 11-2. An Autodelta Alfetta engine with sliding-plate throttles. This type of 2-liter engine is capable of about 200 hp.

Fig. 11-3. *A streetable race car is the dream of many enthusiasts. This coupe attended 1991 Alfa club convention in San Diego.*

fuel pump, increased compression ratio, forged pistons, large aluminum sump, and lower rear axle ratio. Following the Veloce example, Alfa engine modifications need to be multifaceted rather than limited to one or two areas.

One needs to be circumspect about what the Veloce modifications produced. There was a definite penalty in driveability in exchange for the engine's virtually unbreakable willingness to run at electric-motor speeds. It started hard. Its lack of torque forced the driver to treat every stop light as a dragstrip Christmas tree since anything below 3000 rpm would probably stall the engine coming away from a stop. Finally, the Veloce came into its own only at patently illegal road speeds.

Which brings me to the first important point: modifications always involve a trade-off, somewhere. For emission-controlled Alfa engines, the most significant trade-off is a failure to pass smog inspections. For emission-free engines, the trade-off is typically fuel economy, driveability, or reliability. That does not mean that a modified Alfa engine is a flashbulb. The inherent strength of the Alfa engine makes it superbly reliable even when heavily modified for all-out racing; but the reliability is reduced, nonetheless, and the overeager owner can break a modified Alfa engine where a stock engine will keep running.

Most performance improvements made to an engine center on the head. Getting fuel in and exhaust out is the name of the game, and the more deep-breathing

Fig. 11-4. *Other early Alfa hot rods. Front car is a 1900 Corta Gara (short wheelbase racing) from the mid-'50s. Car at rear is another 1900 coupe, also for racing, with body by Zagato. Middle car is a Giulietta Sprint Zagato, with dual Weber carburetors, circa 1961.*

the engine, the more power it can produce. The intake and exhaust passages are smoothed and enlarged to increase the volume of air-fuel mixture that the engine digests, and metal is shaved from the head to reduce the size of the combustion chamber and increase the compression ratio of the engine. Different camshafts are used to hold the valves open longer and improve the engine's breathing.

The technical term for this is "volumetric efficiency." An engine is, basically, an air pump. The more efficiently it processes volumes of air, the more power it puts out. An engine with a displacement of 100 cubic inches, if it is an efficient design, draws in about 80 cubic inches of air; otherwise, it might draw in only 50 or 60 cubic inches. The reason an engine is not 100% volumetrically efficient is that there are a number of forces working to keep air from being drawn in: the basic restriction of the carburetor venturi itself; the throttle plate; the intrusion of the valves into the intake and exhaust paths; friction and turbulence of the moving gasses against the curving manifold walls; inadvertent blockages in the intake path from mismatched ports; and backward-moving pulses of gasses caused by the rhythmic opening and closing of the valves.

Most of the restrictions can be reduced in some way but virtually none of them can be eliminated entirely. The single most effective way of improving volumetric efficiency is to fit larger-capacity carburetors. That is what side-draft Webers are all about.

The mechanical preparation of an engine for racing requires a complete tear-down and inspection. Attention to detail and meticulous cleanliness are essential to assure reliability. All moving parts should be checked for cracks that indicate stress and will eventually prop-

agate to cause the part to fail. Iron and steel parts are checked for cracks using a magnetic field; nonmagnetic parts (aluminum or an alloy) are checked by being sprayed with a fluorescent dye and then inspected under ultraviolet light.

Alfa engines have proven very reliable under racing conditions. Given their heritage, that should not be surprising. While some kinds of engines have to be rebuilt after every other race, Alfas will typically run a whole season—or more—with absolute reliability.

Some racing classes do not permit any modifications to the engine, so read the regulations carefully before having at your engine. Even in those classes, however, there is no rule against careful assembly and "blueprinting" to assure reliability and the maximum horsepower permitted from the stock engine.

1.1 Head

The head, more than any other part of the engine, is responsible for the level of power the engine develops. The Alfa head is not especially strong torsionally. It warps easily and must be checked for flatness every time it is removed from the engine. Milling the head to increase compression reduces its torsional strength even further. A machinist's straight-edge and a feeler gauge set are required Alfa equipment for the racing mechanic. The factory-specification limit for warpage is 0.004 in. For racing purposes, I'd recommend milling to the limits of the machine's accuracy, which will be on the order of 0.002 to 0.001 in.

There are two machining operations typically performed on a head to improve its efficiency: milling it to increase compression and grinding away some of the combustion chamber so that each chamber holds exact-

Fig. 11-5. *Some modifications to this 101 Spider make a lot of sense. An alternator has been fitted for improved electrical system performance. Brake booster has been prettied up and air horns fitted. Webers and headers add gobs of go.*

Fig. 11-6. *Original Giulietta engine, seen here in its lowest-power configuration, left plenty of room for modifications. Single-throat Solex carburetor was appropriate for a Volkswagen, but not for Alfa. This is Barry Frantz's beautifully-restored sedan.*

Fig. 11-7. *A Giulia coupe engine bay shows what a serious race car looks like: spare, but very efficient. The missing oil filler cap says the owner's about to top up. The brake reservoirs have a heat shield wrapped around them.*

ly the same volume liquid (cc'ing the head). The purpose is to make each power stroke as uniform as possible. To the same end, some engine builders also index the spark plugs so that the gaps on the spark plugs all point towards the intake valve. This is a trial-and-error process: use a marker on the insulator to mark the

Fig. 11-8. *Not a casual effort: the PMA Milano Verde shows sponsor decals and is clearly equipped "for off-road use only."*

direction of the gap, then try a plug in a plug hole until you find one that lines up when torqued down. You may need more than four (or six) plugs to have a set that indexes correctly.

The GTA Alfas were equipped with dual-plug heads. If you can race with a dual-plug head and just have to win, then the GTA head is a worthwhile investment. Finding one is only a bit harder than paying its asking price, however, so be prepared.

For 2-liter cars, the current twin-spark head may be of interest. This is a stock Euro-spec Alfa item. Several have been imported into the U.S., primarily for racing. The twin-spark heads can be made to fit the 2-liter motor, so they are an attractive alternative. Similarly, there was supposed to be a prototype 3-valve head going into production any day now.

Camshafts

There are a lot of aftermarket cams available for Alfa engines. Carbureted Giulietta engines can be set up with hot cams and valve trains that will allow the engine to run quite reliably at speeds near 10,000 rpm. The larger displacement engines shouldn't be run so high simply because the inertial masses are greater. Generally, the larger the displacement, the lower the maximum permissible revs.

The big aftermarket cam suppliers who specialize in Alfa have thoroughly researched the question of power versus streetability and their recommendations should be carefully considered. If you're only going to drive your Alfa on the street, don't invest in a "full-race" camshaft because you'll quickly learn to hate what it does to driveability. A prominent authority on Alfa camshafts is John Shankle. Save yourself a lot of headaches and money by following his recommendations. Alfa cams vary in "heat." You can elect to stay with stock Alfa parts and still get improved performance. If you have a stock Giulietta and can obtain Giulietta Veloce cams, you will have obtained probably the wildest cams practical for street use. Remember, though, to follow the lesson of the original Veloce and make modifications to the entire engine. Veloce cams by themselves will not provide a significant increase in power. Giulia owners should look for Giulia Veloce cams for street use, or GTA cams for all-out performance.

Again, if you do fit hotter cams to a carbureted Alfa, it's imperative that you modify the intake and exhaust systems to provide improved gas flow capacity.

The original Veloce Alfas used 2-mm thick washers under the valve springs to increase spring pressure.

You can do the same thing on any of the later Alfa engines to provide an extra margin of safety from valve float. Don't use more than 2 mm without carefully checking that the springs don't close up all the way (bind) when fully opened by the cam lobe.

Owners of SPICA- or Bosch-injected cars should follow Joe Benson's recommendations for cam timing (see the SPICA Fuel Injection modification section below) or fit European-specification cams. Generally, wild cams don't work on fuel-injected cars. I'm told that camshafts for the twin-spark heads can also be used to good advantage.

Alfa is the only engine I can think of that allows an owner to change its camshaft timing for little more than a Sunday afternoon's tinkering. If you have the urge for just a little more power but no cash for a new set of camshafts, then just toddle out and increase the valve overlap a few degrees and advance the ignition timing until the car pings on heavy full-throttle acceleration. As long as you don't keep pedal to the metal all the time, you'll get a satisfying sense of having done something to improve performance and you'll not have to pay a reliability penalty, probably, for the extra power.

Valves

If you've got an Alfa engine torn down, my advice is to replace its valve guides. The same reasoning holds

Table a. Alfa Cam Characteristics
(In order of increasing performance/decreasing driveability)

Part Number	Application	Lift (mm)	Degrees Overlap
101-000-320-000	101 Normal	8.6	44
101-060-320-010	101 Veloce	9.0	64
105-020-320-000	Giulia	9.1	43
101-210-320-000	Giulia Veloce	9.5	59
105-020-320-001	Early 1750	9.5	67
105-480-320-001	2000 Europe	10.1	76
Shankle 6205		10.1	76
Shankle 5417		10.7	75
Shankle 5418		10.5	91
Shankle 5410		10.5	93
Shankle 5413		12.6	119
Shankle 5414		12.6	116
119-000-320-025	V-6 (Right)	9.0	61
119-000-320-125	V-6 (Left)	9.0	61
Shankle 5416	V-6	9.5	74

true for bearings: it's easy now, compared to later, if they fail.

Beginning with the Giulia, Alfa exhaust valves are sodium cooled and require special handling. Don't ever hit a sodium-filled valve with a hammer or grind on it. In a sodium-cooled valve, sodium metal is sealed in the hollow valve stem to help distribute heat away from the valve head. The tip of the valve is welded on to seal the hollow stem. If the sodium is exposed to air, it will spontaneously ignite and may cause very serious burns.

WARNING —
Use factory-approved methods for disposing of sodium-filled valves. Consult a factory repair manual or your local authorized Alfa dealer.

Though the head size of stock Alfa valves is already generous, it is possible to fit larger valves and seats. Of course, you have to seat the valves anew when you replace their seats, so that's reason enough to refinish them. A three-angle valve and seat is the standard for a racing engine. Alfa valves and seats are ground to 30°. The 750 Alfa had valves with 8-mm stems. Later cars all have valves with 9-mm stems. While it's quite possible for the enthusiast to replace valve guides, seat replacement is better left to a shop that is experienced and has the proper tool for extracting the seat from the head. Valves and seats from GTA Alfas are an obvious choice; if you can obtain an entire twin-plug GTA head, you'll have an optimum setup.

If you'd rather do some of the work on your heads yourself, you can have a shop put a three-angle cut on the valve and its seat (remember that Alfa valves faces are 30°) and then you can taper the intake valve guides a bit where they intrude into the inlet passage.

The V-6 changed to cast-iron valve guides and modified valves, and these later styles may have been retrofitted to other engines during an overhaul. Replacing these cast-iron guides is more difficult than bronze guides since the entire head must be heated before installing the new guides. Guides should be frozen overnight to shrink their outer diameter, then handled very quickly during installation before they expand again. Always refinish the valve faces after changing guides.

An Alfa multipiece factory tool is available to assure that the guides are driven to the exact depth. You can achieve the same effect with a simple driver just by driving the guides in carefully and checking them with a vernier caliper to assure they've been installed to the proper height. A basic valve guide driver can easily be turned on a lathe from a solid piece of steel stock. Make the pilot portion of the driver the same diameter as a

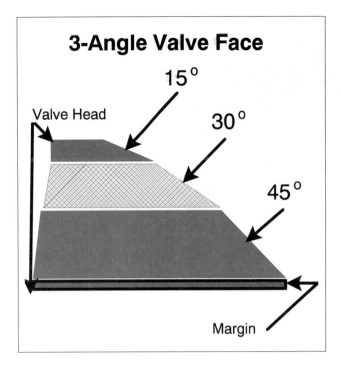

Fig. 11-9. *In a a "three-angle valve cut," face of valve is optimized for best compromise between gas flow past the valve, and size of the valve where it contacts the seat (which affects heat transfer from the valve).*

valve stem and be sure that the shoulder of the driver that bears on the shoulder of the guide is perfectly square.

Porting and Polishing

The reason for porting is a belief that stock inlet passages have too small a diameter for the actual breathing needs of the engine. While this may be a fact for many production engines, it is not demonstrably true for Alfas. Unless you have significantly modified the Alfa engine so it would clearly benefit from larger-diameter ports, leave well enough alone.

If you do decide to do some porting, the goal is to achieve a passage of uniform diameter with no sharp edges. A rotary grinder is used for the rough cut and that is followed, eventually, by hand polishing the surface to a mirror finish. The belief in polishing the passages is that it reduces surface friction between the flowing gasses and the rough surfaces of the casting. But research indicates that a highly polished port may actually reduce flow; the surface imperfections serve to break up the air flow (laminar flow, to get technical) near the outer edges so that the central flow of air is faster.

Though most engine rebuilds should not include porting, you should always carefully match the port openings between head and manifold to remove any

*Fig. 11-10. Top: A rotary tool makes it easy to match port in head to manifold opening. **Bottom**: When matching ports, work carefully and slowly. A lot of aluminum can be cut very quickly with a heavy hand. This is a first cut, which will have to be tapered back into the port to help gas flow.*

shoulders and obstructions to gas flow. Trace the outline of the manifold gasket on both the head and manifold mating surfaces, then grind away to the outline so that both passages match. As a finishing touch, sand the passages in both the head and manifold smooth with 400 wet/dry paper. The V-6 benefits especially from smoothing and matching the ports all along the intake path.

I've seen articles on reshaping the inlet passage near the valve guide to improve breathing. It may be possible to improve on the flow characteristics of the Alfa head, but such an undertaking requires a flow bench if it is to be at all effective. I advise against doing more

than a polish job unless you have a flow bench, and understand then that you might scrap several heads before you achieve any improvement over the stock Alfa design. If you do go at the inlet ports with a grinder, there is always the danger of cutting through the port into a water passage, in which event the head is immediately scrap.

Most racing engines have the intake manifold path enlarged or reshaped to the owner's specifications. Again, porting is not absolutely necessary on an Alfa, and you'll quickly find that different builders have very different opinions about the proper shape of Alfa intake ports. My personal advice is to select one builder and stick with his recommendations, ignoring all others.

A final thought on heroic hot-rodding measures. In the GTA, Alfa found that smaller inlet passages gave more power than larger ones. In mid-1967 Alfa reduced the size of the inlet ports on the Giulia. Simply enlarging the ports on these cars may result in a power loss.

Fig. 11-11. A GTA ready to go. Note the adjustable exhaust muffler, a necessary item for tracks that have noise limitations.

1.2 Headers and Mufflers

The original Veloce header was a hand-fabricated set of tubes that certainly set some kind of record for the most sophisticated "stock" exhaust system ever offered the public. The headers fed into a genuine Abarth exhaust system to provide a truly unrestricted exhaust flow; no baffles anywhere in sight.

Actually, most nonsmog Alfa exhaust systems (early normal Giuliettas and some sedans excepted) are true headers, pairing cylinders 1 and 4, and 2 and 3 for maximum efficiency, and the stock exhaust is a direct descendent of the no-baffle design of the original Abarth system. Alfa headers are cast, in contrast to the original fabricated Veloce manifold, but they are very efficient. Because of Alfa's excellent header design, there is a limited aftermarket for headers to fit pre-'75 cars.

Fig. 11-12. Stock Alfa manifold is actually a header system, with cylinders 1 and 4 going into one pipe and cylinders 2 and 3 into another. This is a complex casting which is prone to break, unfortunately, on number 4 runner.

Fig. 11-13. These powder-coated Euro-style headers add power but are not smog legal in many states. A pristine car.

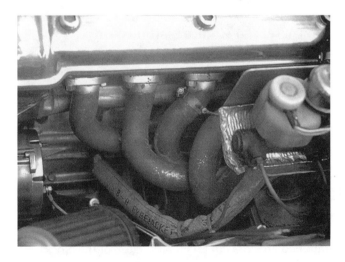

Fig. 11-14. Aftermarket large-bore headers take up a lot of room and give off a lot of heat. Note how brake line is insulated.

Pollution-controlled Alfas have a less-exotic exhaust system than the earlier cars, and headers will show a performance improvement when installed on post-'75 models. Shankle has fabricated headers that will provide in the neighborhood of a 10-hp increase over the rev range. The catalytic converter by its very nature is a significant restriction in the exhaust system. Removing it, however, significantly increases the pollution you add to the air I breathe.

Though stock Alfa exhaust systems are expensive, don't try to save money by installing a generic baffled muffler. I have found that a very serviceable exhaust system for the older cars can be fabricated using two long glasspacks and 2-in. straight pipe. The first glasspack is placed amidships and the other is located beside the fuel tank. The pipe must pass under the axle and will reduce ground clearance somewhat but the result is both quiet and efficient. If you're a stickler for such things, you can have a shop bend the pipe to pass over the top of the axle as in the original design. In this case, a two-piece assembly is required because a single continuous pipe can't be threaded over the axle.

1.3 Air Cleaners

It used to be fashionable to run without air cleaners, especially with Weber-carbureted engines. Ram tubes on air-cleanerless Webers are a beautiful sight, to be sure, and there was a pervasive feeling that any air cleaner (even an Alfa air cleaner) represented a restriction to the intake system that could be easily avoided.

Living in Los Angeles, one is impressed with the desirability of an air cleaner not only for the engine but for one's private intake system as well. The problem for an engine is no better even in smog-free environs. Airborne particulate matter is an abrasive that will prematurely wear rings and valve guides. For that reason, an air cleaner is truly a necessity.

Alfa's air cleaners have quite a large capacity, considering the diminutive displacement of the engines. If you feel the stock system is too restrictive, there are several low-restriction aftermarket units that have been used extensively and successfully on racing engines.

John Shankle feels that the stock fuel-injected air cleaner does, in fact, either detune the system or offer enough restriction to lower performance. If you're after a maximum-performance engine using SPICA fuel injection, then John's Quadraflow filters are recommended.

Note that you cannot change the air cleaner design on Bosch fuel-injected cars (i.e., most Alfas since 1981), though you can probably install an aftermarket high-flow filter from K&N and other suppliers.

Most air cleaner systems on older Alfas have the cold-air inlet tube missing. On all fuel-injected cars and true Veloces, there is an inlet tube from the front grille area of the car to the air cleaner box. This tube is there to

Fig. 11-15. *A Giulietta Sprint Zagato long-tail with short "Sebring" exhaust showing just below door. This was a cast-alloy, oval megaphone that helped extract exhaust gases with-* *out minimizing ground clearance. The sound was especially ear-splitting.*

provide oxygen-rich cold air to the intake system. When the inlet tube is removed, the engine breathes warmer air from the engine bay. The warm air is less dense and therefore contains less oxygen (to burn with fuel) than cool ambient air. For this reason alone, you should always try to provide an ample source of fresh

Fig. 11-16. *The cold air tube: if it's not there, it should be, for best performance.*

air to the intake system. In addition, most of Alfa's cold-air intake systems are fairly tightly sealed and I suspect that they may provide a slight ram effect at high speeds. Alfa used ram intake air on several of its older race cars and I wouldn't be surprised if there is some slight ram effect designed into the system.

1.4 Increased Compression Ratio

Alfa's robust crankshaft and rods will support a significant increase in compression without sacrificing significant reliability. The limiting factor for increased compression is now gasoline quality. Probably a compression over 10.0:1 will cause pinging on any pump fuel you can find, unless your car is equipped with a knock sensor. Don Black, the well-known Alfa engineer and enthusiast, is running an Alfa 4-cylinder engine at 14.5:1 using methanol fuel.

Increased compression can make a dramatic difference. In the early '60s, a Giulietta owned by Paul Tenney had its head shaved right to the valve seats, and Paul had an extra cut taken off the top of the block and cylinders for good measure. As I recall, he estimated the final compression ratio to be around 14:1. The car went like a rocket and when I first drove it I presumed it was a modified Veloce. You can imagine my surprise to find a single 35APAIG Solex carburetor bolted to the engine.

The stock thickness of Giulietta and Giulia heads is 4.4094 in. Although the factory asks for a minimum head thickness of 4.3898 in., a Giulietta head can be safely shaved to within about 1/16 in. of the valve seats, but the resulting head must be regularly and carefully torqued to keep the head gasket from leaking and special care must be used in handling the head to keep it from warping. Extra washers should be used under the head bolts to assure that the bolts don't foul the ends of the normal threads. If you've overdone the milling operation, you can lower the compression ratio of the engine by fitting a thick copper-clad gasket available from several Alfa stores.

The preferred method for obtaining increased compression, however, is to fit higher-compression pistons rather than shaving, and thereby reducing the resistance to warpage of, the head. Most Alfa stores can obtain pistons with compression ratios as high as you dare. Note that Alfas built beginning in winter 1971 have their wrist pins offset in the piston. As a result, these pistons should be installed paying special attention to "front" and "back" indications stamped on the piston crown. The offset is counter to crankshaft rotation, giving a slight mechanical advantage to the piston at top dead center.

If you have a true Giulietta Veloce, don't ever fit cast Giulietta "normal" pistons. The Veloce forged pistons have a lower compression height than stock Giulietta pistons, but the Veloce head is thinner by about 0.040 in.

Any change in compression should be accompanied by rejetted (richer) carburetors and an ignition advance with a steeper curve than normal.

1.5 Block

Typically, the block is not modified, except to drill the second and fourth main bearings for direct oiling. It is necessary to verify that the cylinder liners stand adequately proud of the block deck, but beyond cleaning the block and freeing its "wet" area of any coolant deposits, nothing needs to be done to its excellent design.

In the stock Alfa, each cylinder liner is separate. The 1750 GTAm was fitted with liner sets cast together as one to provide a little extra rigidity between cylinders and, as well, to increase displacement to 2 liters. This "monosleeve" effort was probably overkill. There's no evidence that the stock liners are not adequately rigid as designed, even in racing applications.

It is desirable to glass-bead the rods (versus shot peening) to enhance reliability. Similarly, meticulous balancing of all moving parts will help ensure reliability. The standard checks to assure crank bore accuracy, crank journal parallelism, and freedom from stress cracks are all very necessary when dealing with highly stressed racing engines. Since this is not a shop manual, I can safely skip the details of these standard checks.

A lightweight flywheel improves engine response and can actually aid control when cornering under power at the limit of adhesion since the engine responds more quickly to throttle inputs. The goal for a racing engine is to lighten the flywheel just short of weakening it to the point that it disintegrates under a shock load at maximum rpm. While you can machine the stock flywheel so it loses about six pounds, an aluminum flywheel will give you the same weight loss with much less effort and a larger safety factor.

High-capacity oil pumps are available for Alfas from Autocomponents, as are auxiliary oil coolers. A racing Alfa probably needs an oil temperature gauge. If your Alfa's oil is running too hot, then any number of oil coolers can be used to lower running temperatures. The stock Alfa oil system is already fully race-capable, however, and auxiliary units are probably not necessary for the weekend racer.

There are dry-sump conversions for Alfa, also from Auto components and other suppliers. Some of the factory cars ran dry sumps, but the expense of such a system and the excellent design of the stock system makes a dry sump conversion of questionable value for most applications. Early Giuliettas with stamped-steel oil pans will benefit from the addition of a windage tray to assure a steady oil supply. All cast-aluminum Alfa sumps have windage and cooling provisions built in, but some racers still elaborate on their designs by adding extra baffles, doors, or screens.

Increased Displacement

An early modification to Giulietta engines was to fit a 1500 cc big-bore kit that takes advantage of Alfa's removable liners. The conversion requires boring the block to accept the oversize liners, and that's a machining operation that requires absolute accuracy. It's unlikely that any of these kits are still available.

A second big-bore conversion that permits a standard block is simply to bore out the stock liners to obtain very nearly 1400 cc and then fit custom pistons by Jahns or some other aftermarket piston source. This is actually a very neat modification if the cost can somehow be kept in line, and a similar approach can be used to save a Giulietta or Giulia engine when its existing liners are worn oval and no new liners are available.

In those instances when the liners are good and the only thing wrong with the Giulietta piston is excessive top-groove clearance, have a machinist cut a wider top groove and use Simca 2-mm compression rings to replace the stock 1.5-mm top rings.

It's not possible to use larger-displacement stock Alfa liners in smaller engines unless the block is bored to accept the new liner and the heights of the liners are reduced.

2. Fuel Delivery Modifications

2.1 Weber Carburetors

There is not an Alfa owner alive who, bereft of proper Weber carburetors, does not covet a set. Twin 40DCOEs on an Alfa is one of the rare intimations of automotive nirvana.

Of course, Webers have been associated with high-performance Italian machinery for very many years and the desirable little Giulietta Veloce with its twin DCO carburetors indelibly etched a Weber desire in the true Alfista. The fact that the carburetors are almost the same size as the engine itself suggests enormous power.

Webers certainly do make a difference over a single-Solex setup. Bolting twin Webers to a Solex-equipped Alfa makes the engine wonderfully deep-chested, even without a change of cams. In comparison, the Solex carb seems to begin wheezing in the mid-5000 rpm range. The reason for this transformation is simple: the venturi area added by the Webers is roughly twice that of the Solex, and the Weber intake path is certainly the most efficient possible. If you have a Solex-equipped Alfa, you should want a pair of Webers. For a period of time that achievement was not difficult. When the Giulia was current, Webers were relatively plentiful, and there are probably still a lot of Weberized Giuliettas masquerading as Veloces. But, in the last few years, the 40DCOE Weber has become harder and harder to find. The scarcity was exacerbated for about a year because Weber simply stopped making them.

From an Alfa standpoint, the real source of 40DCOE scarcity was the introduction of the SPICA fuel injection system and Alfa's insistence on a complete absence of service parts for it. If your SPICA pump failed, you could purchase a set of used Webers for about half the cost of replacing the SPICA pump. Many Alfa owners converted from SPICA to Weber defensively, and virtually all of those converting reported a real increase in performance.

Fig. 11-17. Six-cylinder 2600 cars of the '60s sported three Solex side-draft carburetors and respectable performance, though engines were not really highly stressed. A Weber-carburetor option improved performance.

Fig. 11-18. A Giulietta Veloce with its plugs removed. An air cleaner should be located on the firewall. Air path, however, is correct: from driver's side, across engine, to carburetors. Pristine engine compartment and breather tube from oil filler cap suggest that this car is raced regularly. Note headers.

Fig. 11-19. The ultimate road-going hot rod, this Tubolare Zagato Alfa features a highly modified Giulia engine in a chassis that includes independent rear suspension. TZ is one of the most desirable Alfas, ever.

Subsequent experience indicates that there is no real performance increase gained with the conversion. The probable cause of the claim was that the owner was replacing a poorly tuned fuel injection system with a properly tuned Weber set-up. For whatever reason, Weber's lowered production and the greed of SPICA-phobes has effectively dried up the Weber supply line.

There are alternate carburetors. The dellOrto and Mikuni are both Weber work-alikes that are fairly easily obtained. The dellOrto is a simple conversion since it is the carburetor used on European Alfas. Used sets of dellOrtos run about $400 to $700 with manifold. New sets are about $900. Though I have never personally seen a Mikuni setup on an Alfa, there is no reason those carburetors should not work very well. Entire Weber conversion kits are available. DellOrto carbs are available from Alfa Ricambi and Mikunis are available at stores that specialize in Toyota racing supplies.

Since European Alfas were equipped with dellOrtos, carbureting a fuel-injected engine amounts to little more than returning the car to its Euro specs using stock Alfa parts. Sources such as Tom Zat used to obtain entire used assemblies directly from Europe. If you're putting carburetors on a SPICA fuel-injected engine, it's important to blank off the oil feed line to the fuel injection pump with a very sturdy aluminum plate, for a thin plate will be deformed by the oil pressure and eventually leak. Several aftermarket suppliers stock a predrilled blanking plate and can also supply any of the bits and pieces you didn't get with the carbs and manifolds. I also recommend replacing the high-pressure fuel pump near the gas tank with a low-pressure aftermarket unit. There is no sense in running the high-pressure pump unnecessarily. It is an expensive item better saved if not absolutely needed, and using a low-pres-

sure pump eliminates the need for a pressure-reducing valve somewhere in the fuel line to the carburetors.

A brand new set of carbs with adapter plates for the SPICA manifold is available. If you have a serviceable set of carbs, you can buy just the adapter plates and fasteners needed to complete the job. A Weber conversion is illustrated in the author's book on Weber carburetors.

Racing Carburetion

While you may want—even long for—Weber carburetors on your racing Alfa, the class rules may not allow it. Whether you're running Solexes or Webers, however, you can most certainly rejet for maximum performance, though you may not be able to change the size of the venturi. In general terms, you want to run as rich as possible, just short of fouling the plugs. A rich mixture keeps combustion temperatures lower, helps cool the exhaust valve by dumping unburned fuel past it, and generally improves reliability. If you're running a catalytic converter, however (say, on a true dual-purpose car), the rich mixture will also quickly incinerate your converter. Lean mixtures, on the other hand, rob power and increase combustion temperatures.

Also, you may find that whopping big venturis, and plenty of them, will not give you the performance you expect, especially if you're racing on short, tight circuits where the engine runs low in its rev range. This is especially true of cars used for gymkhanas and similar parking-lot heroics. A lot of venturi is good only if your engine plans to live at wide-open throttle. If you typically run between 2000 and 4000 rpm, then smaller overall carburetion will give you improved low-end response and better fuel atomization. Modern theory suggests that it is better to run multiple small venturis rather than a single, large one to keep the air/fuel velocity high even at low engine speeds. Certainly, a higher velocity in a carburetor gives a stronger signal and improves throttle response.

The Solex carburetor fitted to Giuliettas and Giulias is a very satisfactory unit, though you may be able (check the rules!) to exchange it for a Weber. The Solex/Weber swap used to be very popular. The Alfa owner and Fiat 1200 owner were both certain that the replacement carburetor improved performance.

The rules may not allow you to replace a fuel-injection system with a carburetor for racing. If you can, then dual 40 DCOE Webers are fine for most applications. The stock venturi in a 40 ranges from 28 mm for the Giulietta, 30 mm for the Giulia, and 32 mm for the 2-liter. If you're running a 2-liter at consistently wide-open throttle, you might explore 45 DCOEs, but the 40s will probably be fine with slightly larger venturi (try 34 mm).

2.2 SPICA Fuel Injection

The remainder of this section is the last of Joe Benson's series on the SPICA system originally published in the *Alfa Owner*. Though it does not limit itself to modifications of the intake and fuel injection systems, it's included in its entirety here because those interested in modifying a SPICA system for improved performance

also need to know this information. Don't get your hopes up. We won't be resurfacing the 3-dimensional cam. Joe's position is that the stock settings for the SPICA system are perfectly adequate for most requirements. What he offers is much more valuable in practical terms: what you can reasonably do to get maximum performance from a fuel-injected engine.

You can increase the breathing of a SPICA fuel-injected car by enlarging the venturis and fitting larger throttle plates. With this modification, I'd suggest readjusting the long link to maintain mixture strength. Lengthening it enriches and shortening it leans the mixture strength.

Engine Fan

Believe it or not, the largest single performance gain is to be had by modifying the engine fan. You gain about 7 hp at 5000 rpm, and the change doesn't affect any emission equipment. Remove the fan from the water pump pulley (usually requires removal of the radiator), clamp it in a vise, and cut off all but two symmetrically opposite blades. File down the rough stubs to the hub with a coarse file. Leave the radiator fan shroud on to assist cooling. If the car overheats, replace the fan with a proper six-bladed one immediately or you run the risk of a warped head and a blown head gasket.

Fig. 11-20. Coupes do very well in Alfa club events. Here a GTV6 corners hard.

Ignition

Set the ignition timing at idle at the "P" mark (TDC) rather than at the retarded, specified "F" timing mark. To be conservative, you might set it about 1° retarded from TDC. For street usage, you'll notice crisper response and starting and a boost in gas mileage. However, this change does increase your octane requirement and I would recommend the use of premium fuel. If you get any pinging, either switch to better gasoline or retard the ignition closer to the "F" mark or you'll melt pistons. This especially applies if you plan to compete under track conditions of full throttle and high rpm.

In my own experience, I have had excellent success with Champion N6Y spark plugs as an all-around plug for any fuel injected Alfa. In fact, I have "cured" quite a few Alfas of rough idle simply by installing N6Y plugs. I would also like to stress the importance of maintaining the entire ignition system in good condition. Pay particular attention to maintaining tight mechanical contact of the cable terminals in the distributor cap and on the plugs, make sure the plug caps are firmly threaded on (use a pliers) and maintain a high degree of cleanliness, especially on both the inside and outside of the distributor cap.

Cam Timing

True precision cam timing requires the use of a dial indicator to measure valve lift, but since very few have the necessary equipment I'll present a simpler but effective method for performance timing the camshafts.

Since it causes the most confusion, I'd like to begin by defining the terms advance and retard. When standing in front of the car the engine crankshaft and camshafts normally run in a clockwise direction. Advancing either the intake or exhaust cam means just that—turning it further in a clockwise direction from its normal timing. For retarding, you turn the camshaft counterclockwise from its normal setting. See Fig. 11-21.

For performance timing, we want to increase the valve overlap by advancing the intake cam and retarding the exhaust cam. Again viewing from the front of the car, this means the intake cam is rotated further clockwise and the exhaust cam is rotated further counterclockwise from its stock timing marks (Table B).

> *CAUTION —*
> *I recommend stuffing rags under the sprockets while timing cams to catch any parts you may drop. Then, remove the rags and roll the engine slowly in gear before cranking to ensure there is no piston-to-valve interference.*

By inspection of the timing marks, adjusting the cams should appear to have turned both cams inward toward the engine centerline. Each vernier hole step rotates the cam by 1.5° or about half the width of the cam timing marks. The five-hole change should move the

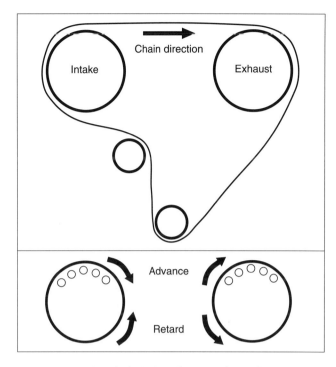

Fig. 11-21. Camshaft timing advance and retard.

cam timing mark about 1/8 in. from the stock timing mark on the cam bearing, or 7.5°. If the cam timing marks are further apart than this, you did something wrong—try again.

Camshaft Selection

Before running out and buying "hot" cams for your car, I urge you to first try the modifications listed above. The 2000 especially is quite satisfactory even with the mild stock cams and will give gas mileage that shames most so-called "economy" cars. But for the power hungry, here's what will work.

Three cams are suitable for use in fuel-injected cars: the stock 1750 cam (P/N 105020320001), the stock 2000 cam (P/N 105200320000) and the European 2000 cam (P/N 105480320001). The European cam is the hottest cam for the injected engines. It is so effective with injection because it combines the moderate duration of the 1750 cam with increased lift. This cam will noticeably increase acceleration through the gears and allow you to rev right up to engine destruction in fifth gear. (Several suppliers have additional hot cams for SPICA fuel-injected cars.) The 1750 cam will significantly pep up a 2000 engine at less expense. Modifications work both ways, too, and I can envision an economy-conscious 1750 owner trading cams with a 2000 owner who's trying to keep up with his neighbor's 911 Porsche.

The old 1600 Veloce sport cam will not work well with SPICA injection. The long duration of this cam is apparently incompatible with the stock timed injection and the result is an overall performance loss.

Table B. SPICA Fuel Injected Cam Timing Changes

Model	Amount and Direction		Result
	Intake	Exhaust	
1750	+ 1	− 1	Performance
2000 up to 1974	+ 4	0	Stock 1750 specs
1974 2000	+ 5	− 1	Performance
1975 2000	+ 4	− 4	Stock 1750 specs
Alfetta	+ 5	− 5	Performance
Listed values refer to the number of vernier holes camshaft should be rotated from stock. "+" means advance, "−" means retard and "0" means no change.			

SPICA Pump Modification

Once inside the pump, its workings suggest other possibilities for increasing performance for those who use their cars off-highway where modifications are still allowable. There are a few ways the SPICA unit can be modified to provide improved fuel delivery for modified engines:

- Over-richen the main richness adjustment at the cold-start solenoid
- Substitute a different injection pump 3-dimensional cam
- Lengthen the long link to uncalibrate throttle plate information to the pump
- Readjust the pump linkage to simulate a higher-flow fuel cam

The first is the most frequently-used and least desirable solution. Over-richening can provide the higher rpm flow required, but the midrange suffers and idling will tend to foul the plugs. Oil dilution and cylinder wall wash-down can also result.

The second procedure is a satisfactory but limited alternative since the hottest fuel cam available is the 1974 production unit. To reach this conclusion, I investigated using a dial indicator to read the movement of the pump linkage that directly controls the position of the richness rack in the pumping section. Linkage position was recorded for every 5° of pump throttle control lever rotation at the low-, mid-, and high-rpm cross sections of the fuel cam.

To date, I've only checked the 1969, 1974, and 1976 pumps. From the curves, it is obvious that the 1974 cam has about 14% more high-rpm, full-throttle fuel delivery than the 1976 cam, while it is quite similar in the other ranges.

Happily, all pump cams are interchangeable, with the exception of the 1969 unit.

I've included adjusting the long link in Joe's list simply because it's an easy way to get a bit more fuel without tampering with anything else. (If it doesn't work for you, it's easy to reset the link to the proper specifica-tions.) Lengthening the long link will make the pump think that the throttle plates are open more than they actually are. Put another way, it rotates the 3-dimensional cam slightly toward a richer setting. This does move the richness curve out of synch with real engine needs, and if you make the link too long, you'll probably lose the fuel cut-off function because the switch arm will never find the fuel cut-off hump on the cam.

Joe's other approach, changing the barometric sensor, also fools the pump but keeps the fuel curve congruent with stock engine needs. If, however, you've modified the engine, its real needs will be different from what the pump is set up for anyway, so either adjustment is worth trying. Under any circumstance, I'd work slowly and use a pyrometer to keep from frying the engine.

The last technique, readjusting the internal pump linkage, is the truly elegant solution—the dial-a-horsepower approach. There are several points inside the pump where pivot point relocations or link length changes could alter the flow characteristics, but such an approach requires pump disassembly, precision machining, and any errors would be magnified by the geometry.

The desired procedure is an easily performed external adjustment that affects primarily top-end fuel flow. The answer is so simple and obvious that once discovered it appears almost trivial: misadjust the barometric sensor bellows to fool the injector into thinking it's at a lower elevation. (This change requires removal of a catalytic converter if one is fitted.)

Note that the altitude specification at sea level for all post-1969 pumps requires the spring clip (viewed through the rear inspection port) to sit in the seventh notch from the top of the vertical notched lever.

Spend a little time observing the operation of this vertical lever—specifically that it rotates fore and aft about its base pivot and that a rearward rotation pulls the horizontal compensator link and spring clip rearward with any pump input that calls for increased richness.

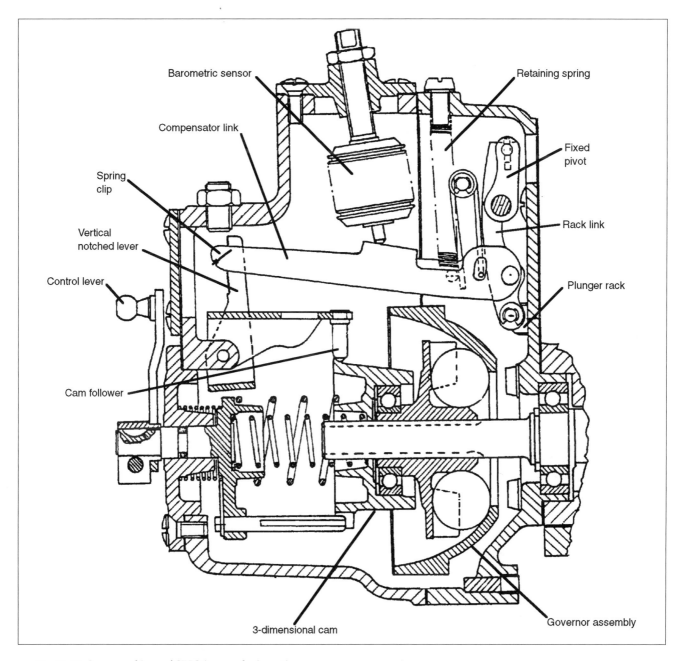

Fig. 11-22. *Inner workings of SPICA pump logic section.*

Consider the longer stroke of the compensator link that results from the spring clip setting in the highest notch in the vertical lever. A given fuel cam can only provide a certain maximum rotation of the vertical notched lever, but by raising the spring clip higher on the lever you achieve a longer lever arm and thereby greater plunger rack movement and increased fuel flow.

The vertical notched lever may also be considered a form of variable mechanical amplifier since it translates a small rise in fuel cam profile into a larger movement of the richness or plunger rack. Our modification simply adjusts it for maximum gain.

Finally, a few observations to facilitate pump recalibration and assist in final setup:

• Adjust the altitude compensator so the spring clip sits in the top or next-to-top notch of the vertical notched lever. Check adjustment by manually running the pump through a sequence in which automatic altitude compensation normally occurs.
• The external stop screw directly above the vertical notched lever may have to be readjusted to prevent the forked arm of the compensator link from hopping off the tip of the vertical lever. Avoid binding the arm.

CHAPTER 11

- The temperature compensator lever feature of the older pumps is a direct bolt-on substitution for the later pumps and provides a quick, external adjustment to the fuel curve when properly set up.
- The optimum fuel cam and vertical lever notch position for a specific application is best determined with a blend of intuition and experimentation. Spark plug analysis, a CO meter, and the Color Tune spark plug tool would be good aids.
- Even more upper end enrichment is available by extending the length of the vertical notched lever. A friendly machinist, and pump disassembly, is required.
- With a little inventiveness, the injection pump fuel flow could be made responsive to boost pressures when a turbo system is installed.
- All other injection linkage and gaps should be set absolutely stock for best results.
- Any time the temperature compensator lever is turned to change the fuel curve, the main mixture must be readjusted for best results.
- Be aware that the bellows is still subject to changing with barometric pressure as well as altitude, so don't be surprised if your settings shift slightly.

2.3 Bosch Electronic Fuel Injection

To understand the Bosch unit, I'd like to begin by comparing three ways Alfas have been fed fuel: carburetor, SPICA mechanical injection, and Jetronic Bosch electronic injection.

A carburetor monitors the engine's operating condition only passively, and really uses only air flow through the venturi to understand what's needed (chokes and accelerator pumps are only slightly less passive exceptions). Because it works so passively, the carburetor requires an intricate series of circuits to provide proper fuel metering. Switching from one circuit to another depends solely on fluid dynamics. A carburetor is inefficient in part because it senses engine needs so poorly, and also because its fuel-metering accuracy is compromised to achieve simplicity.

Carburetor Logic

Input	Processor	Output
Air flow Throttle motion (accel. pump) Engine temperature (choke)	None	Air-fuel mixture

The SPICA system measures several engine parameters independently. It translates the measurements into the fore-aft motion of a rack inside the fuel injection pump. This rack controls the volume of fuel injected into the engine. The SPICA logic is completely mechanical. Input to the SPICA unit includes atmospheric pressure (barometric capsule), coolant temperature

(thermostatic actuator), engine speed (governor assembly in the logic section), starter-engaged (cold-start solenoid), and ambient temperature (season lever on the top of the injector pump).

SPICA Logic

Input	Processor	Output
Throttle position Engine temperature Engine speed Barometric pressure Temperature range Starter engaged	Mechanical	Fuel to injectors

Like the SPICA system, the Bosch EFI measures several engine parameters. Instead of using mechanical sensors and logic, it uses electrical sensors that deliver signals to an electronic control unit (ECU) for processing. The ECU processes the information and outputs a pulse to each fuel injection solenoid, releasing pressurized fuel into the intake manifold.

Bosch EFI Logic

Input	Processor	Output
Battery voltage Ambient air temperature Throttle position Engine speed Engine coolant temperature Starter motor signal Intake air flow Exhaust oxygen	Electronic Control Unit (ECU)	Injectors open and close; fuel pump power

Only a glance at the three charts shows that the electronic system gathers much more information (input) about the engine's operating condition. The fact that the information is processed electronically means that no allowances are needed for inertia or friction in the logic section. By monitoring exhaust oxygen, in fact, the system is self-correcting in setting mixture strength.

Diagnostic and failure-mode routines are built in to help troubleshoot the system as well as to get you home if something does fail.

It's important not to get thrown off base by the fact that you're confronting an electronic system. It still does the same thing (feed fuel) that a carburetor does. Thus, there are some carryover skills you can use to help in diagnosis. If the trouble with your car only occurs when the engine is cold, then the problem is clearly located in the sensors and solenoids that take the place of the choke circuit to help the engine run when it is cold. If the problem is a high-speed one, then the full-throttle switch is suspect.

The ECU is a small but very expensive black box. It's natural to leap to the conclusion that it's bad. Provided you haven't fried any chips by incautious probing, it's every bit as durable as your Walkman. And its connec-

Fig. 11-23. Nitrous oxide injection, here on a Milano, is absolutely cheapest way of adding gobs of power to your car. It's also a guaranteed way to destroy the engine if you have no restraint.

With Bosch injection, nitrous is generally set to inject under full throttle. Note nitrous pressure regulator at lower left, and injector installed in intake bellows just before throttle valve.

tors are keyed, so it's very difficult to hook them up wrong. Relax. It's probably not the ECU.

The inputs and outputs of the EFI system are all that the novice can test. The logic of the ECU belongs to Bosch alone, and its internal workings are proprietary. However, there are some things about it that are generally known. Most ECUs are designed to accommodate several engines. The item in the ECU that customizes it for a particular application is a Programmable Read-Only Memory (PROM) chip and it is replaceable. By reprogramming the PROM, the fuel-delivery characteristics of the system are changed. The contents of the PROM are available to anyone who is experienced in "burning" chips and it is quite possible, through the PROM, to reprogram a custom fuel delivery curve. Reprogrammed PROMS are available for many EFI units, but they are somewhat hard to find. Alfa ECUs with modified PROMs are available from some aftermarket suppliers such as Alfa Heaven, and from Tom Zat on a replacement basis. With the Zat unit installed, the rev

limit is raised to approximately 7200 rpm and extra fuel enrichment begins at about 3500 rpm.

The oxygen sensor forces the Bosch unit to provide the proper air-fuel ratio over a very wide range of engine operating parameters, so the stock Bosch system can be used on turbocharged engines. Since the amount of fuel delivered in the Bosch system depends in part on its delivery pressure, increasing system fuel pressure will enrich the air-fuel ratio overall providing the oxygen sensor is disabled. This is a make-do fix, but it may be adequate for some special cases.

For the average Alfa owner, however, the Bosch unit is about as user-serviceable as a desktop computer. That is, there is little you can do except verify that all the connectors are tight and replace suspected components with known good ones. However, Charles Probst's book on Bosch Fuel Injection, published in 1989 by Robert Bentley, explains Alfa's L-Jetronic system in sufficient detail that you should feel confident working on the system.

2.4 Turbo/Supercharging

The love affair between superchargers and Alfa Romeo dates back to 1923 when Merosi designed the P1 Grand Prix car with a Wittig vane-type supercharger. A Rootes-type supercharger was standard equipment on the most sporting of Jano's passenger car engines up to about 1937.

A supercharger is simply a mechanical pump that crams more combustible mixture into the engine than it could pull in on its own. In effect, a supercharger is nothing more than a method of artificially increasing the engine's displacement by forcing its volumetric efficiency to exceed 100%. The drawback of the supercharger is that the energy needed to turn it is subtracted directly from the engine's output. Thus, though a supercharged engine develops more horsepower than a nonsupercharged one, it does not develop nearly as much as is theoretically possible.

Because of the expense of supercharging, the practice of supercharging passenger cars has been followed only occasionally, and then on "pacesetter" models. If you want to supercharge your Alfa, there is a roots-type supercharger kit, and the Fiat Volumex unit can be made to fit.

The modern jet airplane engine uses a new generation of metallurgy. These improved metals have provided engineers with a method of supercharging an engine using the considerable energy of the engine's exhaust gasses. In fact, turbochargers were first used on airplanes to overcome the low-oxygen conditions of high altitudes. The turbocharger places durable turbine blades in the path of the exhaust gasses and uses them to drive a similar turbinelike compressor to supercharge the intake of the engine. The turbine turns at speeds around 100,000 rpm and withstands exhaust temperatures that would quickly melt lesser metals. The turbocharger has the additional advantage of taking very little power from the engine, since it is not mechanically driven off the crankshaft as is the supercharger.

The biggest single drawback of the turbocharger goes directly to its turbine characteristic: some time is required for the turbine blades to accelerate to speeds

Fig. 11-24. *A turbo engine in the New Giulietta, not available in the U.S. This would have been a great hit, with 175 hp compared to 130 for normally aspirated engine.*

Fig. 11-25. *A few of the 1980 GTA cars were turbocharged, the turbos driven by oil, not exhaust pressure. This gave advantage of more consistent boost. Large fabricated plenum contains Weber carburetors.*

Fig. 11-27. *Engine bay of Ashton's Junior Zagato. Turbo is mounted to exhaust manifold and its outlet snakes in front of the engine.*

where they become an efficient compressor. This is the well-known "turbo lag," the time between flooring the accelerator and the onset of intake system pressurization, or "boost." A turbocharger is not an ideal attachment for a variable-speed engine.

Well, nothing about a car's engine is ideal anyway. In the mid-'80s, turbocharged Alfas were all the rage. The three primary sources of turbocharged Alfas were John Shankle and Jafco's Gary Fortner on the West Coast and Callaway on the East Coast, which has gone on to make quite a name fitting twin turbochargers to Corvettes. All these systems were very serviceable and overcame the dreaded turbo lag with varying degrees of success. The intake systems were boosted by 6–12 psi. Fortner's installation retained the SPICA fuel-injection system and Shankle's fitted a DCOE Weber.

I have driven cars turbocharged by both Shankle and Fortner. Both offered a truly significant performance increase and were still docile enough for every-

Fig. 11-26. *Jim Ashton's Junior Zagato is a regular sight at West Coast events. It's a formidable competitor, with a Jafco-turbocharged engine.*

day driving. The Fortner car was running a boost of about 7 psi and developed about 150 hp from the 2-liter engine.

John Shankle's turbocharged GTV was running a boost of about 12 psi and offered about 200 hp. It was an absolute rocket and still proved outstandingly reliable over the many thousand miles John logged with the car, proof of the Alfa engine's essential reliability. The higher boost pressures required a significant change in driving habits, however. At low engine speeds, opening the throttle quickly would stall the engine. A slow, steady foot was required to get under way. Once under way, however, I had more power at my command than in any other Alfa I have ever driven.

Most turbo systems include water injection to help control pre-ignition.

A modern turbocharger kit with intercooler and EFI is available from Jim Steck at Autocomponents, and you can still obtain the set-up used on European 1.8-liter Alfa 75 turbos. From discussions with both Shankle and Fortner, I can assure you that the selection of a properly sized turbocharger for a custom installation is a difficult process that involves quite a bit of experimentation. The proper location of the turbo in the engine bay is a point of contention. Should it be on the exhaust or intake side? The relation between the sizes of the two turbines is ultimately critical. In addition to that, some method must be devised to provide a suitably enriched air/fuel mixture to the engine. Finally, if the turbo installation is truly successful, you will find yourself needing a different rear axle ratio to enjoy its advantages fully.

It is quite easy to cobble up a turbo system that will turn your engine into a grenade at first full throttle. I advise against the effort unless you can obtain a preengineered kit such as the Alfa 75's turbo system.

I need to note that Alfa did create the GTA-SA, a twin-supercharged race car based on a GTA coupe. Instead of being driven by exhaust gasses, however, the two compressor turbines were powered by a variable-displacement wobble-plate pump. The pump oil volume was controlled by a SPICA pump on early cars but Webers were used later on. Water injection was used to reduce combustion temperatures. The car ran in Group 5 in England.

3. Transmission/Differential

Aside from picking from the few alternate ratios available for Alfa, there is little you can do to enhance either the transmission or the differential for racing. Close-ratio transmission gearsets are available from aftermarket suppliers. For most racers, the stock gears will do.

You can lower the oil resistance in the gearbox by using 40-weight engine oil. If the lower viscosity makes you nervous, then a can of STP and/or a tube of molybdenum disulfide will add an extra margin of lubricity.

I recommend against using a lightweight oil in the differential unit. Indeed, a 140-weight lubricant seems appropriate for racing. The Montreal Alfa used a stock Giulia rear axle with an extra oil sump bolted in place and larger-diameter axle shafts. It's always seemed to me a perfect modification to fit a Montreal sump to a racing Alfa's rear axle.

4. Wheels/Tires

Generally, amateur racing allows a choice of wheel/tire combinations so long as the new shoes are the same approximate size as stock. Since there is a choice, there is an active rumor market about hot wheel/tire combinations.

If there are no regulations, then wider is better unless you're running in rain or snow, when narrower is better. On some Alfas the rear wheel wells will restrict

Fig. 11-28. Alfa GTAs swept July 8, 1970, Italian Championship race at Magione, taking first five places. This is a 1300 cc GTA Junior.

Fig. 11-29. *Derek Bell Takes first overall at 1982 Donnington Park weekend. This is cornering at the limit: note lift of inside front wheel.*

your maximum tire size. Measure carefully and remember that the tire moves up and down during a race!

There is a difference in tires. Modern speeds are wholly dependent on tire technology, and the top speeds of race cars have increased more from tire technology than any other single source. Of course, tire manufacturers know very well when they have produced a superior tire and they invariably charge more for them. It is absolutely true that in a tire you get exactly what you pay for.

If you're confused by claims, just notice what the winners are wearing.

It's common to run tires with treads approximately half-worn. New-tread depths allow the tread to deform or "squirm" too much for the accurate steering that's wanted in racing. If you're dedicated, you'll have your tread shaved at the tire shop to about 50% wear. This means, as a bottom line, that you're throwing away half of those expensive tires you just bought. If the tread is soft enough to make a good race tire, it isn't going to last long at racing attitudes anyway. The biggest cost of a race car is keeping it shod.

Fig. 11-30. *You can't get on a proper race course without a roll bar, helmet and seat belts. This Giulietta is touring the course at an Alfa club national convention.*

Let's say you have a weekend of testing at some track. On the first day you'll sort out carburetor jetting, suspension settings, and such. On the second day, you'll sort out tire pressures. On a properly setup race car, tire pressure changes can be measured in seconds per lap. Don't take the label on the door jamb seriously.

You'll spend a lot of time determining the tire pressures that will produce the lowest lap times for your car.

5. Suspension

Never run a 750 or 101 Giulietta or Giulia with damaged or missing front limit straps. Similarly, the rear limit straps on all solid-axle Alfas must be in place if you want to race safely.

Rules permitting, the lower and stiffer the better, because most races are run on relatively smooth surfaces. An easy way to lower Giuliettas by shortening the limit straps, front and rear. This approach, suggested by Paul Tenney, maintains spring rates. If you're satisfied with a slightly stiffer spring, then you can cut a coil from the stock spring.

Fig. 11-32. A popular modification is to add a Panhard rod to help locate the rear axle. The rod takes the place of the sliding block used on factory race cars.

Fig. 11-31. A proud owner-operator in the process of sorting out the rear suspension.

Generally, an antisway bar stiffer than the stock one is desirable for racing. It's fair to peek at what others are doing, but few competitors would appreciate your measuring the diameter of their antisway bar.

On Giulias and later models with the 105 suspension, you can fit an adjustable upper link that allows easy decambering. This link can be fabricated, or purchased from Shankle.

The 750 and 101 chassis has a weak bracket where the trailing arm attaches to the rear axle tube. For racing, weld plates front and rear to box this bracket. The front plate will be shorter than the rear one, of course, to clear the trailing arm.

Pay frequent attention to all the rubber bushings in the front and rear suspension attaching points on a race car. The bushings can tear or deform, reducing handling near the limit, and probably should be replaced with LS-series Heim uniballs if you're really serious.

The front suspension of 105-series cars can be adjusted by fitting an adjustable upper transverse link,

and this is a very popular modification even on street Alfas. The upper trailing link can also be lengthened to provide additional caster, but the allowable adjustment is limited because the upper transverse link will eventually foul against the body cutout.

On Alfettas it's popular to add a suspension brace that runs between the shock towers and across the top of the engine.

Many racers modify the rear axle of solid-axle cars to lower the roll center and improve rear-axle traction. On the GTA, this modification was achieved with a sliding block between the body and chassis. Adding a Panhard rod does about the same thing. It should be attached at one end of the axle by a joint near the brake backing plate or coil spring support, and at the other end to a joint attached to the body. This provides a large-radius link with a roll center lower than the stock setup.

Shock Absorbers

It is not true that a stiff shock absorber improves traction or reduces body lean in a corner. The shock absorber controls the quickness with which a body recovers from a bump or with which the suspension responds to an irregularity in the road. Body lean is a function of spring rate and suspension geometry and traction is a function of wheel size and suspension geometry. Clearly, if the shock absorbers are so bad that the wheels dribble down the road like a basketball, there can be little traction or control.

In summary, a stock Alfa is a superior handling car on the street. On the track, a stock Alfa can distinguish itself, but probably not win races. To win requires experience that can only be gained in a race. You can shorten the time to learn racing techniques by attending a commercial driver's school. Mechanically, very few things are required to race an Alfa. Many of the modifications have more to do with increasing safety and reliability than improving performance, but they are essential modifications nevertheless.

> According to Don Fuller in his report for *Motor Trend* magazine, John Shankle prepared a Milano suspension for sporting use. Wheels were Simmons 15x7 in. shod with General XP2000H 195/60HR15 tires. Front torsion bars were set to reduce ride height about 1.75 in., the same reduction realized by fitting Shankle 4704 rear springs. Spax adjustable shock absorbers were fitted front and rear and set to the seventh click from full-soft. The front torsion bars were 25.4 mm in diameter, compared to the stock 23 mm, for a 54% rate increase. Rear antiroll bar diameter was 25.4 mm.

6. Chassis

Many serious racers spend some time strengthening the chassis of racing Alfas. Generally, the consensus is that the coupe chassis is stronger than the Spider, not only because of the strengthening sheet metal that forms the roof, but also because the front end is marginally stronger.

The first step in chassis strengthening is to seam weld all the panels to each other. During manufacturing, Alfa spot welded panels to mate them. This causes stress points along the panel as it flexes, rather than a smooth distribution of stress. Seam welding (running a continuous bead along panel joints) spreads the stress, but also creates a great deal of heat that can warp the chassis. Continually check body shell alignment when doing any seam welding.

High-stress areas, such as suspension attachment points, can be further strengthened by welding an extra thickness of sheet metal at the attaching point. Gussets are also a valuable addition wherever possible.

On 105-series cars there is a tendency for the front sub-chassis to crack. A little extra sheet metal welded up where the chassis mates to the body and suspension will help keep things together.

I want to stress that all of the above chassis modifications are "gilding the lily" on all but out-and-out racing Alfas. Under normal use the stock Alfa chassis is very sturdy and fine for occasional racing.

7. Racing

Inherent in every car with sporting pretensions is a rather macho dream called racing. The dream includes driving home late Sunday night with a passenger's seat filled with trophies, if not a worshipful companion. In

the early days, it was quite possible to drive a car during the week and, on the weekend, race it. The dream was very nearly attainable.

Fig. 11-33. *Herb Wetson's ride back in 1971—and the way he got there. This is a modest effort in today's terms.*

Fig. 11-34. *The modern way. This 1989 photo shows two very important competitors on the West Coast. Barber's shop is home to many Alfas, racing or not. In the U.S., Sperry Valve Works is where many serious racers have their head work done. Both are extensive efforts with even their own car transports. This two-day event at Laguna Seca was sponsored by the Southern California Chapter of the Alfa club.*

Those days, largely, are gone. The real racers today are steely-eyed strategists who have lost their dreams somewhere between a flow bench and the dynamometer. Even the Sports Car Club of America, long a bastion of "we won't race for pay" has embraced commercialism. Money breeds a virus that turns fun things into serious matters of life and death.

In some ways, that's not wholly inappropriate. Racing can be a matter of life and death. It is more threatening to the dreamer than the technician simply because the technician's attention is focused on things in the real world. Brakes, suspensions, and tires are very real world.

I want to be somber here simply to dispel up front any suspicion that mere Alfa ownership qualifies one to

Fig. 11-35. Alfa club offers Walter Mitty's dream: drive your car on a real race course. Alfa club national conventions are once-a-year events that attract cars from all over the U.S. This Alfa is prepared for serious racing. Site is 1984 gymkhana event at Seattle.

win races. The real racing cars are to street cars as military fighters are to Piper Cubs. Some race series foster the fiction that Chevies and Fords are out there racing, box stock. While the cars may look stock, their frame, engine, gearbox, suspension, and rear ends have been radically modified to enhance performance and reliability.

Racing is, first of all, a school where people go to learn. What they learn is a great deal about making a car go fast, about changing the suspension in subtle ways so it grips better, and about approaching that edge of speed beyond which the car flies off the track and crashes. Racers also learn something of themselves, for racing successfully depends on concentration and attitude as much as a well-prepared car. If you can regard racing as a school, then the inevitable failures won't hurt so much, and victories can be put in proper perspective.

No one would dream of covering a college degree in one chapter. Neither can I hope to treat racing in more than a cursory manner. Like college, there is no single curriculum in racing; the total experience depends entirely on your expectations and aspirations. Much of the effort in racing, as in college, seems directed to pedestrian details that have little relation to going faster. That appearance hides a crucial difference: while it may be possible to go through college inattentively, racing demands your best concentration. Inevitably, it is some pedestrian detail that makes the difference between winning and losing a race. Innate brilliance, superior strategy and physical prowess can't overcome what a loose bolt will do.

7.1 Regulations

What you can do to your car is severely limited by the regulations under which you'll run. Most impor-

Fig. 11-36. More mods to another Giulietta spider. Upturned air horns are necessary because of the small space between carbs and bodywork. Spin-on oil filter is located for easy maintenance. Cam covers are painted body color.

tantly, these regulations are designed to assure that your car is safe to race. A helmet, rollbar, three-point seat belts, and proper flame-retardant clothing are common requirements. Some regulations also require an onboard fire-suppression system. Regulations also help to equalize competition by grouping similarly-capable cars together and, within those groupings, defining how much you can do to your car and still be "legal." Regulations keep Alfas from having to compete for trophies against Corvettes and Ferraris. To a certain extent, regulations keep the costs of racing in line and assure that no one can buy or cheat his way to the winner's circle. Before you decide to drop a big-block Chevy engine in your Alfa, check the racing regulations.

Race regulations have created an entire political environment with all the intrigue of a CIA/KGB thriller. One of the things you'll learn in racing's "school" is how easily an avocational pastime can become very serious business, indeed.

7.2 Safety

Race clothing is designed to give your body freedom of motion and still provide protection in an accident. Helmets are mandatory in virtually every kind of racing. Most regulations specify helmets approved by a research institution such as Snell. Whatever money you spend on a helmet is more than worth it. The same is true of seat belts and roll bars.

Fire-retardant clothing is required in some classes. We've all seen the complete-coverage undersuits worn by Formula One drivers. Those suits are designed to reduce the danger of being burned in an accident.

A racing seat is designed to hold your body in place against the centrifugal forces of hard cornering. As a result, they are also somewhat hard to get in and out of. They should be light and tight. Giulietta sprint seats are too flat for racing; Spider seats can be padded to hold your lower body in a viselike grip. I owned a '66 Veloce Spider with seats that held me like a suction cup (I think its bolster padding had been increased by a previous owner). The later GTVs were equipped with a variety of seat styles, some of which are too two-dimensional to be really suitable for hard driving. My favorite GTV seat is the 1969 style that offers good support as well as ventilation. My favorite Alfa seat of all time was in my 1900CSS Zagato. The fact that the '69 GTV seat resembles the Zagato seat has a lot to do with my affection for the style.

Most of the Alfa sedan seats are not contoured "deep" enough for racing (a sedan must be easy for the kids to fidget in). If you're racing a sedan, and the regulations allow, get a lightweight fiberglass seat. Remember that your seat belt is going to help hold you in place. In most instances, a stock Alfa seat will work just fine.

While stock Alfa seat belts may be legal for racing, I'd not venture on a course with serious intent unless I had an equally serious five-point racing harness around me. That almost presumes a roll-bar for upper-belt attachment, and that's fine, too. Well, to be honest, you have to picture yourself unconscious in your car as it is busily rolling itself into a ball. Inside is where you want to be, and staying there can't depend on gripping the wheel or bracing yourself against the door.

Which brings us to roll bars. While you can buy fabricated roll bars, you can also have one welded up to your sanctioning body's specification or, if you're not going to race, to your own specification. Any welding shop is competent to fabricate a roll bar, but not necessarily competent to design one. It's important that the roll bar is high enough to protect your head in a rollover and strong enough not to collapse along with the roof. The roll bar must attach to the body at three points and chrome moly tube is the preferred material. The best way to find a local shop that will make a proper roll bar is to ask another competitor. As with so many things about racing, asking around at the event is the best single source of information.

Fig. 11-37. Alfas make great club racers. Here's a perfectly streetable car working hard in a turn and loving it. Sturdy roll bar is low enough to allow the top to be raised. Roll bars are hard to *remove and replace, so the design has much to recommend it for the weekend racer who also drives his Alfa to work.*

A roll bar limits interior room, so if you use your Spider for both transportation and racing you may wish to fit an easily removed unit. Coupes and sedans, generally, will have permanently installed roll bars.

Some fire suppression equipment should be on every car that competes, even if it is nothing more than a small fire extinguisher. The optimum setup is a halon system with stainless-steel braided tubing—expensive but effective.

In summary, much of what you must learn about racing is strictly hands-on. You can abbreviate the learning time by attending a school. Much of what you can do to your car will be limited by the regulations you'll run under. Talking to other racers is actually your most important source of information, whether you're discussing driving techniques or where to get a roll bar. Finally, safety is absolutely the first consideration of racing. All your training and the regulations you'll encounter ultimately go to making racing a safer sport.

Finally, I want to relate a true story. The driver who tells this story is an accomplished racer. During a race, the universal joint on his Giulietta's driveshaft broke. As the shaft whipped around under the car, it smashed against the brake line, rupturing it. At something over 100 mph, this driver suddenly found himself with neither engine nor brakes to slow his speed. The fact that

he survived the inevitable crash is a testimony to his skill and the essential strength of his Alfa. It is also a fine illustration that there is no substitute for meticulous preparation.

Fig. 11-39. You can have fun in an Alfa. And survive. This car rolled four times and then went through a fence at 100 mph. The front and rear of the car crumpled as designed, the doors stayed closed, and the passenger compartment is still relatively undamaged. The driver was securely belted in and stayed inside the car.

Fig. 11-38. In addition to good competition, there's camaraderie at any Alfa event, as well as some interesting and valuable Alfas. A Sprint Speciale and Giulietta SZ gather admiring glances.

7.3 The Hot Setup

What follows is a one-day, gen-u-ine college degree (as I mentioned in the section above on racing). I need to stress that the moment you modify a car, it becomes unique, and anything you do to it subsequently depends completely on what you've done to it before. As a result, there simply can't be a cookbook approach to race-car preparation. Having said that, there are some large generalizations that can guide you in setting up your Alfa for street or autocross. These suggestions are based on the broad, real-world experience of my friend Joel at International Auto Parts.

Hot Street

Engine:
- 40 mm dellOrto carb conversion with matched intake manifold
- Large-valve cylinder head
- Pre-1975, 4-into-2 stock cast-iron exhaust manifold
- Ansa Freeflow exhaust
- 10.0:1 piston and liner set
- 11 mm calibrated performance cams, 300° duration
- Marelli Plex ignition
- Lodge HL plugs

Suspension:
- Performance springs
- Koni sport shocks
- Rear antisway bar
- Adjustable front upper control arms
- 15 x 6 alloy wheels
- 195/60-15 tires

Autocross

Same as Hot Street with the following changes:
- 45 mm dellOrto carbs
- Tubular exhaust header
- 10.4:1 pistons
- 11 mm autocross cams, 312° duration
- Lodge 2HL plugs
- Carrillo connecting rods
- 15 x 7 alloy wheels

A Serious Effort

Al Leake loves to race, and you can find his team at most West Coast events in the U.S. One of his cars is an Alfa single-seat Giulietta known as the "Monoposto." He also fields several other cars, including a Milano Verde. Al's effort represents the high end of enthusiast participation, and includes an enclosed trailer that carries spares and other goodies.

Body panels are easily removed on serious race cars to ease maintenance. Though fenders of stock Giulietta Spider are structural members, here they have been cut away to provide fast access to suspension parts.

Giulietta Monoposto carries a removable metal tonneau over the passenger compartment. Driver's compartment has room for driver and spare tire.

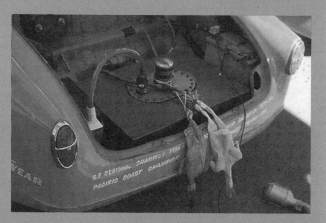

Point of racing is, after all, to have fun: Leake's team is not devoid of humor.

APPENDIX 1: TOOLS

Tools

Author's note: I'm indebted to Tom Tompkins for writing this chapter. Tom's knowledge of Alfa special tools is unsurpassed, as is his enthusiasm for the marque.

Part of the fun of owning an Alfa, especially a classic Alfa, is being able to work on it. To do so, a set of basic tools is essential.

Since Alfa is a metric car, your toolbox should be stocked with the usual metric hand tools: open and box-end wrenches ranging from 7 mm to about 25 mm; a socket set of the same range that includes a spark plug socket; Allen wrenches, especially a 14-mm Allen wrench for the cam cover bolts; and assortments of screwdrivers and pliers. Before spending hundreds of dollars for a set, however, keep in mind that the majority of Alfa fasteners are either 8, 13, 14, 17, or 19 mm and the other sizes will go largely unused.

The possible exception is a 1-1/16 in. socket for the oil drain plug and a 1-1/2 in. socket for the crankshaft pulley, which has to be removed on 2-liter Alfas to replace the toothed drive belt for the SPICA pump. I give these two measurements in inches because those two sockets are hard to find in their metric equivalents.

In addition to the basic hand tools, you'll want a set of feeler gauges and a micrometer for valve adjustment, a torque wrench measuring in lb. ft., a high-impedance (read expensive) volt/ohm multimeter (VOM), and a bright strobe light for timing the engine.

Most Alfa owners have the popular Haynes or Chilton workshop manual for their car, and a few have the official Alfa publications. The first thing you'll notice in an official factory publication, and in some sections of the popular shop manuals, is the need for special tools. The official factory manuals are the basis for the popular ones and if you compare the two you'll see that they are very similar. The official shop manuals are written presuming that the reader is an Alfa mechanic in a fully equipped Alfa shop, so reference to special tools is of no concern. One of the goals of the popular manuals is to shield the owner from having to use special tools. Whenever a screwdriver will work instead of special tool X9934378, the popular manual will show a screwdriver.

Most mechanical components on an Alfa are very forgiving, so repairs requiring the use of special tools can frequently be accomplished without them. In addition, there are whole categories of repair, such as cylinder head refurbishing, which are usually given over to specialist shops, thus obviating the need for the owner to have the special tools.

But sometimes you have to have the special tool, which was designed by the same engineer who designed the part you're working on. Alfa special tools fall under the following headings:

Convenient and Handy Tools

These tools are nice to have, will speed up your work and impress your friends, but you can get along without them. Tools in this category include the prop for holding the Al-

fetta's de Dion triangle in its tipped-down position during rear-end work.

Most owners like to adjust their own valves. Alfa has an exotic tool set that includes a dial indicator for the task when a micrometer and set of feeler gauges will work just fine. The Alfa cam turning tool, A50103, is a highly recommended "convenience" because it eliminates the necessity of having to use Vise-grips, which chew up the camshaft and drop steel shreds into the oil pan.

Another convenient hand tool that is still available from Alfa is the 10-mm distributor locknut wrench, A50213. Its long shaft and "T" handle make adjusting the distributor a pleasure.

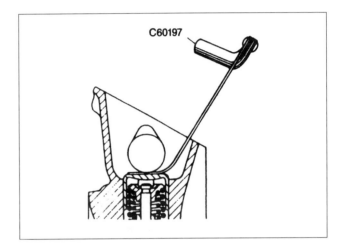

C60197: A curved metric feeler gauge set will be hard to find in aftermarket. You must not try to use a straight feeler when measuring intake valve clearance on a V-6.

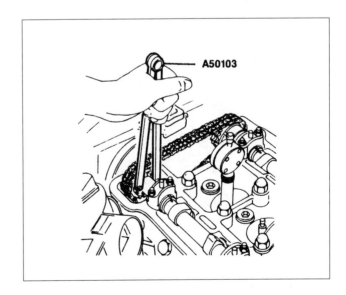

A50103: Camshaft rotating tool makes valve checking easy.

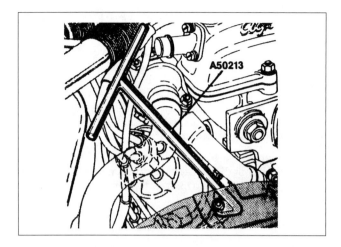

A50213: The 10 mm distributor wrench gets to a locknut not otherwise easily accessible. A 10-mm crowfoot socket also works.

Essential Tools

This kind of tool can't be replaced by a screwdriver or a hammer, or cobbled up from some scrap metal very easily. A lot of drivers for bearings and seals fall in the essential category. A similar tool for a domestic car might work with a little modification, but unless you get something like the official tool, you can't do the job.

Some essential tools are surprisingly easy to fabricate. An especially useful tool for a cylinder head stuck to the block is the Alfa tool A20451. This tool is still available from Alfa and most aftermarket suppliers. But one clever home mechanic removed the porcelain from two old spark plugs, brazed bolts into the shells, and then used them with a hoist and chain to lift the head free.

Another essential tool, no longer available, is the wrench required to remove the lower Weber carburetor mounting nuts from a Giulietta Veloce. A very bent 14-mm box-end wrench works after a little trial and error (check first that the nuts are not replacement 13 mm). To prevent breaking the wrench, bend it using a torch to heat the metal.

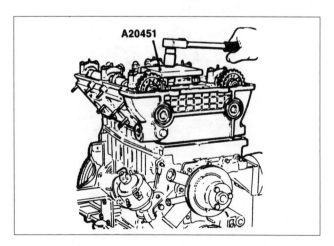

A20451: Nothing is more unforgiving than a stuck cylinder head. This tool secures in a spark plug hole and presses against tops of cylinder head studs.

For the adventurous soul willing to tear down an engine with the intent of replacing the main bearings and wet liners, Alfa has several special tools that make the job easier. With a little more time, care, and difficulty, the job can be done without any of these official tools. Typically, seal drivers, the most demanding of the essential tools, can be replaced by proper-sized sockets or lengths of PVC water pipe.

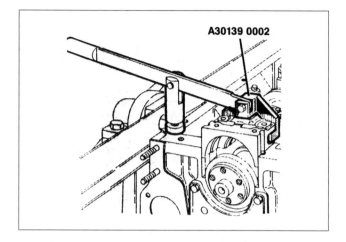

A30139-0002: Rear main bearing cap on a 4-cylinder engine is almost impossible to remove without scratching things. This tool slips into cutouts on either side of bearing cap and attaches to a puller arm.

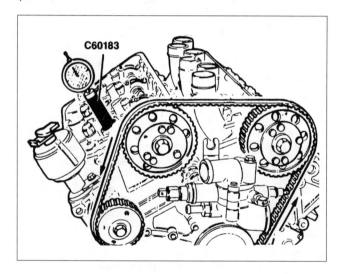

C60183: This is an adapter for a dial indicator to be used when setting top dead center on a V-6.

Another frequently performed procedure is the replacement of the clutch assembly. Don't try centering the driven plate (clutch disc) without a tool. For this task, Alfa tool A40103 is used. A similar tool is available from many tool shops, but be sure you get the proper size for the Alfa clutch hub, which is 21 mm at the splines and 16 mm at the pilot bushing for most Alfas (check yours first to be sure).

A transmission involves probably the most complex component of the car. Most of its special tools are pullers, drivers, and a number of plates and half-plates used to engage bearings and gears on a hydraulic press. Many of the tools

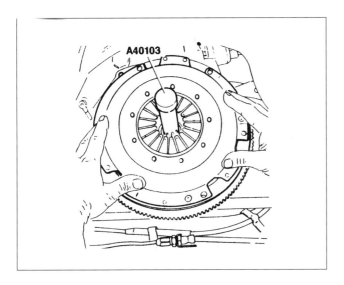

A40103: This is clutch disc centering tool.

are available from sources other than Alfa, but it's uneconomic to buy them for just one rebuild. This applies especially to the quantity of tools required to rebuild the transaxle on an Alfetta, GTV6, or Milano.

The driveshaft donuts should be removed with a clamp, Alfa tool A20315, but a very large radiator hose clamp works as well.

There are a number of special tools for the suspension. The two most used are tool A30156 for removing upper ball joints and tie rod ends, and tool A30157 for removing lower ball joints. An aftermarket tool, commonly known as a "pickle fork" is a brutal substitute. The front-axle grease seal driver, tool A30192, can be improvised or obtained from Alfa aftermarket sources. On later torsion-bar cars, the torsion-bar removing tool A30374 and the forcing screw tool A30374/0001 can both be fabricated, once you understand how the bars remove.

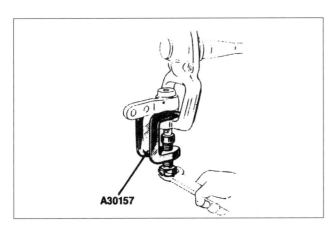

A30157: This is a puller similar to A30156, but for lower ball joint only.

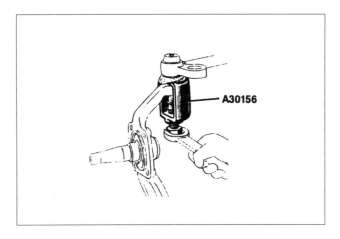

A30156: This is a puller for upper ball joint and tie rod ends. This is a hardened-metal tool and you probably shouldn't try to fabricate one. If desperate, use a "pickle fork" instead.

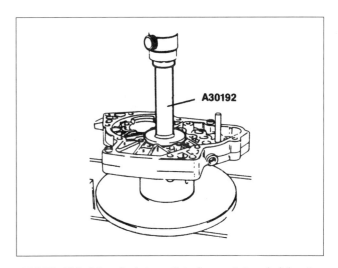

A30192: This driver for intermediate flange pinion shaft bearing also works as a Spider front axle grease seal driver. Clearly, any driver with correct dimension will work.

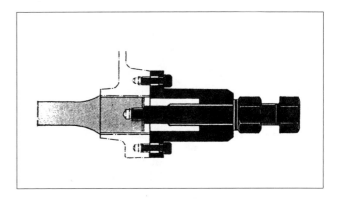

A30374: This is a push/pull tool for working with torsion bars. Large center bolt screws into bar itself while tool mounts up solidly to car with two smaller bolts. Use lead plumbing pipe to fabricate.

Most owners are willing to replace their own disc brake pads. Shop manuals show a brake piston positioning template that is used to set the notch on the piston to its proper rotation. The template for the front brakes is A20160 and the rear is A20149. These templates can be copied in cardboard for occasional use, and they are also available from Alfa aftermarket sources.

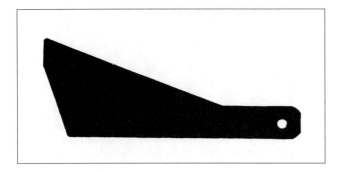

A20149: This is template for setting disc brake piston. Angles shown here can be copied to cardboard.

Rear Alfetta/V-6 brakes are a bit more involved to replace because they are inboard and have mechanical adjusters. The wrench A50194 makes the job much easier.

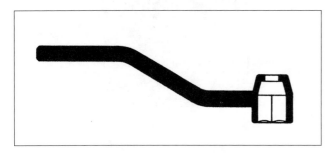

A50914: Inboard rear brakes on transaxle are easier to adjust with this wrench.

Most any mechanical task on an Alfa can be undertaken confidently with simple hand tools and a willingness to fabricate a few essential tools. Most important of all is a little ingenuity in improvising a tool. One enthusiast, needing to replace the rear motor mount on his Alfetta, marched down to the local hardware store and purchased several different sizes of inexpensive PVC fittings, one of which worked perfectly for pressing in the bushing. Another, when replacing pistons, fabricated a ring compressor from an old wet liner. He cut it to about 70 mm length and beveled the top edge to ease ring compression.

Critical Tools

Only the Alfa tool will work. Fortunately, there are very few critical Alfa tools, and the average owner is not likely to need them. The gauges for setting up a ring and pinion are examples of critical tools.

While it is certainly possible to rebuild a differential without the Alfa special tools, the job is much easier and more precise with them. Most shops do not have the tools (or the motivation to make them) and use a hit-and-miss technique that depends on coating the gears with Prussian Blue or Zinc Chromate and then "reading" the gear pattern. If you are very demanding of precision, you'll have to hunt even for an Alfa shop that has the proper tools for a proper differential rebuild.

Fuel Injection Tools

The SPICA fuel injection system inspires the most awe because of the many critical protractors, gadgets, and calculations required by the official manual (the aftermarket manuals won't go near the subject). Only two tools are actually needed to make the adjustments. The first is the dummy thermostatic actuator. This tool is no longer available from Alfa, but an adjustable actuator is available in the aftermarket. It is possible to fabricate a dummy actuator using a bolt, a large washer, and several nuts, but the low price of the aftermarket part makes fabrication hardly worthwhile.

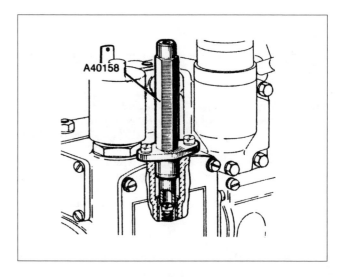

A40158: Standard dummy thermostatic actuator for all 1969–74 cars (shown) has a projection 27.0 mm from lower surface of mounting flange. For 1975–76, 1980–81, and 49-state 1977 cars projection is 27.8 mm. For 1977 California and all 1978–79 cars projection is 29.0 mm.

The second critical tool is the mixture locknut tool that fits over the fuel cutoff solenoid and engages a crown-headed locking nut. Tool A50177 is for 1970–79 cars and tool A50244 fits the 1980–81 models. Hint: a work-alike for tool A50177 can be fabricated from a very heavy U-bolt if its ends are bent and then ground down to fit the slots in the locking bolt.

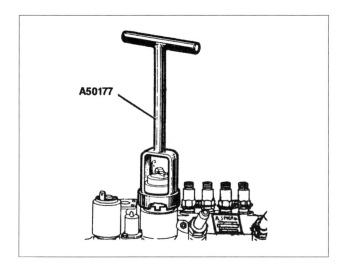

A50177: *Mixture locknut wrench for straddling fuel cutoff solenoid. This may be a critical tool if locknut was last tightened by a gorilla; made-up substitutes won't be strong enough to work.*

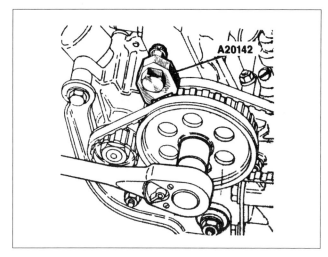

A20142: *This lock for fuel injection pump pulley is about the only way you can remove center nut without tearing up toothed belt.*

A50224: *Similar to A50177, but for 1980–81 cars.*

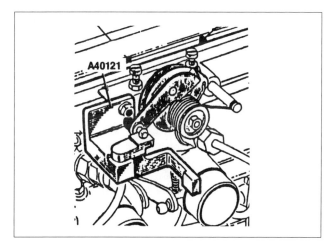

A40121: *Probably the most-coveted tool for SPICA owners, this is two-position stop setter for throttle bellcrank.*

There are two other fuel injection tools that are nice to have, but can be easily improvised. One is the fuel injection pump pulley tool, A20142 which is replaceable by a small aftermarket puller. The other is the Alfa relay crank travel and position setting tool A40121. This tool appears to be critical, but a plastic protractor and the proper SPICA manual will work instead.

Special V-6 Tools

The 6-cylinder Alfas have a set of special tools unique to them. Valve adjustment on the V-6 is slightly more involved and requires several additional special tools that are all available from Alfa or on the aftermarket. Tool A20361 is a universal wrench for the distributor, oil pump, and camshaft drive pulleys. Puller A30521 is used with tool A20361 to remove the camshaft drive pulley. The locking pin tool A20363 can be made up with a little effort, but its use is critical to setting the timing belt tension. An exotic exhaust-valve adjusting tool, reminiscent of the tool used on the classic prewar Alfas, is A50220.

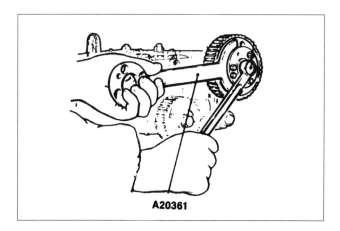

A20361: *This is a "universal" V-6 wrench for distributor/oil pump and camshaft drive pulleys.*

APPENDIX 1

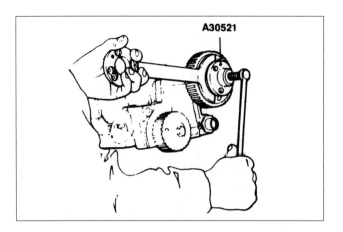

A30521: This tool is used with handle A30521 to remove camshaft drive pulley flange on V-6.

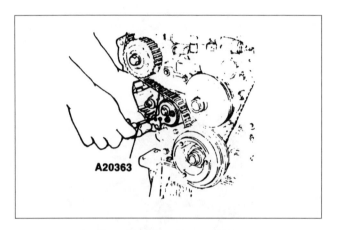

A20363: This is detensioner unit locking pin for setting proper drive belt tension.

Bosch EFI

The Bosch electronic fuel injection system is a quantum leap in complexity from the mechanical SPICA fuel injection system. The Bosch unit requires some elaborate test equipment, but according to Charles Probst, in his book *Bosch Fuel Injection and Engine Management*, most fuel injection problems can be solved with a digital VOM and a six-bar fuel pressure gauge, which costs about $200. If you're even mildly interested in Bosch EFI systems, then Probst's book, also published by Bentley, is essential reading.

APPENDIX 2: RESOURCES

Annotated Bibliography

Over the years, as a writer, I've collected over forty books on Alfa. For most owners who do not care to make a profession of researching the marque, one or two core books on Alfa will supply virtually all the information one is likely to need, and I will list these core books first.

There are a few caveats that preface my reviews. The most important is that several of the books are not written in English. Those in Italian may be worth getting simply because you'll enjoy the pictures. The second warning is that a large library of Alfa books can quickly become expensive and not be awfully informative, since most titles add only a very little to the core volumes that I'll discuss first.

The mother lode of Alfa photographs is the archive at Arese and almost all have been placed in the public domain. Anyone who tries to write a book on Alfa is obligated to mine this primary source. As a result, you will find a duplication of photographs in Alfa books that reason and copyright laws would otherwise prohibit. Buying another title may add only a very few new photographs for your reference, and no new information. I hope my reviews here will save you the considerable expense of obtaining Alfa books that don't really help you learn what you want to know.

If you want to know all about Alfa and would enjoy reading a fine book, then *Alfa Romeo*, by Peter Hull and Roy Slater, is really the only resource you'll ever need. It's the standard work on the subject, readable, engaging, and thoroughly authoritative.

If you don't care to read narrative but want to get directly to the heart of the matter, and absolutely have to know everything there is to know about every Alfa ever built, Luigi Fusi's book *Alfa Romeo—All the Cars Since 1910* is the ultimate source. This is not a book so much as a catalog. It is chronologically arranged by model so that, by fanning its pages, you can piece together your own history of the marque. Fusi's book went through two editions before going out of print. While the first edition had a few phrases translated into English, it was basically in Italian. The second edition of Fusi includes a roughly parallel English text. To get all the information the second edition contains, however, a reading knowledge of Italian is still required. Though it's hard to find, Fusi's book (either edition) is worth having. In the fall of 1991, Fusi's book was rumored to be scheduled for reprinting, but while a limited number of signed books was released, no publication date has been set. You can get the same information at about four times the price in the Alfa Catalog Rasionne, listed below.

Fusi worked at Alfa during the halcyon years and kept his own records. He was an Alfa employee until 1982, when he retired. He continued to act as a consultant until 1989 when he ceased consulting because of poor health.

In fact, it was Fusi who saved all the historic artifacts that contributed to his work, and the museum at Arese would never have been accomplished were it not for his dedication. It is currently not popular at Alfa to credit Fusi so effusively, but the cars in the museum and virtually all historic documents at the Arese archives have survived because of this one man's dedication.

Significantly, the information that Fusi had to obtain from official Alfa records, most notably the serial numbers of the postwar cars, is much less accurate than his own note-keeping.

I need to recommend one other book by a personal friend, Joe Benson. If you're in the market to buy—or have just bought and want to find out what you've done—then Joe's *Alfa's Buyer's Guide* will make rewarding reading while giving you a dollars-and-cents history of all the Alfas you're ever likely to see in person.

Finally, I should recommend my own book on the Giulia (with Jim Weber), and John Tipler's work on the same subject.

To find many of these books you'll probably have to order from a specialist book store. Try your local book dealer first. You'll find that most booksellers are eager to serve you. Many of the books are available from Classic Motorbooks, which advertises in virtually every car magazine you're likely to pick up. Albion Scott is another general source. If you prefer slightly more personal attention, I will recommend Phil Lampman of Shelby's Dad (Issaquah, Washington) and Brian Hernon of Automod Atlanta (Georgia), both of whom will chase out-of-print books for you. Tom Warth, founder of Motorbooks International, now has his own rare book concern, T.E. Warth, which publishes a catalog of automotive titles.

The order of reference is: title, author, place, date, publisher.

Core Books

Le Vetture Alfa Romeo Dal 1910
Luigi Fusi
Milano, 1965, Editrice Adiemme

Luigi Fusi was an Alfa employee from 1920 to 1983. When Jano arrived from Fiat in 1923 to help on the troubled Grand Prix project, it was Fusi who penned the original drawings for the P2 car that would win the championship in 1925. It was also Fusi who saved the Alfa archives, conceived, fought for, and organized the museum. If there is a Mr. Alfa in the world, the title goes to Fusi hands down.

This is the first edition of his work, done in Italian with a very brief polylingual appendix. When it appeared, it was absolutely revolutionary, providing information not otherwise available. It is not a reading book so much as it is a catalog of all the cars from 1910 to 1965 (ending with the 2600 Sprint Zagato). Thus, all the important modern cars are missing, and so is all the color of Alfa's history. This book is out of print and not at all available. I prefer it over the second edition only because it is slimmer and thus more portable for my briefcase.

Tutte Le Vetture Alfa Romeo Dal 1910
Luigi Fusi
Milano, 1978, Emmetigrafica

This is the second edition of Fusi, which adds some parallel English text and covers the cars until 1977. The main catalog is still exclusively Italian but quite manageable simply because its so repetitive in terminology. This book too, is

APPENDIX 2

out of print, though two thousand signed copies were released in 1992. No Alfa library is complete without a copy.

Alfa Romeo
Peter Hull and Roy Slater
London, 1964, Cassell

These two English enthusiasts have created a book that is
engagingly narrative and most authoritative, especially
on the older cars. Its date makes it of limited interest to
owners of Milanos and 164s, however. It is out of print.

Alfa Romeo
Peter Hull and Roy Slater
London, 1982, Transport Bookman

Reprinted in 1983 with a different format and some new
information. If you must have only one Alfa history book,
this is it. Cars post-AlfaSud are not covered.

Alfa Romeo, the Legend Revived
David Styles
London, 1989, Dalton Watson

This is a new work that matches the readability of the
Hull/Slater work and adds the advantage of dealing with
the newest cars, including the ES30. It is slightly more cursory than Hull/Slater, but filled with pictures and information not found otherwise (a list of the cars at Arese, for
instance). If you can't find Hull/Slater, or have it and want
just one more title, choose this one.

Alfa Romeo Catalog Raisonne
Bruno Alfieri, Editor
Cremona, 1982, Automobilia

This two-volume work is written in parallel columns of
Italian, English, and French. Fusi contributed to the work,
along with Giovanni Lurani, so its pedigree is impeccable.
Besides, Alfieri is a writer of the first rank himself. The first
volume is textual and approaches the Styles work for pre-
sentation, while the second volume is a Fusi-like catalog of
the various models. At $250, though, you can get almost
the same information for less; I'd recommend Hull/Slater
and Styles.

General Histories

Alfa Romeo
Evan Green
Beacon Hill, NSW, 1976, Evan Green

The Australian origins of this work make it both notable
and refreshing. While Green has mined the Arese archives,
his unique photos of Alfas in Australia brings something
important to the history of the marque. The many color
pages show that Green put his wallet where his pen was.
Even though cursory, this is a brave publication and one
worth owning if you've already got the core books.

Alfa Romeo
Peter Hull
New York, 1971, Ballantine

This is a paperbound overview of Alfa's history that is out
of print. One in a series of auto histories by the publisher, it
serves as a summary that can be read in a single evening.

Alfa Romeo Pocket History
Gonzalo Alvarez Garcia
Cremona, undated, Automobilia

This is one in a series of pocket auto histories published in
Italy. The translation by Mary Trotter is fluent and the numerous illustrations are well served by glossy paper, an
advantage Peter Hull's overview doesn't enjoy.

Great Marques—Alfa Romeo
David Owen
Secaucus (USA), 1989, Chartwell Books

Another series title, this is a large-format book filled with
original and crisp color photography. Owen is an Alfa authority and this work offers both coherence and perspective. The text is short and the pictures, though seductively
gorgeous, do no always do the work of a thousand words.

Alfa Romeo—Milano
Michael Frostick
London, 1974, Dalton Watson

Publishers are convinced that pictures sell books. Frostick's work is the logical extreme of this approach and his
text is reduced almost entirely to photo captions. The significance of the photos—indeed, the marque itself—is left
largely to the reader's imagination.

Museo Alfa Romeo
Akira Fujimoto
Tokyo, 1979, Car Styling

This is a book in both Japanese and English, with a short
color section and many exceptionally well-reproduced
black-and-white photographs. The English is fluent and
the photos are gathered from the archives, other contemporary sources, and new photography. The book is a fine
visual resource, but the three-page history sandwiched between the color and black-and-white sections is certainly
too brief for the novice who wants to know about Alfa.

Alfa: Immagini e Percorsi
Tito Anselmi and Carlo Pirovano
Milano, 1985, Electa

This is almost an official Alfa book, published for the Alfa
Romeo exhibition at the Triennale of Milan in 1985. It is in
both English and Italian, and gives a fine overview of the
marque using photographs from the archives.

The Alfa Romeo Tradition
Griffith Borgeson
Kutztown, 1990, Automobile Quarterly

This is a very important work about Alfa, but it is not so
much about the cars as about the people who made them.
There are several ways to chronicle a business enterprise.
You can talk about the product, the organization itself, or
the people who make up the organization. The three approaches demand incremental familiarity; even casual
readers can appreciate the features of a car when it is as
special as Alfa. On the other hand, business-school-like
studies of the growth of the Prototype Section of the Engineering Department are a bit restricted in their appeal, and
an in-depth dissection of an ngineer, even if he is a heavy
hitter in the organization, can all too easily become sopo-

rific. This is the only book that deals directly with the players who collectively made Alfa what it is. Thankfully, Borgeson's meticulous research and fine style keep it readable. His approach illuminates the first 30 years or so of Alfa history exceedingly well. It does less well for the faceless corporation that Alfa became in the 1970s.

Era- or Model-Limited Works

Illustrated Alfa Romeo Buyer's Guide
Joe Benson
Osceola, 1983, Motorbooks International

Joe was the club's technical guru for years and this overview of Alfa models is perfect if you want to look at completely original pictures, get the most important facts, and then decide whether or not to buy the car. It is one in a series of practical buyers' guides written to the same format. As a result, it doesn't try to be a complete history, and Joe begins with the 6C2500 cars. It does, however, establish a relative value of the many Alfa models, something no other book does so explicitly. The prices quoted in the original edition are laughably low now, and a new edition has just been released that covers contemporary Alfas and prices.

Le Alfa di Merosi e di Romeo
Luigi Fusi
Milano, 1985, Editrice Dimensione S

Le Alfa di Vittorio Jano
Luigi Fusi
Roma, 1982, Edizioni di Autocritica

These two books are the stuff of a postgraduate seminar, and then only if you read Italian. The first covers the cars designed by Merosi, that is, from 1910 to 1923. The second covers Jano's work from 1923 to 1937. Since Fusi and Alfa History are virtually synonymous terms, the selection of photos and their explication reflects ultimate wisdom. If you're really interested in the old Alfas, these two books are reason enough for a few semesters spent learning the language.

The Immortal 2.9
Simon Moore
Seattle, 1986, Parkside Publications

This award-winning book covers a single model Alfa with a fineness of detail that almost defies belief. One is left wondering what caused Moore to omit the family trees of all the individuals who designed, manufactured, or even touched these important cars. Any marque would be honored to have such a work of scholarship, which, in its medium, is every bit as masterful as the objects it chronicles are in theirs. With the possible exception of Pomeroy, Moore's is the finest car book I know.

Simon is preparing a similar work on the 8C2300 Alfas that will also be published by Parkside.

Alfa Romeo 8C2300
Angela Cherrett
Dorset, 1992, Veloce

Alfa Romeo Tipo 6C
Angela Cherrett
Veloce

Angela serves as secretary for the Alfa Romeo section of the British Vintage Sports Car Club. Her credentials are impeccable and these are essential books if you want to know about Vittorio Jano's 6- or 8-cylinder cars. Large photos, literate writing and plenty of detail make a superb book. Much technical detail is included, so you have a real sense of exactly what these cars were about. A list of race victories is appended.

The 6C1750 Alfa Romeo
Luigi Fusi and Roy Slater
London, 1968, MacDonald

Roy owned a 1750 when he and Peter Hull (owner of an RLSS at the time) wrote the major history of Alfa. Roy later moved to Italy with his wife, Edna, and I suspect that this book was a happy collaboration for both Roy and Luigi Fusi. While the Cherrett book is more of an overview of the model, this book is really an expanded owner's manual, complete with a section on maintenance. You could not ask for more impeccable credentials nor a more authoritative work. A list of known cars is included.

Alfa Romeo Tipo A
Luigi Fusi
Milano, 1982, Emmetigrafica

Everyone knows the Tipo B, but the Tipo A was certainly more curious. This is a book about a single car, created in 1931 and recreated under Fusi's supervision in 1973. The Tipo A used two 1750 supercharged engines side by side, complete with two gearboxes and two driveshafts. Perhaps appropriately, the text is in parallel columns of Italian and English. The book gives an inside look at Alfa's racing efforts in the early 1930s as well as Fusi's passion, in the '60s and '70s, to create an Alfa museum.

Alfissimo
David Owen
London, 1979, Osprey

Owen is a British journalist and one of his works on Alfa has already been noted (above). This book covers only a few postwar models, from the 1900 to the New Giulietta, which was not brought into the U.S. Owen is a writer first and an Alfa enthusiast second; his style is most readable and he ends the work with a chapter on how much he really likes the marque. This is a good book if you're just interested in knowing about the Alfas you're likely to encounter in person. The mix between text and illustration is very good, and the color photo section in the middle of the book is a plus.

Alfa Romeo Alfetta GT
David Owen
London, 1985, Osprey

This book, one in the Osprey AutoHistory series, covers the Alfetta coupe in both 4- and 6- cylinder editions. It is a relatively small book, but then it really only covers two Alfa models. Like his other works, it is readable, well illustrated, and generally interesting.

Annotated Bibliography 261

Alfa Romeo Veloce
Don Hughes/Vito Witting da Prato
Newbury Park, 1989, Haynes

This book hits a small target squarely. It deals with the Giulietta Veloce and its Zagato- bodied variants. I suspect da Prato provided most of the unique photos in this book, while Alfa-club member Hughes's contribution is a very readable text. Don maintained a Zagato register for the club over a number of years. If you have a Giulietta, especially one with a Zagato body, then this is a mandatory work. It includes a list of Alfa wins for the years 1956–63, and a partial list of known SZ cars by serial number and name of first owner.

Alfa Romeo Giulietta
Evan Wilson
London, 1982, Osprey

Evan was editor of the *Owner* preceding me, and this book includes an appendix of *Alfa Owner* coverage of the Giulietta. What is really curious is that the "coverage" is only an index and is useless unless you have access to a collection of *Owners* of the era. If this book has a fault it is that it far exceeds its announced goal. Instead of dealing exclusively with the Giulietta, the work is padded out with an introductory history and a final chapter on the Giulia. There is a lot of meat here, but some fat too. Nevertheless, this is an essential book if you own a Giulietta.

Alfa Romeo Giulietta
Angelo Tito Anselmi and Lorenzo Boscarelli
Milano, 1985 Ed. della Libreria dell'Automobile

This is probably the best work on the Giulietta, but it is in Italian. Anselmi is a respected Italian journalist and Boscarelli is a trained mechanical engineer. Their work is filled with detail, including lists of racing successes and serial numbers of early/significant cars. Clearly, there's no substitute for being there; the book is filled with photos of prototypes that we may never have seen otherwise. Clearly, the scholarship is both meticulous and expansive. The book is worth getting just for the photos.

Alfa Romeo Giulia
Pat Braden, with Jim Weber
Osceola, 1991, Motorbooks International

While the Giulietta is the darling of Alfadom, there were more models of the Giulia than of any other Alfa. This book explores them all, from the Giulia TI to the luscious Tubolare Zagatos. A brief Alfa history puts the series in context. The mechanical details are explored in some depth using numerous photographs.

Giulia Coupe GT & GTA
John Tipler
Dorset, 1992, Veloce

Veloce is a new publisher that has also brought us Angela Cherrett's book on the 8C Alfas. While it is a competing work to my own book, I have no hesitancy to commend this as significant. Because John focuses on the English experience of Giulia ownership, he offers a refreshing counterpoint to the American emphasis of my book. His anecdotes of racing Giulia coupes are especially readable.

Alfa Romeo Duetto
Giancenzo Madaro
Milano, 1990, Nada

If the current Spider is, in fact, the last of the series that was introduced in 1966 at the Geneva show, then this book might have been the definitive coverage of one of the longest model runs in auto history. It isn't. There is a characteristic Italian approach to a car book that is well illustrated here: lots of press-release photos and a slick, somewhat laudatory text. By comparison, British works are filled with nitty-gritty: inch-by-inch race coverage, tables of owners' registers and a real sense of persona that may even include acid criticism. This book has just the facts. For me, it misses the enthusiasm and the behind-the-scenes details that might have made it truly significant. It's in Italian, and there is a companion volume on the Giulia GT (see below).

Alfa Romeo Giulia GT
Brizio Pignacca
Milano 1990, Nada

A companion work to the Duetto book (above) published by Nada. Like the Duetto book, this one breaks no new ground, and gives a comprehensive if rather sanitary recitation of the marque. This book recognizes the rapidly increasing value of the coupes and gives a final chapter on restoration that features body work on a GTAm.

Alfa Romeo 1900 Sprint
Gonzalo Alvarez Garcia and Angelo Tito Anselmi
Milano, 1983 Ed. della Libreria dell'Automobile

The 1900 Sprint is a car that deserves a book, and this one satisfies. As Alfa's first serial-production car, the 1900 is arguably the most significant car Alfa has ever created. The 1900 was a favorite platform for body builders during an era that groped out of the 1930s classicism toward modern aerodynamics. This is much more a picture book than Anselmi's work on the Giulietta, so the Italian text is less of a drawback. The 1900 was a wonderful old car. If you've never owned or ridden in one, this book may not appeal. But if you have been fortunate enough to experience a 1900, this book is mandatory even if you don't read Italian.

Alfa Romeo Spider
Bruno Alfieri
Milano, 1988, Automobilia

Brought to you by the same publisher as the Catalog Rasionne, this is a slick overview of Alfa Spiders, from the 6C1750 to the Vivace, a 1986 show car by Pininfarina. There are a number of original color photos in this book and, I suppose, most of the press photos that Alfa Romeo has released. The emphasis is on the Giulietta/Giulia/etc. Spider series. The book is written in three parallel columns of Italian, English, and French. This is a great book for casual reading or for just decorating the coffee table. If your Spider has left you with funds for only one book, this one would not be a bad choice.

Alfa Romeo Giulia Berlinas 1962–1976
Alfa Romeo Giulia Coupes 1963–75
Alfa Romeo Spider 1966–1990
Brooklands Books Series
R.M. Clarke
Cobham, Surrey, various years, Brooklands Books

These three books are collections of magazine road tests. They make interesting reading, especially if you don't subscribe to British magazines such as *Autocar, Autosport* or *Motor*.

Collected in these three volumes is a microcosm of the world of buff-book road tests. Alfa has for a very long time been the darling of the motoring journalists. The phenomenon comes through loud and clear and we can share the reporters' joys at the prospect of testing an Alfa.

Road & Track on Alfa Romeo
Cobham, Surrey, undated, Brooklands Books

This is a virtuoso performance from *Road & Track* that plays off the Brooklands series reviewed above. There are four titles in this series, covering road tests published in 1949–63, 1964–70, 1971–76 and 1977–1984. Included are extended-use tests, owner survey results, and classic road tests published during the years covered.

Books Published by Alfa Romeo

I'm not exactly sure how one goes about getting any of these books. My impression is that they were somewhat widely distributed through the Alfa organization when new, but I think they probably count as collectors' items now. They are all in slip-covers and some are numbered. Lucio Simonetta is the lead author of all these books and all are published by Alfa Romeo at Arese.

Alfa Romeo History Museum
1979

This book includes many new photos of cars in the museum at Arese and a coherent history of Alfa in fluent English. Like all the other Alfa books, it is large-format, with color on virtually every page. This book is available from the enthusiast bookstores.

Giulia: l'ha disegnata il vento
1982

"Designed by the Wind" is a working translation of this peek behind the scenes during the development of the Giulia. The work begins with a few brief pages of copy in Italian and follows that with over one hundred (unnumbered) pages of uncaptioned photographs showing the various Giulia models in great detail. If you were there, the photos are probably very evocative. Not having been there, some of the photos are cryptic. My favorite photo in the book is a side view of a crash-test Giulia sedan, the front of which is perhaps 50% its original length. The windshield is intact and the fit between the front and back doors remains unchanged.

Alfa Romeo 70 anni di immagini
1980

This is a picture book that brings some new photos from the archives. Brief captions help the novice identify what he's looking at. At the back of the book is a brief chronology of Alfa history in Italian, English, German, and French.

Cento Manifesti
1981

One hundred ads about Alfa, from 1919 to 1980. Actually, not quite a hundred, since some of the "ads" are really photo layouts, covers of magazines, and catalogs, or just photos of the product. Alfa has an ad style: few words and big pictures. It's especially elegant with the old cars.

Effetto Alfa
Franco Nencini

It is perhaps overblown to suggest that Alfa influenced the tastes of several generations, but it is also closer to fact than many may realize. And, if one considers only the Italian culture, then the premise is not overblown at all. This book/record combination captures the milieu of Alfa: prosperity, war, and reconstruction.

Books About Alfa Body Builders

Carrozzeria Italiana
Angelo Tito Anselmi
Milano, 1980, Automobilia

This is a photographic record of a touring exhibition of Italian coachwork, accompanied by a selection of essays by automotive historians. The book reveals that there is virtually no style of bodywork that was exclusive to Alfa, and one may find Alfa look-alikes among all the Italian manufacturers, most especially Fiat. As a reality check for Alfa Euphoria this book is especially valuable. Note: an identical title, by Marchiano, has just been published.

Car Styling Special #19 Bertone
Akira Fujimoto
Tokyo, 1977, Car Styling

This is a paperbound work in Japanese and English with gorgeous color photography. All the Italian body builders indulged in weirdness from time to time and Bertone is no exception. But most of Bertone's work is very main-line. Indeed, the Bertone shop handled the production of several Alfa models, including the Giulietta Sprints and the Montreal.

Touring Superleggera
Carlo Anderloni and Angelo Tito Anselmi
Milano, 1983, Autocritica

Superleggera means superlight, a technique of body building pioneered by Touring in the 1930s. Instead of laying metal over wood, the superleggera technique uses a network of small tubes to form the basic body. Of course, the technique is not unique to Touring, since Zagato also built his bodies over a latticework of metal tubes.

With Bertone, Touring is the body builder of choice for production Alfas. This book offers over three hundred pages of clear photographs of Touring-bodied cars, from the late 1920s to the present. Of all the body builders, Touring is perhaps the least venturesome. His designs have a "rightness" that reflects the median taste of the era most elegantly.

APPENDIX 2

Books About Zagato

The association between Alfa Romeo and Zagato has been both long and honorable. As a result, a book about Zagato is also unavoidably about Alfa Romeo. Looking at these books is especially illuminating because one sees many of the styling techniques, so easily recognized on Alfa, applied to other marques.

Zagato
Gianfranco Fagiuoli and Guido Gerosa
Rome, 1969, l'Editrice dell'Automobile

This book, in Italian, gives a very balanced overview of the products of the Zagatos and a lot of information about the founder, Ugo. It is notable in including fold-out illustrations of some of the most beautiful of the Zagato-bodied cars.

Zagato—Seventy Years in the Fast Lane
Michele Marchiano
Milan, 1989, Giorgio Nada

This book is in both English and Italian and is larger than the Fagiuoli/Gerosa work, both in format and number of pages.

Zagato
Michele Marchiano
Milan, 1984, Automobilia

Marchiano is certainly the name to consult on Zagato. This earlier work is about the same size as Fagiuoli/Gerosa. A chronology at the end is very helpful in sorting out who Zagato worked for and when.

Alfa Romeo Zagato SZ and TZ
Marcello Minerbi
Brescia, 1980, La Mille Miglia Editrice

Arguably the most beautiful Alfa Romeo series of all, the SZ and TZ cars are superbly detailed in this book. A list of known cars is given at the end of the book. It is in Italian and English.

Le Zagato Fiat 8V and Alfa 1900 SSZ
Michele Marchiano
Milano, 1987, Edizioni della Libreria dell'Automobile

This book is in both Italian and English. The Ottovu Z was a 2-liter pushrod V-8, while the 1900 was a 4-cylinder of almost identical displacement, so the conjunction of the two makes for an interesting comparison. But Marchiano never does a comparison, preferring instead to treat each car independently. You either love these Zagatos or you hate the styling. There is no in-between. The same can be said for the current ES30.

Books Containing Significant Alfa References

A lot of Alfa lore is sprinkled throughout automotive literature. Several key Alfa players switched companies. Ferrari is clearly the most famous ex-Alfista (see below), but Jano came from Fiat and went to Lancia, Hruska came from Porsche and Cisitalia, Colombo went on to Ferrari, and Ricart returned to Spain where he designed the Pegaso automobile. While Nuvolari is perhaps best remembered for his Alfa drives, he also managed stints with other German and Italian manufacturers. Fangio, the Argentine Chevrolet driver and five-time world champion, also drove for numerous European manufacturers. Piero Taruffi is perhaps better remembered for his book, *The Technique of Motor Racing*, than his consistent performance as an Alfa world-championship team member.

So important is Alfa that virtually any book about an Italian car manufacturer, or any grand prix driver, will contain interesting Alfa references. I offer the following notes as a guide:

Enzo Ferrari

Ferrari was a large Alfa concessionaire whose organization (Scuderia Ferrari) became Alfa's racing team for a few years in the 1930s. Whenever Ferrari reminisces about his life in the 1920s and 1930s, the odds are he's talking about Alfa. To say there are several books on Ferrari is a gross understatement. Works about him or by him would fill their own library.

Juan Manuel Fangio

The current Fangio book in print is by Carozzo. Another, by Fangio himself, is out of print. Since Fangio was the premiere postwar Alfa driver, his story is interwoven with Alfa.

General-Interest Books

Alfa dominated several Italian race-courses, so books that deal with the Targa Florio and Mille Miglia, or the track at Monza, will certainly have numerous Alfa references.

The two-volume set by Laurence Pomeroy, *The Grand Prix Car*, deserves special mention, even though it is out of print. Pomeroy's work is classic, and his detail on the P2, P3, and Alfetta Alfas has not been matched by any other author.

A further sense of Alfa's standing in the automotive world can be gained by some comparisons. I suggest Dante Giacosa's book *Forty Years of Design With Fiat*, and *La Lancia* by Wim H.J. Oude Weernink as informative studies of contemporary Italian designs. Orsini's book on Maserati will also make interesting reading. Of course, Hugh Conway's book on Bugatti is a classic work detailing Alfa's primary competition both on the race track and in the concours.

Shop and Parts Manuals

Official Alfa shop manuals are available from your local Alfa dealer, and updated lists of the available manuals are occasionally published by Alfa (just in case the guy at the parts counter gives you a blank look). The official Alfa manual is better than a commercial aftermarket manual, usually, because it offers clear photographic reproduction. The official manuals presume that you have a set of special tools, especially pullers and drivers, some of which not even dealers had. While the aftermarket manuals have a tendency to pick up procedures from the official factory manuals, special tools and all, they occasionally reveal alternate procedures, suggesting that a large screwdriver can be used in place of special tool X001234. For that reason alone, you may well wish to get both the official and aftermarket manuals for your car.

Members of the Alfa Romeo Owners Club may obtain some shop manuals through the club.

Both Haynes and Autobooks publish shop manuals for Alfa that are frequently available off-the-shelf at your local book store or Alfa emporium.

This may sound like heresy, but a shop manual has very limited utility, and typically has every bit of information excepting the exact thing you're interested in. Along the same line, most of the pictures in a shop manual are composed to obscure just the part you need to see, or they are reproduced so the critical area is an impenetrable blob of black ink.

The proper use of a shop manual is to jog an experienced technician's memory. The fact that they are almost universally used as primers makes them potentially destructive. A shop manual cannot teach manual dexterity, common sense, or basic safety, three attributes that, taken together, usually obviate the need for a shop manual.

If you have a SPICA-injected Alfa, you should buy the reprint of the factory tuning manual from Classic Motor books in Osceola, WI. It presumes that you have several special tools, but virtually all of those tools can be improvised. The thermostatic actuator test plunger can be made up of a threaded bolt and washer; the throttle bellcrank adjuster can be made up using a protractor; and the manometer used to set the throttle plates can be adapted from an inexpensive motorcycle unit (or made from scratch using plastic tubing).

If your Alfa has Bosch electronic fuel injection, then the book by Charles Probst (published by Robert Bentley) is mandatory.

A parts manual, even an outdated one, is frequently more informative to an experienced technician than a shop manual because the separate parts of an assembly are clearly shown. If you can obtain a parts manual for your car, it's good insurance. Don't forget, though, that part numbers are frequently updated, so follow the advice of your dealer when ordering a specific part.

Appendix 2

Sources

Membership in an Alfa Romeo owners club will put you in touch with knowledgeable, helpful people who can guide you in the unmatched pleasures of Alfa ownership.

All of the listed specialist shops are in addition to the official Alfa dealer organization.

NOTE—
We have endeavored to make this list as comprehensive as possible at the time of publication. Failure to list a supplier is merely an oversight. Listing does not imply endorsement or recommendation of any of the suppliers.

Alfa Clubs

Australia

AROCA (New South Wales) Inc.
Ian Johnson
PO Box R23
Royal Exchange, New South Wales 2000
02-967-2279

AROCA (Western Australia) Inc.
David Canute
PO Box 316
West Perth, Western Australia 6872
09-470-5822

AROCA (South Australia)
Peter Axford
PO Box 355
North Adelaide, South Australia 5006
08-396-3815

AROCA (Victoria) Inc.
Victor Lee
PO Box 216
Camberwell, Victoria 3124
03-879-0400

AROCA (Queensland) Inc.
John Stappleton
PO Box 104
Paddington, Queensland 4064
07-300-5297

Belgium

Il Biscione
Merkenclub voor Alfa Romeo v.z.w.
Kerkstraat 7
8340 Damme

Canada

Alfa Club of Edmonton
PO Box 1484
Edmonton, Alberta T5J 2N7

Sprint Speciale Register
Leslie J. Hegedus
77 Christena Crescent
Pickering, Ontario L1V 2K5

England

Alfa Romeo Owner's Club
Michael Lindsay
97 High St.
Linton, Cambridge CB1 6JT

105-Series Register
Chris Sweetapple
26 Honey Lane
Waltham Abbey, Essex EN9 3AS
0992-760364

The 1900 Register
Peter Marshall
Mariners, Courtlands Avenue
Esher, Surrey KT10 9HZ

The Giulietta Register
Gavin McGuire
Chart House
Moorhouse Road
Limpsfield, Chart
Surrey RH8 0SQ

Junior Zagato Register
Franco Macri
6 Oast House Road
Icklesham, Winchelsea
East Sussex

France

Alfa Romeo Club de France
PO Box 103
92322 Chatillon, Cedex

Germany

Alfa Classic Club
Dr. Stephan Epping
Sauerbruchstraße 31
5657 Haan

Alfaclub e. V.
Siegfriedstraße 73
4800 Bielfeld 1

Alfa Romeo Club Freiburg
Bernd Tritschler
Hurstweg 35
7800 Freiburg

Holland

Dutch Alfa Romeo Owners Club
Ben Hendriks
Ruynemanstratt 56
5012 JH Tilburg

Register Giulia Bertone
(C.J.A.M. De Groot)
Amersfoortsewag 14a
3712 BC Huis ter Heide

Italy

Zagato Car Club
Carrozzeria Zagato
20017 Terrazzano do Rho
Milano

New Zealand

Alfa Romeo Owners Club of New Zealand Inc.
Glen Watson
PO Box 507
Feilding
06-356-3878

Sweden

GTA Register
Anders Ericsson
Vikingagatan 49
S-753 34 Uppsala

Switzerland

Swiss Alfa Romeo Register
Registre Suisse Alfa Romeo
Casa Postale 196
1000 Lausanne 12

U.S.A.

AROC (Alfa Romeo Owners Club)
2468 Gum Tree Lane
Fallbrook, CA 92028

Alfa Romeo Association
546 W. McKinley Avenue
Sunnyvale, CA 94086

Alfa Romeo Association,
San Francisco
28617 Miranda
Hayward, CA 94544

Alfa Romeo Market Letter
7017 SE Pine
Portland, OR 97215

The GTZ Register
George Pezold
73 Bay Avenue
Huntington, NY 11743

The Giulietta Register
Russ Baer
1729 Linden Avenue
Baltimore, MD 21217

The Giulietta SZ Register
Donald E. Hughes/UMC
810 12th Avenue South
Nashville, TN 37202

Montreal Register
Gene Ross
PO Box 3925
Visalia, CA 93287

Puget Sound AROC
Terry Gates
8719 Green Avenue North
Seattle, WA 98103

Suppliers

England

Alfa II
Ramesh Bharadia
Unit 6 Bowman Trading Estate
Westmoreland Road
London NW9 9RL
081-206-2075
Sales and servicing

Alfetta Racing
Piero Pesaro
1A Commerce Road
Brentford
Middlesex TW8 8LE
081-560-6194 or 081-847-3897/8
Sales, maintenance, and performance
tuning

Automeo
Les Dufty
36 Gypsy Patch Lane
Little Stoke, Bristol
0272-695771
Maintenance and performance
tuning

Richard Banks
Commerce House
Wickhambrook
Newmarket, Cambridgeshire CB8
8XL
0440-820291
Sales and restoration

Benalfa
Alan Bennett
19 Washington Road
West Wilts Trading Estate
Westbury, Wilts BA13
0373-864333
Restoration and engine rebuilds

Brookside Garage
Jon Dooley
55 High Street
Wrestlingworth
Sandy, Bedfordshire
076723-217
Restoration, maintenance, and per-
formance tuning

T.A. & J.M. Coburn
Widhill House
Blunsdon
Swindon, Wilts
0793-721501
Upholsterers

EB Spares
31 Link Road
West Wilts Trading Estate
Westbury, Wilts BA13 4JB
0373-823856 fax: 0373-858327
Parts

Brian Hammond
Station Cottage
Thuxton
Dereham, Norfolk
0362-850320
Sales, maintenance, and performance
tuning

The Highwood Motor Company
Chris Sweetapple
26 Honey Lane
Waltham Abbey, Essex EN9 3AS
0992-760364
Restoration and parts

Peter Hilliard & Son
41 High Street
Penge, London SE20 7HJ
081-778-5755
Maintenance and performance
tuning

Huntsworth Garage
Tim Stewart
Boston Place, London NW1 6ER
071-724-0269
Sales and maintenance

Intaparts
Richard Igsar
Unit 15 Vulcan Road
Leicester
0533-519901
Parts

Alwyn Kershaw
Chessingham Park
Dunnington, York YO1 5SE
0904-488778
Sales, maintenance, and performance
tuning

Malcolm Morris
MP Motors
Horton
Devizes, Wilts SN11 3LX
0380-860611
Restoration

Ramponi Rockell
41 Lancaster Mews
London W2 3QQ
071-262-2449
Sales and maintenance

Rossi Engineering
Rob Giordanelli
Sunbury on Thames
0932-786819
Restoration and performance tuning

Spider's Web
Roger Longmate
Westgate Street
Hillborough
Thetford, Norfolk IP26 5BN
07606-229
Maintenance and restoration

John Timpany
20 Jason Close
Brentwood, Essex
0277-211296
Performance tuning

Chris Whelan
Magnolia House
49 Southwick Street
Southwick, West Sussex BN42 4TH
0273-593963
Engines

Italy

AFRA
Via Caracciola, 24
Milano, Italy
38.17.06
The major Italian source of
obsolescent Alfa parts

U.S.A.

APE (Alfa Parts Exchange)
2436 Whipple Road #2
Hayward, CA 94544
510-471-7132

Alfa Heaven
111 Zagato Lane
Aniwa, WI 54408
Restoration

Alfa Ricambi
6644 San Fernando Road
Glendale, CA 91202
800-225-2532
Performance parts

Autocomponents
8906 Dark Country Line Road
Brookville, OH 45309
513-884-5144
Turbos, Giulietta parts, race engineering

Automotive Systems Group
6644 San Fernando Road
Glendale, CA 91201
818-956-7933
Combines Alfa Ricambi and Shankle
Engineering

Barber's Shop
116 18th St.
Sacramento, CA 95814
916-448-6422
Service and aftermarket parts

Black Bart's Auto Emporium
Bob Bartel
2315 South Calhoun
Fort Wayne, IN 46807
219-456-3435
Aftermarket parts, cloisonne em-
blems

Appendix 2

Bobcor Motors
243 W. Passaic St.
Maywood, NJ 07607
201-587-9000
Old Alfa shop, changed management
several years ago

Centerline Alfa Products
PO Box 1466
Boulder, CO 80306
303-447-0239
Engine/driveline parts

DiFatta Brothers
5928 Belair Road
Baltimore, MD 21206
410-426-7524

Eastwood Company Tools
Box 296
Malvern, PA 19355
1-800-345-1178
Tools, general restoration

Ereminias Imports
3000 S. Main St.
Torrington, CT 06790
203-496-9800
Enthusiastic Alfa dealership, large
inventory

EuroParts
1425 Gardena Ave. Unit 7
Glendale, CA 91204

Exoticar
Al Cortes
1412 Mansel Ave.
Lawndale, CA 90260
Specializes in older Alfas, especially
the 1900

HRE Used Cars and Parts
Box 293
Chico TX 76030
Used parts

Wes Ingram
309 South Cloverdale St.
Suite D-5
Seattle, WA 98108
206-938-2558
SPICA injection

International Auto Parts
Rt. 29 North
Charlottesville, VA 22906
804-973-2892
Large parts inventory

Jon Norman
1221 Fourth Street
Berkeley, CA 94701
Racing enthusiast

Quattroruote
Carter Hendricks
3336 Washington Blvd.
St. Louis, MO 63103
314-531-3088
Service and restoration

Stewart Sandeman
22692 Granite Way
Laguba Hills, CA 92653
Service, some used parts

Sperry Valve Works
2829 Gundry Ave.
Signal Hill, CA 90806
310-988-5960

Paul Spruell Alfa
3320 LaVenture Drive
Chamblee, GA 30341
404-458-8458

Gary Valant
13551 Method
Dallas, TX 75243
214-423-8204
Service

Ward and Deane Racing
115 N. Oak St
Inglewood, CA 90301
213-754-6008
Performance parts, especially
suspension

Wefco Rubber
1655 Euclid Ave.
Santa Monica, CA 90404
Extruded rubber moldings

Tom Zat
111 Zagato Lane
Aniwa, WI 54408
715-449-2141
"One-of-everything" used parts

Index

INDEX

> **WARNING** —
> • *Automotive service, repair, and modification is serious business. You must be alert, use common sense, and exercise good judgement to prevent personal injury.*
>
> • *Before beginning any work on your vehicle, thoroughly read all the Warnings and Cautions listed at the front of this book.*
>
> • *Always read a complete procedure before you begin the work. Pay special attention to any Warnings and Cautions, or any other information, that accompanies that procedure.*

Glossary

American Term	English Term	American Term	English Term
A-Arm	Wishbone (suspension)	Knocking/Pinging	Pinking
Allen Screw	Grub Screw	Lash	Clearance, Free-play
Antenna	Aerial	Light	Lamp
Axleshaft	Halfshaft	Lock Washer, Tang	Tab Washer
Back-up	Reverse	Lug Nut	Wheel Nut
Barrel	Venturi	Metal Chips	Swarf
Block	Chock	Misses	Misfires
Bushing	Bush	Muffler	Silencer
Coast	Freewheel	Multimeter	VOM (Volt-ohmmeter)
Convertible	Drop Head	Open Flame	Naked Flame
Cotter Pin	Split Pin	Parking Brake	Handbrake
Counterclockwise	Anti-clockwise	Parking Light	Side Light
Countershaft	Layshaft (gearbox)	Pinging/Pre-ignition	Pinking
Dashboard	Facia	Piston/Wrist Pin	Gudgeon Pin, Small End
Denatured Alcohol	Methylated Spirit	Pitman Arm	Drop Arm
Dome Lamp	Interior Light	Power Brake Booster	Servo Unit
Drive Axle	Driveshaft	Primary Shoe	Leading Shoe (of brake)
Driveline	Transmission	Prying	Levering
Driveshaft	Propeller Shaft	Repair Shop	Garage
Fender	Wing, Mudguard	Replacement	Renewal
Firewall	Bulkhead	Ring Gear (Differential)	Crownwheel
Flashlight	Torch	Rod Bearing	Big End Bearing
Float Bowl	Float Chamber	Sedan	Saloon
Freeway/Highway	Motorway	Shock Absorber	Damper
Frozen	Seized	Snap-ring	Circlip
Gas Pedal	Accelerator Pedal	Soft-top	Hood
Gas Tank	Petrol Tank	Spacer	Distance Piece
Gearshift	Gearchange	Spark Plug Wires	HT Leads
Generator	Dynamo (DC)	Spindle Arm	Steering Arm
Ground	Earth (Electrical)	Stumbles	Hesitates
Header	Exhaust Manifold	Throw-out Bearing	Thrust Bearing
Heat Riser	Hot Spot	Tie Rod	Trackrod (steering)
Hood	Bonnet	Transmission	Gearbox
Idle	Tickover	Troubleshooting	Fault Finding
Installation	Refitting	Trunk	Boot
Intake	Inlet	Turn Signal	Indicator
Jackstands, Safety Stands	Axle Stands	Valve Cover	Rocker Cover
Jumper Cable	Jump Lead	Wheel Cover	Roadwheel Trim
Keeper (valves)	Collet, Split Cotter	Windshield	Windscreen
Kerosene	Paraffin	Wrench	Spanner

About the Author

In 1961, Pat Braden became the first Vintage Editor of the *Alfa Owner*, the national club publication. At the time, he owned a 1929 1750 Gran Sport Alfa Romeo. Since then, he has owned over 50 Alfas, including 8C2300, 6C2500, and 1900 CSS Zagato models, as well as the expected gaggle of Giuliettas, Giulias, 1750s, 2-liters, 2600s, 2000s, Alfettas—and a GTV6. Beginning in 1979, he became Editor of the *Owner*, a post he held for six years. He is an honorary life member of the Alfa Romeo Owners Club and still contributes a monthly article.

This is Pat's sixth book and his second on Alfa Romeo. He is an instructional designer, creating audiovisual training programs for a client list that includes most import and domestic marques. In his off-hours he is an automotive journalist, shade-tree mechanic, husband, and father.

Acknowledgments

This book started out as a collection of reminiscences and grew into much more of an encyclopedic work than I ever anticipated. That is because some good friends helped me understand that there is a need for a general, single-source guide to Alfa ownership.

Publishers usually get a perfunctory thanks, probably more often than not to assure that the next manuscript finds a warm reception. But David Bull and John Kittredge deserve first credits for guiding this work. David, more than anyone else, saw the possibilities.

My team of technical advisors has included Bill Pringle, Phil Lampman, Fred Di-Matteo, Paul Tenney, Tom Zat, and John Hertzman, all of whom are distinguished Alfista as well as good friends. Joe Benson gets special thanks for permission to include his work on the SPICA fuel injection system, and Jim Weber has my gratitude for his continued review and encouragement during the book's long gestation.

Don Black deserves singular recognition, not only for his profound knowledge of Alfa Romeo and long stewardship of the marque, but as well for an incisive review of the manuscript which has kept me faithful to the facts.

Finally, a word of gratitude to my wife, Cheryl, whose love and unbounded patience has made this work possible.

Front Cover: Front grille detail of a Sprint Speciale, photo by David Gooley.

Above: A GTV stretches its legs, photo by Ian Dawson.

Back Cover: Clockwise from top: a) Alfa Spider, photo by Martyn Goddard; b) European GTV6, photo by Ian Dawson; c) a race-prepped GTA 1300 Junior, photo by Jean Taylor; d) 164 in the rain, photo by Martyn Goddard; e) a pristine Giulietta Veloce engine compartment, photo by the author.